PLANT GROWTH SUBSTANCES

PLANT GROWTH SUBSTANCES

Principles and Applications

Richard N. Arteca

The Pennsylvania State University

CHAPMAN & HALL

New York • Albany • Bonn • Boston • Cincinnati • Detroit • London • Madrid • Melbourne
Mexico City • Pacific Grove • Paris • San Francisco • Singapore • Tokyo • Toronto • Washington

Cover design: Andrea Meyer, emDASH inc.

Copyright © 1996
Chapman & Hall

Printed in the United States of America

For more information, contact:

Chapman & Hall
115 Fifth Avenue
New York, NY 10003

Thomas Nelson Australia
102 Dodds Street
South Melbourne, 3205
Victoria, Australia

Nelson Canada
1120 Birchmount Road
Scarborough, Ontario
Canada, M1K 5G4

International Thomson Editores
Campos Eliseos 385, Piso 7
Col. Polanco
11560 Mexico D.F.
Mexico

Chapman & Hall
2-6 Boundary Row
London SE1 8HN
England

Chapman & Hall GmbH
Postfach 100 263
D-69442 Weinheim
Germany

International Thomson Publishing Asia
221 Henderson Road #05-10
Henderson Building
Singapore 0315

International Thomson Publishing-Japan
Hirakawacho-cho Kyowa Building, 3F
1-2-1 Hirakawacho-cho
Chiyoda-ku, 102 Tokyo
Japan

1 2 3 4 5 6 7 8 9 10 XXX 01 00 99 97 96 95

Library of Congress Cataloging-in-Publication Data

Arteca, Richard N.
    Plant growth substances : principles and applications / by Richard N. Arteca.
        p.   cm.
    Includes bibliographical references and index.
    ISBN 0–412–03911–7
    1. Plant growth promoting substances.   I. Title.
QK731.A76   1995
581.3'1—dc20   94–43486
                                                          CIP

British Library Cataloguing in Publication Data available

To order this or any other Chapman & Hall book, please contact **International Thomson Publishing, 7625 Empire Drive, Florence, KY 41042.** Phone: (606) 525-6600 or 1-800-842-3636.
Fax: (606) 525-7778, e-mail: order@chaphall.com.

For a complete listing of Chapman & Hall's titles, send your requests to
Chapman & Hall, Dept. BC, 115 Fifth Avenue, New York, NY 10003.

To my wife, Jeannette for her dedication to the successful completion of this book and the countless hours spent preparing figures, editing, organizing, and whatever else was necessary.

CONTENTS

CHAPTER THREE

CHEMISTRY, BIOLOGICAL EFFECTS, AND MECHANISM OF ACTION OF PLANT GROWTH SUBSTANCES

CHAPTER 4

SEED GERMINATION AND SEEDLING GROWTH

CHAPTER FIVE

ROOTING

CHAPTER SIX

DORMANCY

CHAPTER SEVEN

JUVENILITY, MATURITY, AND SENESCENCE

CHAPTER EIGHT

FLOWERING

CHAPTER NINE

ABSCISSION

PHYSIOLOGY OF FRUIT SET, GROWTH, DEVELOPMENT, RIPENING, PREMATURE DROP, AND ABSCISSION

TUBERIZATION

CHAPTER TWELVE

MANIPULATION OF GROWTH AND PHOTOSYNTHETIC PROCESSES BY PLANT GROWTH REGULATORS

CHAPTER THIRTEEN

WEED CONTROL

PREFACE

It has been well documented that extremely low concentrations of plant growth substances have the ability to regulate many aspects of plant growth and development from seed germination through senescence and death of the plant. More than 60 years ago auxins, which are the first class of plant growth substances, were discovered. Since then four additional classes of plant growth substances have been recognized, namely, gibberellins, cytokinins, abscisic acid, and ethylene. Most recently brassinosteroids, salicylates, and jasmonates are beginning to gain acceptance as the newest classes of plant growth substances. Numerous advances in the use of plant growth substances on a practical scale along with basic research at the biochemical, physiological, and molecular levels have been made. The use of these compounds in agriculture has a great deal of potential in regulating many, if not all, plant physiological processes, and with time their commercial use for these purposes may become a reality. However, at the present time the use of plant growth substances is in its infancy stage of development. This becomes quite evident when reviewing the literature in the area of plant growth regulating compounds where one can find that a given class of plant growth substances promotes, inhibits, or otherwise has no effect on a specific plant process. Although there is a great deal of variation in the literature some general theories have been established and, for the most part, can now be applied to many situations. These general theories and concepts along with the relationship between plant growth substances and different physiological processes will be discussed in this book, which will be organized as follows: It starts with historical aspects and fundamental terms/concepts. This is followed by methodology for extraction, purification, and quantification of plant growth substances in Chapter 2; chemistry, biological effects, and mechanism of action of plant growth substances are covered in Chapter 3. The first three chapters will provide a firm foundation on how these plant growth substances work, thereby making many of their practical uses evident. In subsequent chapters the involvement of plant growth substances in seed germination, seedling growth, rooting, dormancy, juvenility, maturity, senescence, flowering, abscission, fruit set, fruit growth, fruit development, premature drop, ripening, promotion of fruit drop, tuberization, photosynthesis, and weed control will be covered. Throughout this book practical uses of plant growth substances in agriculture will be presented and, when possible, examples will be given showing how basic laboratory research has translated into increased production and profits to the grower. This text is designed primarily for upper-level students and will provide an excellent resource book for people entering this exciting field of research.

CHAPTER 1

Historical Aspects and Fundamental Terms and Concepts

GENERAL HISTORICAL ASPECTS FOR PLANT GROWTH SUBSTANCES

In the early 1900s F. W. Went made the profound statement *"Ohne Wuchstoff, kein Wachstum"*, translated, "without growth substances, no growth." It is now generally accepted that plant growth substances have an important regulatory role throughout the plant kingdom. Since the pioneering work of Went, research in the area of plant growth substances has made considerable strides. At the present time plant growth substances are used in agriculture for purposes such as delaying or promoting ripening, induction of rooting, promotion of abscission, control of fruit development, weed control, size control and many other responses. Although they are currently used in agriculture, there are many questions which remain to be answered in order to maximize the true potential of plant growth substances.

Duhamel du Monceau (1758) concluded from his experiments that sap movement controlled the growth of plants with one sap moving upward and the other downward. He showed that the sap moving downward from the leaves controlled nutrition of the roots, and when the downward movement of sap was interrupted by girdling both callus and root formation would occur above the girdle. In 1880, Julius von Sachs, the father of plant physiology, presented evidence revising du Monceau's theory. Von Sachs (1880) was the first to propose that organ-forming substances were produced by the plant and that these substances moved in different directions controlling growth and development. He further postulated that the organ-forming substances were produced in response to environmental conditions such as light and gravity. Beyerinck (1888) wrote the first article on organ-forming substances. He studied willow gall, which is caused by the leaf wasp *Nematus capreae*. When the wasp deposits its eggs a large gall develops. He suggested that "growth enzymes" moved out of the eggs and caused a gall to form. It was not until 40 years later that indole-3-acetic acid (IAA) was identified as the cause of the gall. For many years researchers assumed that nutritional factors rather than growth substances were involved in regulating plant growth and development. One of the most commonly used examples of the nutritional approach is the classical work of Kraus and Kraybill (1918) establishing the nutritional theory. They studied changes in the carbohydrate-to-nitrogen (C/N) ratio in relationship to growth of tomato plants. They found that normal fruiting and vegetative growth occurred when high carbohydrate and low nitrogen levels were present within the plant, whereas high levels of nitrogen and low levels of carbohydrate promoted vegetative growth and reduced fruiting. This work was misinterpreted to suggest that this ratio had a direct effect on flowering. Subsequent work showed that although a particular carbohydrate-to-nitrogen ratio may be associated with a certain type of growth, this ratio is in no way a direct cause of flowering. Today it is generally accepted that plant growth substances are involved in the regulation of plant responses rather than the C/N ratio and this is called the hormonal theory.

Bayliss and Starling (animal physiologists) (1904) coined the word *hormone* and defined it as arousing to activity. Today in animal systems the term *hormone* is used to define a compound synthesized at a localized site, and transported via the bloodstream to a target tissue where it regulates a physiological response based on concentration in that tissue (Trewavas 1981). Fitting (1910) first introduced the term *hormone* into plant physiology, and it has been used since to define specific naturally occurring organic substances with regulatory functions in plants (Fitting 1910). However, its use has historically caused a great deal of confusion in the plant world and still remains unclear today. In 1951, Dr. Kenneth V. Thimann, who at the time was the President of the American Society of Plant Physiology, appointed a group of scientists to establish uniform nomenclature for plant growth substances. This committee, in conjunction with 200 other scientists from all over the world interested in chemical regulation of plant growth and development, established nomenclature on regulators in plants (van Overbeek et al. 1954). The definition of *plant hormones* (*phytohormones*) was established by this group to be "regulators produced by plants, which in low concentrations regulate plant physiological processes. Hormones usually move within the plant from a site of production to a site of action."

DISCOVERY OF PLANT GROWTH SUBSTANCES

Auxins

Although Darwin is better known for his theory of evolution, he is considered to be the scientist responsible for initiating modern plant growth substance research. In his book entitled, *The Power of Movement in Plants* (Darwin 1880), he described the effects of light on movement of coleoptiles from etiolated *Phalaris canariensis* (canary grass) seedlings. The coleoptile is a specialized leaf in the form of a hollow cylinder that encloses the epicotyl and is attached to the first node. It protects the growing tip of the grass seedling until the more rapidly growing leaf emerges above the ground. When the coleoptile was supplied with unidirectional light it resulted in bending toward the source. If the tip of the coleoptile was covered with an aluminum foil cap and the lower part was unilaterally illuminated, curvature typically did not occur. However, when the lower part of the coleoptile was covered and the coleoptile tip was exposed to light, bending did occur. These experiments suggested that the tip, of the coleoptile perceived the light and produced a signal which was transported to just below the tip, resulting in bending. This was further confirmed by an experiment which showed that when unidirectional light was given to a coleoptile with its tip removed, bending did not occur (Figure 1.1).

Salkowski (1885) discovered indole-3-acetic acid in fermentation media. It was not until many years later that this substance was found in plant tissues, and today it is known to be the major auxin involved in many physiological processes in plants. Rothert (1894) confirmed and extended Darwin's experiments showing that the phototropic signal resulting in bending is conducted in the parenchyma of the coleoptile.

Fitting (1907) evaluated the effect of making unilateral incisions in *Avena* coleoptiles in an environment saturated with moisture so the cut surfaces did not dry before or after they were pressed together. He showed no effect of lateral incisions on the growth rate and phototropic response regardless of their positions with regard to light (Figure 1.1). He concluded that the stimulus is transmitted through living material and goes around incisions. He further

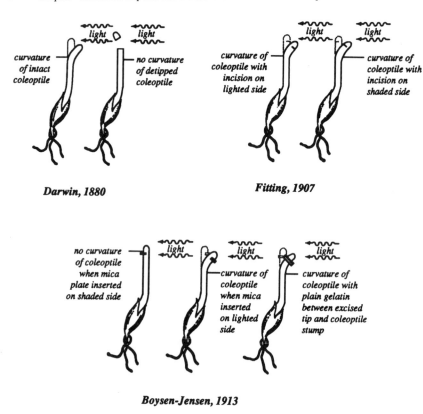

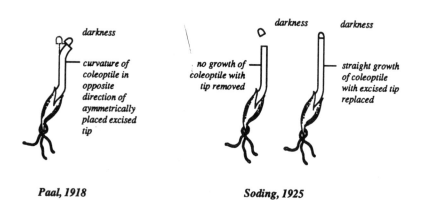

Figure 1.1. Historical studies on the involvement of auxin in coleoptile curvature from Devlin and Witham (1983)

speculated that the positive phototropic response occurred because the light sets up a polarity in the cells of the tip and that the stimulus is transmitted from the cells of the unilaterally illuminated tips to the cells in the darkened basal portion. Unfortunately, his observations were incorrect because a barrier to translocation was never formed.

Boysen-Jensen (1913) showed that the light-induced stimulus could be transported through nonliving material. He provided proof for this by cutting the tips off of *Avena* coleoptiles and then inserting a block of gelatin between the tip and the severed stump. When he illuminated only the tip, curvature was obtained in the basal portion below the gelatin (Figure 1.1).

In another experiment, Boysen-Jensen made incisions on different sides of the coleoptile and inserted a piece of mica to assure that a barrier had been achieved, unlike Fitting's earlier work. He showed that when the incision was made on the shade side of the coleoptile there was no curvature; however, when it was made on the light side of the coleoptile curvature occurred (Figure 1.1). This work showed that the signal was transported down the shaded side of the coleoptile, promoting curvature.

Paal (1918) confirmed Boysen-Jensen's findings and provided additional information showing that a diffusible substance produced in the tip controls the growth of the *Avena* coleoptile. He found that if the tip of a coleoptile was removed, then placed on one side of the cut surface in the dark, negative curvature (away from the tip) was induced (Figure 1.1).

Soding (1925) extended Paal's work using the straight growth test, which is based on elongation of the *Avena* coleoptile in the dark. He showed that if the tip of the coleoptile was removed there was a reduction in elongation; however, when the tip was removed then put back on the coleoptile stump in the original orientation straight growth resumed (Figure 1.1).

Went (1926) published a report describing how he had obtained an active chemical substance from the *Avena* coleoptile tip by placing coleoptile tips on blocks of agar, waiting for a period of time, removing the tips, and cutting the agar into smaller blocks (Figure 1.2). He found that agar blocks containing the diffusate from excised tips promoted renewed growth of the coleoptile when placed on decapitated stems.

Went (1928) also developed a method for the quantification of plant growth substances present in a sample (Figure 1.3). This test indicated that the curvatures were proportional, within limits, to the amount of active plant growth substance. Went's findings greatly stimulated plant growth substance research. In fact, much of our current knowledge regarding IAA is based on results obtained with the *Avena* curvature test, which is still commonly used today.

Kögl and Haagen-Smit (1931), starting with 33 gallons of human urine, performed a series of purification steps, and after each step they tested biological activity using the *Avena* curvature test. Their final purification step yielded

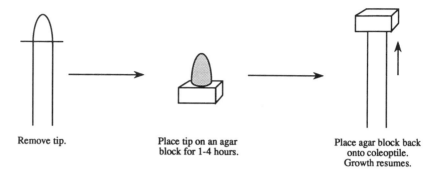

Remove tip. **Place tip on an agar**
block for 1-4 hours.

Place agar block back
onto coleoptile.
Growth resumes.

Figure 1.2. Early study by Went (1926) showing that a chemical from excised coleoptile tips promoted renewed growth of decapitated coleoptiles in the dark.

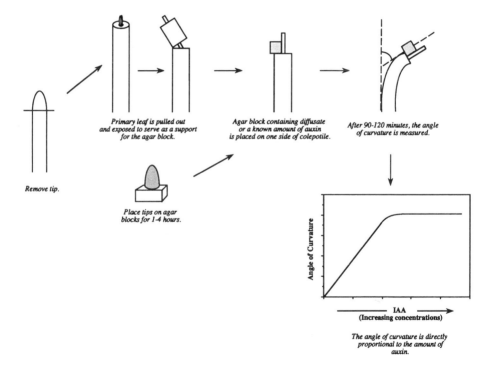

Figure 1.3. Description of the *Avena* coleoptile bioassay developed by Went (1928).

40 mg of purified compound which they called *auxin A* (*auxentriolic acid*). Later Kögl et al. (1934a, 1934b) once again analyzed human urine and found a compound similar in structure and activity to auxin A and gave it the name *auxin B* (*auxenolenic acid*). Also contained in this extract was *heteroauxin* (*other auxin*), now known as indole-3-acetic acid, which had been initially discovered and characterized by Salkowski in 1885.

Kögl and Kostermans (1934) isolated IAA from yeast and Thimann (1935) isolated IAA from cultures of *Rhizopus suinus*. It was not until 1946 that Haagen-Smit and others isolated pure IAA from the endosperm of immature corn grains showing that IAA was found in a higher plant (Haagen-Smit et al. 1946). Vliegenthart and Vliegenthart (1966) presented evidence that auxin A and B are not natural plant products; however, IAA has since been isolated from numerous plant species and has been shown to be ubiquitous in the plant kingdom. Went's auxin was probably mostly IAA, however, other growth promoters may have been in his initial diffusion studies on the phototropic response, possibly IAA derivatives. The term *auxin* is derived from the Greek *auxein*, which means "to grow" and was proposed originally by Kögl, Haagen-Smit, and Went to designate a particular substance which had the property of promoting curvature in the *Avena* coleoptile curvature test.

Gibberellins

Many years ago Japanese farmers observed that certain rice plants were taller earlier in the season, suggesting that growth was better and the crop was going to produce high yields. However, as the season progressed the plants became spindly, chlorotic, and, more importantly, sterile and devoid of fruit. Instead of a bumper season, 40% crop losses were typically found annually. This disease received many names from Japanese farmers based on the symptoms observed, some commonly used terms follow: *bakanae* (foolish seedling), *ahonae* (stupid seedling), *yurei* (ghost), *somennae* (thin noodle seedling), and so on. The most commonly used term is *bakanae seedling*. In 1898, the first scientific paper was published by Hori demonstrating that the *bakanae* disease was caused by infection by a fungus belonging to the genus *Fusarium* (Hori 1898). Sawada (1912) was the first to suggest that elongation of rice seedlings was due to the stimulus of the fungal hyphae. Kurosawa (1926) presented evidence that a substance secreted by the *bakanae* fungus was responsible for the elongation. Following the initial report by Hori there was a lot of controversy over the nomenclature of the *bakanae* fungus. This problem was resolved by Wollenweber in 1931 when he named the imperfect stage (asexual) *Fusarium moniliforme* (Sheldon) and the perfect stage (sexual) *Gibberella fujikuroi* (Saw.) Wr. (Takahashi et al. 1991). The purification of the *bakanae*-producing substance was hindered by the presence of a growth-inhibiting material, fusaric acid (5-n-butylpicolinic acid). However, in 1935 Yabuta isolated an active crystalline material from *Gibberella fujikuroi* sterile culture filtrate (Yabuta 1935). This substance was found to stimulate growth when applied to the roots of rice seedlings and was called *gibberellin A*, this was the first time that the term *gibberellin* was used in scientific literature. Yabuta and Sumiki (1938) were successful in crystallizing gibberellin A and *gibberellin B*; however, due to the war, gibberellin research was placed on hold. In the 1950s there was intensive research by English, American, and Japanese workers on the growth-regulating properties of gibberellic acid and identification of gibberellins in fungal extracts, and they were finally discovered in higher plants. In 1954 English researchers (Brian et al. 1954) identified the plant growth-regulating properties of gibberellic acid in a product from the fungus *Gibberella fujikuroi*. In 1955 American researchers (Stodola et al. 1955) identified what they called gibberellin A and *gibberellin X* from *Gibberella fujikuroi* sterile culture filtrates. Also in 1955, Japanese researchers (Takahashi et al. 1955) found that gibberellin A contained three distinct compounds, which they called GA_1, GA_2, and GA_3. It is now generally agreed upon that gibberellin X, gibberellic acid, and GA_3 are all the same compound, in fact, today gibberellic acid and GA_3 are synonymous. Radley (1956) identified substances similar to gibberellic acid in higher plants, and since this time gibberellins have been shown to be ubiquitous in higher plants (Takahashi et al. 1991). Takahashi et al. (1957) identified GA_4 from *Gibberella fujikuori* and showed that GA_1 was the same as Stodola's (Stodola et al. 1955) gibberellin A; however, there were no counterparts for GA_2 or GA_4. MacMillan and Takahashi (1968) proposed assigning numbers from gibberellin $A_1–A_X$ irrespective of their origin. This procedure is still used today for the more than 90 gibberellins currently known.

Cytokinins

Haberlandt (1913) showed that phloem diffusates had the ability to stimulate cell proliferation in potato tuber tissue. Almost 30 years later van Overbeek et al. (1941) showed that a naturally occurring substance found in coconut milk (liquid endosperm) had the ability to promote cellular proliferation in young *Datura* embryos. Van Overbeek et al. (1944) reported that unpurified extracts of *Datura* embryos, yeast, wheat germ, and almond meal promoted cell division in *Datura* embryo cultures, thereby showing that these substances were widespread in their occurrence. Haberlandt's work was extended by Jablonski and Skoog in 1954; they showed that vascular tissue cells contain materials that stimulate cell division in tobacco plants. Miller et al. (1955b) were the first to report the isolation and identification of kinetin (6-furfurylaminopurine), which was obtained from aged or autoclaved herring sperm deoxyribonucleic acid (DNA). They named the compound *kinetin* because of its ability to promote cell division or *cytokinesis* in tobacco pith tissue (Miller et al. 1955a). Hall and deRopp (1955) showed that kinetin could be produced by autoclaving a mixture of adenine and furfuryl alcohol, thereby showing that kinetin could be formed from DNA degradation products. Miller (1961) reported the identification of a naturally occurring kinetin-like compound in maize, this compound was later found to be zeatin. Letham (1963) published a report on zeatin as a factor inducing cell division from *Zea mays* and later described its chemical properties (Letham 1964). Working in another laboratory Shaw and Wilson (1964) confirmed the identity of zeatin structure and showed that it was not an artifact. After sorting through the literature in this area it is clear that the credit for the discovery of zeatin should be given to both Letham and Miller in 1963. Since the discovery of zeatin, numerous additional cytokinins have been discovered and shown to be ubiquitous in the plant kingdom.

Abscisic Acid

Liu and Carns (1961) isolated a substance from mature cotton fruit and found that it stimulated abscission of cotton petioles. The structure of this compound, which was called *abscisin I* was never determined. In 1963, Addicott's group in the United States isolated a substance from young cotton fruit and found that it also caused abscission of cotton petioles (Ohkuma et al. 1963). They partially characterized this compound showing that it was a 15-carbon compound and called it *abscisin II*. Almost simultaneously with the reports of abscisin II, Wareing's group in England isolated an inhibitory substance from birch leaves exposed to short days. They showed that when this substance was applied to birch seedlings it inhibited growth of the apical bud (Eagles and Wareing 1963). This led them to suggest that the compound was a dormancy inducer and thus the uncharacterized chemical was termed *dormin*. In 1965, workers in Addicott's laboratory proposed the chemical structure for abscisin II (Ohkuma et al. 1965; Ohkuma 1965). Again almost simultaneously Wareing working in collaboration with Shell Researchers Ltd. in England, showed that dormin and abscisin II were the same compound (Cornforth et al. 1965a, 1965b). To minimize problems with nomenclature, plant scientists active in this area of research agreed on the name *abscisic acid* and reported their recommendation at the Sixth International Conference on Plant Growth Substances in Ottawa in 1967; the name was approved and the decision published by Addicott et al. (1968). Today, the terms *abscisin I, abscisin II* and *dormin* have been dropped and only appear in the early pioneering studies. Since its discovery, abscisic acid has been shown to be widely distributed in the plant kingdom and has a variety of effects, in addition to dormancy and abscission.

Ethylene

Practical uses for ethylene started many years ago with ancient Egyptians, who would gash figs in order to stimulate ripening (Galil 1968), and the Chinese, who would burn incense in closed rooms to enhance the ripening of pears (Miller 1947). Girardin (1864) showed that leaks of illuminating gas (produced from combusting coal) from street lights caused premature shedding of leaves near the source. It was also shown that illuminating gas would cause stunting, twisting, and abnormal horizontal growth of shoots (Molisch 1884). It was not until much later that the Russian scientist Neljubow (1901) showed that ethylene was the active component in illuminating gas. His research showed that ethylene, a component of illuminating gas, caused a triple response in etiolated pea seedlings. This information was used to develop the first bioassay for ethylene based on its ability to suppress stem elongation, increase radial expansion (lateral expansion), and promote bending or horizontal growth in response to gravity (Neljubow 1901). Doubt (1917) discovered that ethylene promoted abscission. A method which used combustion products for stimulating ripening in citrus was patented by Denny (1923). Denny (1924) demonstrated that ethylene was the active component in combustion products causing ripening. The first suggestion that fruits release a gas which stimulates ripening was by Cousins (1910), who, in an annual report to the Jamaican Agricultural Department mentioned that oranges stored with bananas caused premature ripening. It was more than 20 years until Gane (1934) proved that ethylene is synthesized by plants and is responsible for speeding the ripening process. Shortly after this finding Crocker et al. (1935) at the Boyce Thompson Institute proposed that ethylene was the fruit-ripening hormone which also acted as a plant growth regulator in vegetative plant organs. This hypothesis was supported by many other researchers (Hansen 1943; Kidd and West 1945). Biale et al. (1954), using available but insensitive assay procedures for ethylene, found that fruits did not produce sufficient amounts prior to ripening to induce the process. This work put the hypothesis that ethylene affects fruit ripening on hold for about five years. In 1959, gas chromatographic methods for the detection of ethylene in plant tissues were developed which increased the sensitivity over existing methods by approximately one million fold (Burg and Stolwijk 1959; Burg and Thimann 1959). With this technique available, the work in ethylene research went full speed ahead. Ethylene is now generally accepted to be a plant growth substance which has many effects in plants from seed germination through senescence and death of the plant.

Brassinosteroids

In the 1960s Mitchell et al. at the U.S. Department of Agriculture (USDA) Research Center began screening pollen in search of new plant growth substances. It had been known for many years that pollen is a rich source of plant growth-regulating substances, thereby making it a logical choice for screening. Nearly 60 species of plants were screened, and about half caused increased growth in the bean second internode bioassay. The greatest growth increases were obtained from alder tree (*Alnus glutinosa L.*) and rape plant pollen (*Brassica napus L.*). Extracts from these two pollens caused such rapid growth in the bean second internode bioassay that the stem would split above the second pair of leaves (Figure 1.4). The USDA workers proposed that this was a new class of lipoidal hormones, which they termed *brassins* (defined as a crude lipoidal extract from rape pollen). Mitchell and Gregory (1972) showed that brassins could enhance crop yield, crop efficiency, and seed vigor. Milborrow and Pryce (1973) believed that brassins were a crude extract containing gibberellins and other compounds rather than endogenous lipids. In an effort to isolate the active components of brassins, 500 lb of bee-collected rape pollen, which was more readily available then alder pollen, was extracted and purified resulting in 10 mg of active crystalline material. Grove et al. (1979) identified brassinolide as the active component in brassins. Brassinolide is the first plant growth substance shown to have a steriodal structure and is the first naturally occurring steroid that has a seven-membered lactone ring as part of a fused ring system. Shortly after the identification of brassinolide by the USDA group it was also identified in *Distylium* extracts (Abe and Marumo

Figure 1.4.   Brassinosteroid-induced splitting of a bean stem caused by rapid growth (Courtesy of N. B. Mandava).

1991). Since this time brassinolide and many other related compounds have been found to be widely distributed in the plant kingdom and have many effects on plant growth and development (Cutler et al. 1991).

Salicylates

The ancient Greeks and American Indians discovered that the leaves and bark of the willow tree cured minor pain and fevers. In 1828, Johann Buchner, working in Munich, Germany, was the first to isolate trace amounts of salicin, which is the glucoside of salicyl alcohol and the major salicylate in willow bark (Weissmann 1991). Raffaele Piria (1838) named the active ingredient in willow bark *salicylic acid* (SA) from the Latin word *Salix*, meaning willow tree. In 1874, the first commercial production of SA began in Germany, and by 1898, *aspirin*, which is the trade name for acetylsalicylic acid, was introduced by the Bayer Co. (Raskin 1992a). There are many references in the literature in which plant scientists have used aspirin and salicylic acid interchangeably in experiments. However, it should be noted that aspirin has not been shown to be a natural plant product. It has probably been shown to be effective because acetylsalicylic acid is readily converted to salicylic acid in aqueous systems. Today, salicylic acid is known to be in a wide variety of plants and is thought by some to be an important plant growth substance (Raskin 1992a, 1992b).

Jasmonates

Demole et al. (1962) were the first to isolate (−)-jasmonic acid methyl ester from the essential oil of *Jasminum grandiflorum*. Today jasmonic acid (−)-JA and its stereoisomer (+)-7-iso-JA are the major representatives of jasmonates, although a number of other structurally related cyclopentane fatty acids have been identified. Initially, jasmonic acid was recognized for its growth inhibitory activity, but now it has been shown to be widespread within the plant kingdom, and renewed interest has focused on its ability to increase expression of specific plant genes, some of which occur in response to wounding (Staswick 1992; Sembdner and Parthier 1993).

FUNDAMENTAL TERMS AND CONCEPTS

Jacobs (1959) developed a simple yet elegant set of rules for assessing the extent and strength of evidence that a given chemical controls a process in an organism. Although these rules were derived for organizing and evaluating evidence and its application to auxins and phototropism, they can still be used today to determine whether a plant growth substance is involved in a given process. The capital letters of the first word of each of the six rules can be used as a memorization tool: PESIGS—presence/parallel variation, excision, substitution, isolation, generality, and specificity. These rules follow:

1. Presence/parallel variation: The chemical in question must be present in the organism and the parallel variation between the amount of the chemical and the relative activation of the process should be demonstrated. The latter requirement should be modified as designated by Davies (1987):

   a. The plant growth substance level should be measured in the exact tissue, cell, or even subcellular compartment where the response is occurring.

   b. The possible changes in sensitivity that may occur during development should be taken into account. One additional factor should be added to this section, the chemical should be found in a wide range of organisms.

2. Excision: Excision refers to the practice of removing the organ, tissue, or organelle known to be the source of the chemical under investigation and demonstrating a subsequent cessation of the process.

3. Substitution: A pure chemical can be substituted for the excised normal source and should lead to restoration of the process.

4. Isolation: One should isolate as much of the reacting system as is feasible and demonstrate that the chemical's effect is the same as in the less isolated system.

5. Generality: Demonstrate that the results obtained by a chemical apply in all similar situations.

6. Specificity: The chemical should be specific.

Today it is clear that the definition of a mammalian hormone cannot be directly applied to plant growth substances for many reasons. Although the term *plant hormone* has been used for many years, it is an inaccurate term, therefore the term *plant growth substance* will be used in this text. The following definition of a plant growth substance will be used in this text:

1. It must be a chemically characterized compound which is biosynthesized within the plant and broadly distributed within the plant kingdom.

2. It must show specific biological activity at extremely low concentrations.

3. It must be shown to play a fundamental role in regulating physiological phenomena *in vivo* in a dose/dependent and/or due to changes in sensitivity of the tissue during development.

This definition includes all of the generally accepted plant growth substances including auxins, gibberellins, cytokinins, abscisic acid, and ethylene, plus brassinosteroids, salicylates, jasmonates, and others by removing the requirement for transport. The requirement for transport in the initial definition has been deleted from the definition contained in this text because although many of the existing plant growth substances may be transported and have their action far from their site of synthesis this is not always true. For example, cytokinins can be produced in the roots and transported to the shoots where they are involved in delaying senescence, whereas ethylene may either be transported or promote changes at the site that it is synthesized. In order to strengthen the initial definition it is important to know the structure of a plant growth substance and that it is widely distributed in the plant kingdom, by including this information researchers will be able to show that a specific compound plays a fundamental role in regulating a response in vivo. By adding the statement that a plant growth substance acts in a dose-dependent manner and/or due to changes in sensitivity of the tissue further strengthens the definition because changes in the concentration of plant growth substances do not always correspond to an actual response. No matter what the definition of a plant growth substance, one should always keep in mind that it is not the individual compound causing a given response, but rather it is an interaction of all known substances plus many others not yet discovered. In addition, it must also be remembered that animal hormones are generally compounds like proteins having high information content, whereas plant growth substances are low-molecular-weight chemicals that only provide a turn-on or turn-off signal by stimulating a cascade of events in the cell leading to a response.

For the remainder of this section, some additional terms which will be helpful in future sections will be defined. The term *plant growth regulator* (PGR) has largely been used by agrichemical companies to designate synthetic regulators. The definition given by van Overbeek et al. (1954) still applies today; plant growth regulators are organic compounds other than nutrients (materials which supply either energy or essential mineral elements), which in small amounts promote, inhibit, or otherwise modify any physiological process in plants. Therefore, in this text we will use the term *plant growth regulator* to designate synthetic compounds (although it is a very broad definition which includes plant growth substances) and the term *plant growth substance* for naturally occurring compounds produced by the plant.

Abscisic acid and related inhibitors naturally inhibit or retard many plant physiological or biochemical processes, however, to use these chemicals to retard growth is not practical for many reasons, cost being one. Different types of synthetic organic chemicals which retard plant growth are used in agriculture today. The definition of a plant growth retardant is an organic compound that retards cell division and cell elongation in shoot tissues and thus regulates plant height physiologically without causing malformation of leaves and stems (Weaver 1972). Plants which are treated with a plant growth retardant typically have darker green leaves and flowering is affected indirectly. The growth of these plants is not completely suppressed but rather slowed down dramatically producing a more compact plant. Some examples of plant growth retardants are Cycocel (2-chloroethyl)trimethyl-ammonium chloride), Paclobutrazol (1-(4-chlorophenyl)-4,4-dimethyl-2-(1H-1,2,4-triazol-1-yl)pentan-3-ol) and Bonzi (2 RS, 3RS)-1-(4-chlorophenyl-4-dimethyl-2-(1,2,4-triazol-1-yl)penten-3-ol). Many plant growth retardants which have been shown to counteract the effects of gibberellins are called antigibberellins.

SUMMARY OF PLANT GROWTH SUBSTANCES

Auxins

The term *auxin* is derived from the Greek word *auxein* meaning to grow. Auxin is a generic term representing a class of compounds which are characterized by their capacity to induce elongation in shoot cells in the subapical region and resemble indole-3-acetic acid in physiological action. Auxins may, and generally do, affect other processes besides elongation, but elongation is considered critical. Auxins are generally acids with an unsaturated nucleus or their derivatives (van Overbeek et al. 1954). See Figure 1.5 for the structure of an auxin.

IAA was first discovered in human urine. Since this time there have been numerous reports indicating that IAA is found ubiquitously within the plant kingdom. In addition, it has been shown that IAA is synthesized in numerous species of nonseed plants including bacteria, fungi, and algae. Even though there have been suggestions' in the literature that there are other naturally occurring auxins which qualify as a plant growth substance, IAA is still the only naturally occurring indole which meets all of the criteria.

It is generally accepted that IAA is synthesized from tryptophan in pollen and actively growing tissues such as shoot meristems, leaf primordia, young expanding leaves, developing seeds, and fruits.

Auxins are involved in many physiological processes in plants. Extensively covering all auxin-mediated responses is beyond the scope of this book; therefore, the following are a list of known responses in plants that have been shown to be regulated by auxins: cellular elongation, phototropism, geotropism, apical

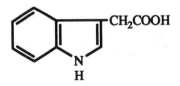

Figure 1.5. Structural formula for the naturally occurring auxin indole-3-acetic acid.

dominance, root initiation, ethylene production, fruit development, partheno-carpy, abscission, and sex expression.

At the present time there are many synthetic auxins which are used for a variety of purposes. Several examples along with their biological effects include: indole-3-butyric acid (root initiation), 2,4-dichlorophenoxyacetic acid (broad-leaf herbicide), and naphthaleneacetic acid (apple fruit thinning).

Gibberellins

Gibberellins are a class of plant growth substances having an ent-gibberel-lane skeleton (Figure 1.6) and stimulating cell division and/or cell elongation and other regulatory functions in the same manner as gibberellic acid (GA_3).

GA_3 was the first commercially available gibberellin. It has historically been called *gibberellic acid* and has been used as a standard in bioassay systems. For these reasons it is the representative structure for more then 90 gibberellins known today (Figure 1.7).

Gibberellins were first identified in the fungus *Gibberella fujikuroi*. Since their discovery gibberellins have been found to be widespread in the plant kingdom including angiosperms, gymnosperms, ferns, brown algae, green algae, fungi, and bacteria (Lang 1970).

It is generally accepted that gibberellins are synthesized via the mevalonic acid pathway in young actively growing shoots and developing seeds.

Gibberellins have been shown to be involved in many physiological processes in plants; however, the genus and/or species plus other factors will determine the specific gibberellin which is most effective in promoting a given response. The following are responses shown to be regulated by gibberellins: stem growth, bolting/flowering, seed germination, dormancy, sex expression, senescence, parthenocarpy, fruit set, and growth.

The substituted phthalimide AC 94, 377 (1-(3-chlorophthalimido) cyc-lohexanecarboxamide) has been shown to act similarly to gibberellins in a number of plant systems (Rodaway et al. 1991).

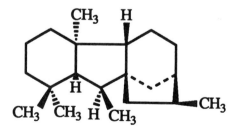

Figure 1.6. Structural formula for ent-gibberellane which is the backbone for all gibberellins.

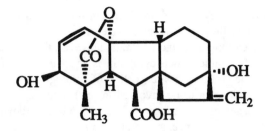

Figure 1.7. Structural formula for gibberellic acid (GA_3).

Cytokinins

Cytokinins are substituted adenine compounds that promote cell division and other growth regulatory functions in the same manner as kinetin (6-furfurylaminopurine).

The first cytokinin was isolated from autoclaved herring sperm DNA and called kinetin (6-furfurylaminopurine) because of its ability to promote cell division or *cytokinesis* in tobacco pith tissue. The first naturally occurring cytokinin was isolated from immature corn kernels and was called zeatin (6-(4-hydroxy-3-methyl-trans-2-butenyl-amino)purine). The most commonly found cytokinin in plants today is zeatin (Figure 1.8).

Cytokinins have been found in almost all higher plants, mosses, pathogenic and nonpathogenic fungi, bacteria, and also in the tRNA of numerous microorganisms and animal cells. At the present time there are more then 200 natural and synthetic cytokinins (Matsubara 1990).

Cytokinins are highest in meristematic regions and areas of continued growth potential including roots, young leaves, developing fruits, and seeds. They are thought to be synthesized in the roots and translocated to the shoots because there are many reports showing that cytokinins are found in the xylem

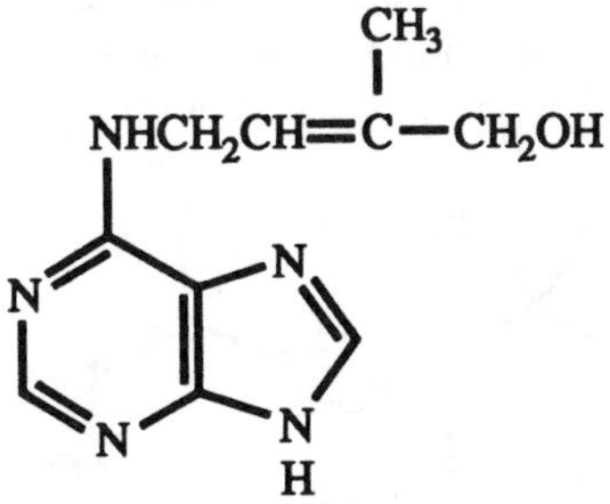

$$NHCH_2CH{=}\overset{\displaystyle CH_3}{\underset{\displaystyle |}{C}}{-}CH_2OH$$

Figure 1.8. Structural formula for zeatin (6-(hydroxy-3-methyl-trans-2-butenyl-amino)purine).

sap of plants. However, cytokinins are found in high levels in fruit and seed tissues suggesting that they may be synthesized there.

There are many cytokinins found in plants, however, the genus, species, and other factors will determine which cytokinin is most effective in affecting a given response. The following are a list of biological responses which cytokinins have been shown to be involved in: cell division, organ formation, cell and organ enlargement, retardation of chlorophyll breakdown, chloroplast development, delay of senescence, stomatal opening and closing, bud and shoot development, and preferential translocation of nutrients and organic substances to cytokinin treated tissues.

There are many synthetic cytokinins known today; however, three common examples are: kinetin (6-furfurylaminopurine), BA (6-benzylaminopurine), and BPA (6-(benzylamino)-9-(2-tetrahydropyranyl)-9H-purine).

Abscisic Acid

Abscisic acid (ABA) is a 15-carbon sesquiterpenoid which is partially produced in the chloroplasts and other plastids via the mevalonic acid pathway. It is a naturally occurring compound which is involved in many aspects of plant growth and development, some of which are inhibitory (growth) and some which are promotive (normal embryogenesis, seed storage proteins). See Figure 1.9 for the structure of abscisic acid.

Abscisic acid is widely distributed throughout the plant kingdom, it has been found in higher plants, mosses, green algae, fungi, and recently in rat brains, however, it has not yet been identified in bacteria.

Abscisic acid is synthesized via the mevalonic pathway in the leaves (chloroplasts and other plastids) and this response is accentuated by stress.

Abscisic acid acts as a signal that a plant is undergoing stress, however, it is also involved in normal physiological processes in plants. The following is a list of responses where ABA has been shown to have an effect: stomatal opening and closing, water stress signal, defense against salt stress and cold stress, normal embryogenesis, induction of seed storage proteins, dormancy, abscission, seed germination, growth, and geotropism.

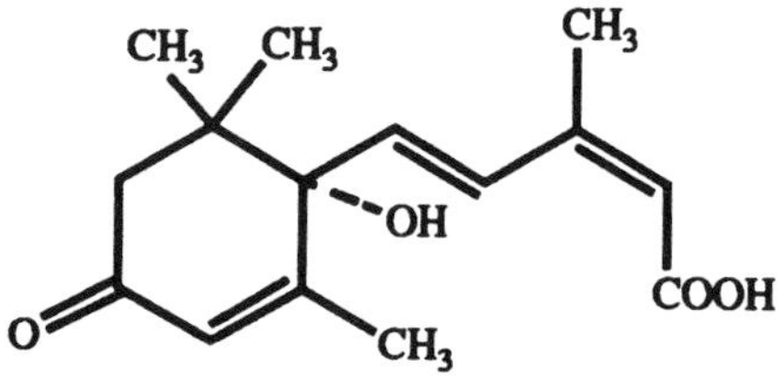

Figure 1.9.   Structural formula for abscisic acid (ABA).

Rapid metabolism and photodestruction limit the usefulness of ABA. While it is important that synthetic derivatives of ABA are produced there are none commonly used at the present time, however, research is ongoing in this area.

Ethylene

Ethylene is a simple unsaturated hydrocarbon which promotes fruit ripening and causes a triple response in etiolated pea seedlings including inhibition of elongation, increased radial expansion, and horizontal growth of stems in response to gravity.

Ethylene has the simplest structure of all known plant growth substances (see Figure 1.10).

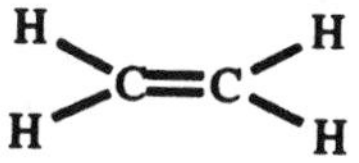

$$H{\diagdown}{\diagup}H$$
$$C{=}C$$
$$H{\diagup}{\diagdown}H$$

Figure 1.10.   Structural formula for ethylene.

Ethylene is produced in all higher plants, however, at the present time it is unclear as to its range of production among lower plants. Ethylene is produced from methionine by essentially all parts of higher plants; however, production varies with the type of tissue and the stage of development.

Ethylene is involved in many physiological processes in plants from seed germination through senescence and death of the plant. The following are effects ethylene has been shown to be involved in: fruit ripening, dormancy, abscission, flowering, senescence, shoot, and root growth and apical dominance.

2-chloroethylphosphoric acid (Ethephon, Ethrel) is a synthetic form of ethylene which in its liquid state at the proper pH does not yield ethylene; however, when the pH is elevated it breaks down to form ethylene.

Brassinosteroids

Brassins are crude lipoidal extracts from rape pollen causing swelling and splitting of bean second internodes. Brassinolide is a steroids which are the active component found in brassins causing swelling and splitting of bean second internodes. Brassinosteroids are a class of steroid compounds having activity similar to brassinolide in the bean second internode bioassay. For the structure of brassinolide see Figure 1.11.

Brassinosteroids have been found in a wide range of plants including dicots, monocots, gymnosperms, and algae. Over 60 kinds of brassinosteroids have been found; however, 31 have been fully characterized including 29 free and

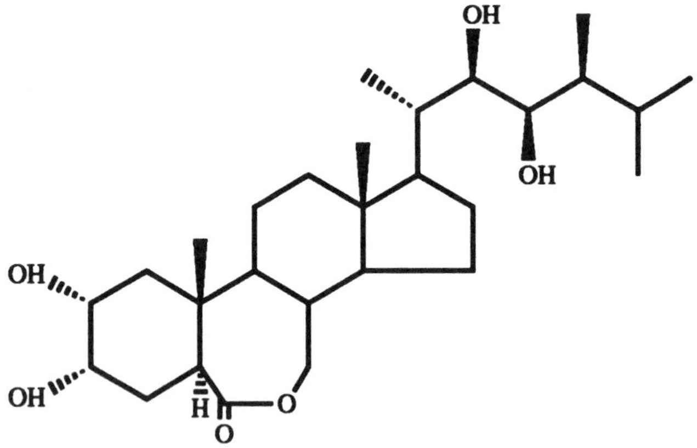

Figure 1.11.   Structural formula for brassinolide (2α, 3α, 22α, 23α,-tetrahydroxy-24α-methyl-B-homo-7-oxa-5α-cholestan-6-one).

two conjugates. A sequential numerical suffix has been established for brassinosteroids occurring in nature: BR_1 is used to designate brassinolide because it was the first discovered and so on. Among the naturally occurring brassinosteroids, brassinolide and castasterone are considered to be the most important brassinosteroids because of their wide distribution as well as their biological activity (Kim 1991).

At the present time it is not clear where brassinosteroids are synthesized in plants; however, they have been detected in many parts of the plant such as pollen, leaves, flowers, seeds, shoots, galls, and stems, but not in plant roots. However, it is likely that future studies will show their presence in roots (Kim 1991; Sasse 1991).

Since the discovery of brassinolide, its biological activity has been evaluated in numerous test systems and has been shown to be involved in the following: enhanced resistance to chilling, disease, herbicides, and salt stress; increased crop yields, elongation, and seed germination; decreased fruit abortion and drop; antiecdysteroid activity; and inhibition of root growth and development (Cutler et al. 1991; Iwahori et al. 1990).

24-epibrassinolide is an analogue of brassinolide which has potential use as a synthetic form of brassinolide because of its high biological activity and relatively easy preparation from brassicasterol (Ikekawa and Zhao 1991).

Salicylates

Salicylates are a class of compounds having activity similar to salicylic acid (ortho-hydroxybenzoic acid) which is a plant phenolic. Phenolics are defined

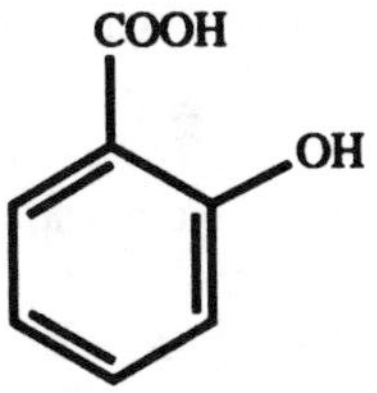

Figure 1.12. Structural formula for salicylic acid (orthohydroxy benzoic acid).

as substances that possess an aromatic ring bearing a hydroxyl group or its functional derivative (Figure 1.12).

Salicylic acid has been found to be widely distributed within the plant kingdom, thus far it has been identified in more then 34 plant species.

Salicylic acid has been identified in leaves and reproductive structures of plants, with the highest levels reported to date in the inflorescence of thermogenic plants and plants infected by necrotizing pathogens (Raskin 1992a, 1992b).

Salicylic acid has been shown to have an effect on a variety of plant processes; however, flowering, heat production in thermogenic plants, and promotion of disease resistance are processes where salicylic acid has its major effect.

Aspirin (acetylsalicylic acid) acts in the same manner as salicylic acid because it is readily converted to salicylic acid in aqueous solutions in plant as well as animal systems.

Jasmonates

Jasmonates are a specific class of cyclopentanone compounds with activity similar to $(-)$-jasmonic acid and/or its methyl ester. See Figure 1.13 for the structure of jasmonates.

Jasmonates have been detected in 206 plant species representing 150 families including ferns, mosses, and fungi suggesting that they are ubiquitously distributed throughout the plant kingdom (Meyer et al. 1984; Sembdner and Gross 1986; Sembdner and Parthier 1993).

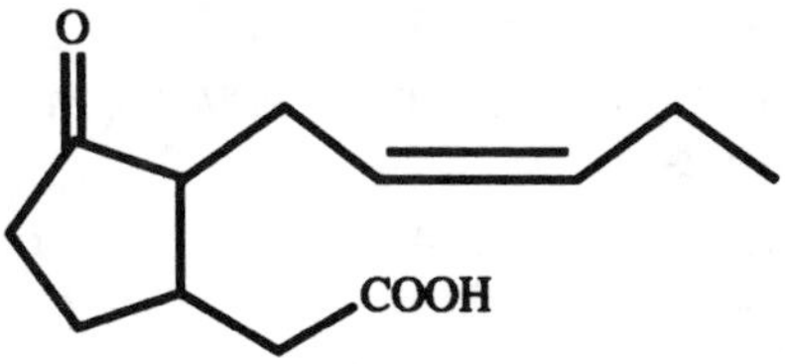

Figure 1.13. Structural formula for jasmonic acid.

Although there is little known about jasmonate biosynthesis evidence has been presented that the stem apex, young leaves, immature fruits and root tips contain the highest levels of jasmonates (Sembdner and Parthier 1993).

Jasmonates have been shown to elicit a wide variety of physiological responses in plants. When jasmonic acid is exogenously applied to plants it promotes senescence, petiole abscission, root formation, tendril coiling, ethylene synthesis, and β-carotene synthesis (Staswick 1992). In addition to its promotive effects, jasmonic acid has been shown to inhibit seed germination, callus growth, root growth, chlorophyll production and pollen germination (Anderson 1989; Parthier 1990; Vick and Zimmerman 1986).

Jasmonic acid has also been shown to induce gene expression in many plant species. Proteins induced by jasmonic acid with a known function are vegetable storage proteins of soybean (Staswick 1990); wound-induced proteinase inhibitors of tomato and potato (Farmer and Ryan 1992), and seed storage proteins and oil body membrane proteins (oleosins) (Wilen et al. 1991).

Endogenous jasmonic acid levels have been shown to increase in response to external stimuli such as wounding, mechanical forces, elicitors from pathogen attack and osmotic stress (Sembdner and Parthier 1993). At the present time there are no synthetic forms of jasmonates commonly used.

REFERENCES

Abe, H. and Marumo, S. (1991). "Brassinosteroids in leaves of *Distylium racemosum* Sieb. et Zucc.: The beginning of brassinosterold research in Japan". In *Brassinosteroids. Chemistry, Bioactivity and Applications*, eds., H. G. Cutler, T. Yokota, and G. Adam, American Chemical Society, Washington, DC, pp. 18–25.

Addicott, F. T., Lyon, J. L., Ohkuma, K., Thiessen, W. E., Carns, H. R., Smith, O. E., Cornforth, J. W., Milborrow, B. V., Ryback, G., and Wareing, P. F. (1968). "Abscisic acid: A new name for abscisin II (dormin)". *Science* 159:1493.

Anderson, J. M. (1989). "Membrane derived fatty acids as precursors to second messengers". In *Second Messengers in Plant Growth and Development*, eds., W. F. Boss and D. J. Morre, Alan R. Liss, New York, pp. 181–212.

Bayliss, W. M. and Starling, E. (1904). "The chemical regulation of the secretory process". *Proc. Royal Soc. (Series B)* 73:310–322.

Beyerinck, M. W. (1888). "Uber das Cecidium von *Nematus capreae* auf *Salix amygdalina*". *Zeitschr. Bot.* 46:1–11.

Biale, J. B., Young, R. E., and Olmstead, A. J. (1954). "Fruit respiration and ethylene production". *Plant Physiol.* 29:168–174.

Boysen-Jensen, P. (1913). "Uber die Leitung des phototropischen Reizes in der *Avena*-koleoptile". *Ber. Deut. Bot. Ges.* 31:559–566.

Brian, P. W., Elson, G. W., Hemming, H. G., and Radley, M. (1954). "The plant-growth promoting properties of gibberellic acid, a metabolic product of the fungus *Gibberella fujikuroi*", *J. Sci. Food. Agr.* 5:602–612.

Burg, S. P. and Stolwijk, J. A. A. (1959). "A highly sensitive katharometer and its application to the measurement of ethylene and other gases of biological importance", *J. Biochem. Microbiol.Technol. Eng.* 1:245–259.

Burg, S. P. and Thimann, K. V. (1959). "The physiology of ethylene formation in apples", *Proc. Natl. Acad. Sci. USA* 45:335–44.

Cornforth, J. W., Milborrow, B. V., Ryback, G., and Wareing, P. F. (1965a). "Identity of sycamore 'dormin' with abscisin II". *Nature* 205:1269–1270.

Cornforth, J. W., Milborrow, B. V., and Ryback, G. (1965b). "Synthesis of ($\pm$)abscisin II". *Nature* 206:715.

Crocker, W., Hitchcock, A. E., and Zimmerman, P. W. (1935). "Similarities in the effects of ethylene and the plant auxins". *Contrib. Boyce Thompson Inst.* 7:231–48.

Cutler, H. G., Yokota, T., and Adam, G. (1991). *Brassinosteroids. Chemistry Bioactivity and Applications*, American Chemical Society, Washington, DC.

Darwin, C. R. (1880). *The Power of Movement in Plants*, Murray, London.

Davies, P. J. (1987). *Plant Hormones and Their Role in Plant Growth and Development*, Martinus Nijhoff Publishers, Boston.

Demole, E., Lederer, E., and Mercier, D. (1962). "Isolement et determination de la structure du jasmonate de methyle, constitutuant odorant characteristique de l'essence de jasmin". *Helv. Chim. Acta* 45:675–685.

Denny, F. E. (1923). "Method of coloring citrus fruits". *U.S. patent 1,475,938*.

Denny, F. E. (1924). "Hastening the coloration of lemons". *J. Agr. Res.* 27:757–769.

Devlin, R. M. and Witham, F. H. (1983). *Plant Physiology*, Wadworth Publishing Co. Belmont, CA.

Doubt, S. L. (1917). "The response of plants to illuminating gas". *Bot. Gaz.* 63:209–224.

du Monceau, D. (1758). *La physique des arbres. Volume I.*

Eagles, C. F. and Wareing, P. F. (1963). "Experimental induction of dormancy in *Betula pubescens*". *Nature* 199:874.

Farmer, E. E. and Ryan, C. A. (1992). "Octadecanoid precursors of jasmonic acid activate the synthesis of wound-inducible proteinase inhibitors". *Plant Cell* 4:129–134.

Fitting, H. (1910). "Weitere entwicklungsphysiologische Untersuchungen an Orchideenbluten". *Zeitschr. Bot.* 2:225–267.

Fitting, H. (1907). "Die Leitung tropistischer Reize in parallelotropen Pflanzenteilen", *Jahrb. Wiss. Bot.* 44:177–253.

Galil J. (1968). "An ancient technique for ripening sycamore fruit in East-Mediterranean countries". *Econ. Bot.* 22:178–190.

Gane, R. (1934). "Production of ethylene by some ripening fruits". *Nature* 134:1008.

Girardin, J. P. L. (1864). "Einfluss des leuchtgases auf die promenaden und strassen-baume". *Jahreb. Fortschritte Agrik.Chemie* 7:199–200.

Grove, M. D., Spencer, F. G., Rohwedder, W. K., Mandava, N. B., and Worley, J. F. (1979). "A unique plant growth promoting steroid from *Brassica napus* pollen". *Nature* 281:216–217.

Haagen-Smit, A. J., Dandliker, W. B., Wittwer, S. H., and Murneek, A. E. (1946). "Isolation of 3-indoleacetic acid from immature corn kernels". *Amer. J. Bot.* 33:118–120.

Haberlandt, G. (1913). "Zur Physiologie der Zellteilung". *Sitzber. K. Preuss. Akad. Wiss.* 318.

Hall, R. H. and deRopp, R. S. (1955). "Formation of 6-furfurylaminopurine from DNA breakdown products". *J. Am. Chem. Soc.* 77:6400.

Hansen, E. (1943). "Relation of ethylene production to respiration and ripening of premature pears". *Proc. Am. Soc. Hort. Sci.* 43:69–72.

Hori, S. (1898). "Some observations on 'Bakanae' disease of the rice plant". *Mem. Agric. Res. Sta. (Tokyo)* 12:110–119.

Ikekawa, N. and Zhao, Y.-J. (1991). "Application of 24-epibrassinolide in agriculture". In *Brassinosteroids. Chemistry, Bioactivity, and Applications*, eds., H. G. Cutler, T. Yokota, and G. Adam, American Chemical Society, Washington, DC, pp. 280–291.

Iwahori, S., Tominaga, S., and Higuchi, S. (1990). "Retardation of abscission of citrus leaf and fruitlet explants by brassinolide". *Plant Growth Reg.* 9:119–125.

Jablonski, J. R. and Skoog, F. (1954). "Cell enlargement and cell division in excised tobacco pith tissue". *Physiol. Plant.* 7:16.

Jacobs, W. P. (1959). "What substance normally controls a given biological process? I. Formulation of some rules". *Dev. Biol.* 1:527–533.

Kidd, F. and West, C. (1945). "Respiratory activity and duration of life of apples gathered at different stages of development and subsequently maintained at a constant temperature". *Plant Physiol.* 20:467–504.

Kim, S.-K. (1991). "Natural occurrences of brassinosteroids". In *Brassinosteroids. Chemistry, Bioactivity, and Applications*, eds., H. G. Cutler, T. Yokota, and G. Adam, American Chemical Society, Washington, DC, pp. 26–35.

Kögl, F. and Haagen-Smit, A. J. (1931). "Uber die Chemie des Wuchsstoffs K. Akad. Wetenschap. Amsterdam". *Proc. Sect. Sci.* 34:1411–1416.

Kögl, F. and Kostermans, D. G. F. R. (1934). "Heteroauxin als Stoffwechselprodukt niederer pflanzlicher Organismen, Isolierung aus Hefe". *Zeitschr. Physiol. Chem.* 228:113–121.

Kögl, F., Haagen-Smit, A. J., and Erxleben, H. (1934a). "Uber die Isolierung der Auxine a und b aus pflanzlichen Materialien, IX Mitteilung". *Zeitschr. Physiol. Chem.* 225:215–229.

Kögl, G., Haagen-Smit, A. J., and Erxleben, H. (1934b). "Uber ein neues Auxin (Heteroauxin) aus Harn, XI Mitteilung". *Zeithschr. Physiol. Chem.* 228:90–103.

Kraus, E. J. and Kraybill, H. R. (1918). "Vegetation and reproduction with special reference to the tomato". *Ore. Agr. Expt. Sta. Bull.* 149:5.

Kurosawa, E. (1926). "Experimental studies on the nature of the substance secreted by the 'bakanae' fungus". *Nat. Hist. Soc. Formosa* 16:213–227.

Lang, A. (1970). "Gibberellins: Structure and metabolism". *Annu. Rev. Plant Physiol.* 21:537–570.

Letham, D. S. (1963). "Zeatin, a factor inducing cell division isolated from *Zea mays*". *Life Sci.* 2:569–573.

Letham, D. S., Shannon, J. S., and McDonald, I. R. (1964). "The structure of zeatin, a factor inducing cell division". *Proc. Chem. Soc.* 230–231.

Liu, W. C. and Carns, H. R. (1961). "Isolation of abscisin, an abscission accelerating substance". *Science* 134:384.

MacMillan, J. and Takahashi, N. (1968). "Proposed procedure for the allocation of trivial names to the gibberellins". *Nature* 217:170–171.

Matsubara, S. (1990). "Structure-activity relationships of cytokinins". *Plant Sci.* 9:17–57.

Meyer, A., Miersch, O., Buttner, C., Dathe, W., and Sembdner G. (1984). "Occurrence of the plant growth regulator jasmonic acid in plants". *J. Plant Growth Reg.* 3:1–8.

Milborrow, B. V. and Pryce, R. J. (1973). "The brassins". *Nature* 243:46.

Miller, C. O. (1961). "A kinetin-like compound in maize". *Proc. Natl. Acad. Sci. USA* 47:170–174.

Miller, C. O., Skoog, F., Okumura, F. S., von Saltza, M. H., and Strong, F. M. (1955a). "Structure and synthesis of kinetin". *J. Am. Chem. Soc.* 77:2662–2663.

Miller, C. O., Skoog, F., von Saltza, M. H., and Strong, F. M. (1955b). "Kinetin, a cell division factor from deoxyribonucleic acid". *J. Am. Chem. Soc.* 77:1392.

Miller, E. V. (1947). "The story of ethylene". *Sci. Monthly* 65:335–342.

Mitchell, J. W. and Gregory, L. E. (1972). "Enhancement of overall growth, a new response to brassins". *Nature* 239:254.

Molisch, H. (1884). "Sitzungsberitche der kaiserl". *Akademie der Wissenschaften (Wein)* 90:111–196.

Neljubow, D. N. (1901). "Uber die horizontale nutation der stengel von *Pisum sativum* und einiger anderen". *Pflanzen Beitrage und Botanik Zentralblatt* 10:128–139.

Ohkuma, K. (1965). "Synthesis of some analogs of abscisin II". *Agr. Biol. Chem.* 29:962–964.

Ohkuma, K., Addicott, F. T., Smith, O. E., and Thiessen, W. E. (1965). "The structure of abscisin II". *Tetrahedron Lett.* 29:2529–2535.

Ohkuma, K., Lyon, J. L., Addicott, F. T., and Smith, O. E. (1963). "Abscisin II, an abscission-accelerating substance from young cotton fruit". *Science* 142:1592–1593.

Paal, A. (1918). "Uber phototropische Reizleitung". *Jahrb. Wiss. Bot.* 58:406–458.

Parthier, B. (1990). "Jasmonates: Hormonal regulators or stress factors in leaf senescence?" *J. Plant Growth Reg.* 9:57–63.

Radley, M. (1956). "Occurrence of substances similar to gibberellic acid in higher plants". *Nature* 178:1070–1071.

Raskin, I. (1992a). "Role of salicylic acid in plants". *Annu. Rev. Plant Physiol. Plant Mol. Biol.* 43:439–463.

Raskin, I. (1992b). "Salicylate, a new plant hormone". *Plant Physiol.* 99:799–803.

Rodaway, S. J., Gates, D. W. and Brindle, C. (1991). "Control of early seedling growth in varietal lines of hexaploid wheat (*Triticum aestivum*), durum wheat (*Triticum durum*), and barley (*Hordeum vulgare*) in response to the phthalimide growth regulant, AC 94,377". *Plant Growth Reg.* 10:243–259.

Rothert, W. (1894). "Uber heliotropismus". *Beitr. Biol. Pflanzen (Cohn)* 7:1–212.

Salkowski, E. (1885). "Uber das verhalten der skatolcarbonsaure im organismus". *Zeitschr. Physiol. Chem.* 9:23–33.

Sasse, J. M. (1991). "Brassinosteroids—Are they endogenous plant hormones?". *PGRSA Quarterly* 19:1–18.

Sawada, K. (1912). "Disease of agricultural products in Japan". *Formosan Agr. Rev.* 36:10.

Sembdner, G. and Gross, D. (1986). "Plant growth substances of plant and microbial origin". In *Plant Growth Substances*, ed., M. Bopp, Springer, Berlin, pp. 139–147.

Sembdner, G. and Parthier, B. (1993). "The biochemistry and the physiology and molecular actions of jasmonates". *Annu. Rev. Plant Physiol. Plant Mol. Biol.* 44:569–589.

Shaw, G. and Wilson, D. V. (1964). "Synthesis of zeatin". *Proc. Chem. Soc.* 231:1.

Soding, H. (1925). "Zur kenntnis der wuchshormone in der haferkoleoptile". *Jahrb. Wiss. Bot.* 64:587–603.

Staswick, P. E. (1990). "Novel regulation of vegetative storage protein genes". *Plant Cell* 2:1–6.

Staswick, Paul E. (1992). "Jasmonate, genes, and fragrant signals". *Plant Physiol.* 99:804–807.

Stodola, F. H., Raper, K. B., Fennell, D. I., Conway, H. F., Sohns, V. E., Langford, C. T. and Jackson, R. W. (1955). "The microbiological production of gibberellins A and X". *Arch. Biochem. Biophys.* 54:240–245.

Takahashi, N., Kitamura, H., Kawarada, A., Stea Y., Takai, M., Tamura, S., and Sumiki, Y. (1955). "Isolation of gibberellins and their properties". *Bull. Agric. Chem. Soc. Japan* 19:267–277.

Takahashi, N., Phinney, B. O., and MacMillan J. (1991). *Gibberellins*, Springer-Verlag, New York.

Takahashi, N., Seta, Y., Kitamura, H. and Sumiki Y. (1957). "A new gibberellin, gibberellin A_4". *Bull. Agric. Chem. Soc. Japan* 21:396–398.

Thimann, K. V. (1935). "On the plant growth hormone produced by *Rhizopus suinus*". *J. Biol. Chem.* 109:279–291.

Trewavas, A. (1981). "How do plant growth substances act? I". *Plant Cell Environment* 4:203–228.

van Overbeek, J., Conklin, M. E., and Blakeslee, A. F. (1941). "Factors in coconut milk essential for growth and development of *Datura* embryos". *Science* 94:350.

van Overbeek, J., Siu, R., and Haagen-Smit, A. J. (1944). "Factors affecting the growth of *Datura* embryos *in vitro*". *Am. J. Bot.* 31:219.

van Overbeek, J., Tukey, H. B., Went, F. W. and Muir, R. M. (1954). "Nomenclature of chemical plant regulators". *Plant Physiol.* 29:307–308.

Vick, B. A. and Zimmerman, D. C. (1986). "Characterization of 12-oxo-phytodienoic acid reductase in corn. The jasmonic acid pathway". *Plant Physiol.* 80:202–205.

Vliegenthart, J. A. and Vliegenthart, J. F. G. (1966). "Reinvestigation of authentic samples of auxin a and b and related products by mass spectometry". *Rucueil* 85:1266–1272.

von Sachs, J. (1880). "Stoff und Form der Pflanzenorgane. I". *Arb. Bot. Inst. Wurzburg* 2:452–488.

Weaver, R. J. (1972). *Plant Growth Substances in Agriculture*, W. H. Freeman and Company, San Francisco.

Weissmann, G. (1991). "Aspirin". *Sci. Am.* 264:84–90.

Went, F. W. (1926). "On growth-accelerating substances in the coleoptile of *Avena sativa*". *Proc. Kon. Ned. Akad. Wet.* 30:10–19.

Went, F. W. (1928). "Wuchsstoff und Wachstum". *Rec. Trav. Bot. Neerland.* 25:1–116.

Wilen, R. W., van Rooijen, G. J., Pearce, D. W., Pharis, R. P., Holbrook, L. A., and Moloney, M. M. (1991). "Effects of jasmonic acid on embryo specific processes in *Brassica* and *Linum* oilseeds". *Plant Physiol.* 95:399–405.

Yabuta, T. (1935). "Biochemistry of the 'bakanae' fungus of rice". *Agr. Hort. (Tokyo)* 10:17–22.

Yabuta, T. and Sumiki, Y. (1938). "On the crystal of gibberellin, a substance to promote plant growth". *J. Agric. Chem. Soc. Japan* 14:1526.

CHAPTER 2

Methodology for the Extraction, Purification, and Determination of Plant Growth Substances

Plant growth substances are involved in nearly all aspects of the plant's life cycle as deduced from the fact that exogenous applications by them profoundly affect plant responses. However, it is difficult to detect plant growth substances within the plant for numerous reasons. At the present time it is not known whether plant growth substances act at certain levels in a dose-dependent manner or if they remain constant, while changes in tissue sensitivity (receptors) are more closely related to a given response. Our knowledge in the area of plant growth substances is still quite limited and the major reason is that the techniques for their analysis are still very cumbersome. There are many methods for extracting, purifying, and quantifying plant growth substances; however, it is not the intent in this section to provide an exhaustive review of the literature in the area. This chapter is intended to be a basic outline of all steps from tissue preparation through final identification and quantification of plant growth substances. However, for detailed protocols see Rivier and Crozier (1987) and Hedden (1993).

EXTRACTION METHODS

Diffusion Method

This method can only be used for fresh intact tissues or plant parts. When using this method the tissue is first dipped in a gelatin solution warmed to 30°C to prevent moisture loss. In order to prevent inactivation of plant growth substances at the cut surface, butylated hydroxytoluene (BHT) (1–2%) is incorporated into the agar where the tips are inserted acting as an antioxidant. The tissue once placed on agar is put in a humid atmosphere from a few minutes to a few hours depending upon the type of tissue used (Figure 2.1).

After a designated period of time the agar block can then be used for the desired bioassay. This is a very good method for evaluating relative plant growth substance levels in different plant parts. It must be noted that this procedure will only give a relative indication of what is present since other promoters or inhibitors at the cut surface can interfere with the bioassay being used. This is a poor method if the compound is to be analyzed chemically because plant growth substances are generally found in small amounts, thereby making separation and quantification a difficult process.

Solvent Extraction

Tissue Preparation.   Tissue handling following harvest of plant samples is extremely important in order to minimize artifacts and loss of plant growth substances being extracted. It must be remembered that these compounds are found in plant tissues at extremely low levels, therefore, any loss due to enzymatic degradation or interconversion could lead to incorrect results. The most effective way to process samples following harvest is to immediately freeze the tissue in liquid nitrogen, freeze-dry and store at −80°C or lower under anhydrous conditions, this will minimize any potential problems. If it is not possible to process the tissue in this manner then modifications can be made, although any changes will reduce storage efficiency.

Extraction.   There are many procedures in the literature for the extraction of plant growth substances, some of which work better for a given compound in a specific tissue than others. Therefore, a general rule of thumb is to evaluate the different methods available and determine which is best suited for your particular situation. In this section some of the most commonly used methods will be summarized in general terms, for more details see Rivier and Crozier (1987) and Hedden (1993).

There are a wide variety of solvents which can be used for the extraction of plant growth substances: methanol or ethanol can be used to extract IAA, ABA, GA, BR, cytokinins, SA, and JA; acetone can be used to extract IAA,

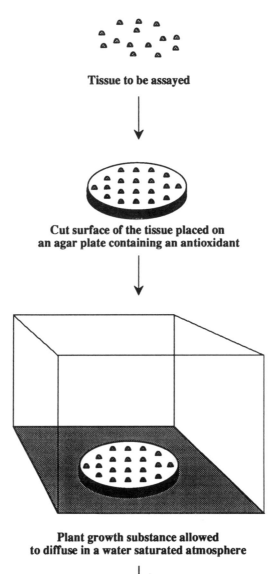

Figure 2.1. Schematic representation for the diffusion method of extracting plant growth substances from fresh tissues.

ABA, and GA; isopropanol or chloroform can be used to extract BR; and the Bieleski medium (Bieleski 1964) which is a mixture of methanol/chloroform/ 90% formic acid/water (12:5:1:2 v/v) or perchloric acid (Laloue et al. 1974), can be used to extract cytokinins.

The most commonly used extraction cocktail for all of the plant growth substances is 80% methanol plus an antioxidant such as butylated hydroxytoluene (BHT). A Waring blender or Polytron homogenizer is commonly used to macerate the tissue in an excess of cold solvent. The sample is spiked with an internal standard to correct for losses encountered during extraction and subsequent purification steps. It is then filtered and the tissue reextracted to assure that maximum recovery has been achieved. Note that the number of extractions varies between tissues and other factors; therefore, the number required to remove the highest percentage of the plant growth substance being analyzed should be tested experimentally. The filtrates should be combined and reduced in vacuo (under vacuum) either to dryness or to an aqueous solution prior to solvent partitioning and/or purification using one or more of the procedures outlined in the next section. Following the removal of organic solvent from the extract a considerable amount of solid debris is often found in the aqueous phase, and following acidic partitioning this debris can dissolve in the organic phase, thereby contaminating it with unwanted compounds. By adding a simple filtration or centrifuge step to remove debris from the aqueous residue, it is possible to reduce the dry weight by up to 80%. It should be noted that following extraction the sample is very crude and contains promoters as well as inhibitors plus a lot of other contaminants. In many cases the extract may show no activity in a bioassay because the action of the plant growth substance may be completely masked by interfering compounds.

PURIFICATION OF EXTRACTS

The first purification step following extraction of plant tissues has traditionally been solvent partitioning, which involves partitioning between an aqueous phase and an immiscible organic solvent (the specific solvent used will be determined by the plant growth substance being extracted, tissue used, and other factors) (Rivier and Crozier 1987). There are many variations on solvent partitioning procedures in the literature for plant growth substances. Instead of explaining the procedure in detail, an example for acidic plant growth substances will be given (see Rivier and Crozier (1987) for more details). For acidic plant growth substances the pH of the water phase from the crude extract should be adjusted to approximately 2.5 and the solution placed into a separatory flask together with equal amounts of diethyl ether and shaken gently. The flask is allowed to stand for a few minutes to enable the phases to separate. If an emulsion forms because shaking was too vigorous the separatory funnel should be placed in the freezer until separation of the two phases is complete. A portion of the acidic plant growth substances will then go from the water phase to the diethyl ether phase. The process is generally repeated about three or four more times to remove all of the acidic plant growth substances from the liquid phase; however, the number of times required to obtain maximum recovery should be evaluated experimentally. The liquid phase is then discarded and the diethyl ether phase taken to dryness in vacuo. Although solvent partitioning has been used traditionally it is probably not the best in all cases, in fact, it is now being replaced with solid-liquid extraction procedures. The use of a solid phase packed into small disposable columns (Powell and Maybee 1985) has led to rapid sample preparation, good recoveries and a requirement for small solvent volume. Procedures using these minicolumns have been developed for most of the plant growth substances (Hedden 1993), thereby providing an excellent alternative to solvent partitioning. Additional forms of purification which have successfully been used for the purification of plant growth substances are paper, thin layer, column, and gas-liquid and high-performance liquid chromatography. The chromatographic procedures previously mentioned can, in some cases, be used alone in order to obtain sufficient purification for quantification, however, most commonly various combinations are required (Rivier and Crozier 1987; Hedden 1993).

QUANTIFICATION OF PLANT GROWTH SUBSTANCES

Following the extraction and purification of a given plant growth substance it must be quantified. Plant growth substances can be quantified with bioassays (Yopp et al. 1986), physiochemical methods (gas-liquid-chromatography or high-performance liquid chromatography), or immunological methods (enzyme-linked immunosorbent assay (ELISA) or radioimmunoassays (RIA)) (Rivier and Crozier 1987; Hedden 1993) each of which has advantages and disadvantages associated with it. It should be noted that it is extremely important that a standard curve is run no matter which method is used; in fact, this step is commonly done before running a sample. It may sound like common sense, but it is a common mistake among researchers when starting in the quantification of plant growth substances.

Bioassays

A bioassay may be defined as a biological system used to test the activity of substance with respect to a physiological response. In order to be a useful bioassay the following criteria is necessary:

1. It must only be specific for the compound being assayed.
2. It must be very sensitive in order to detect the small amounts of a given plant growth substance found in plant tissues.
3. It must be quick and easy to obtain large amounts of uniform plant tissue. After the tissue is obtained the response to a specific plant growth substance must also be quick and easy.
4. The chemical being assayed for or related compounds must be present at very low levels or absent from the plant material.

There are numerous bioassay systems for each of the known plant growth substances; however, it is important to select the bioassay which suits your particular situation. Yopp et al. (1986) has divided bioassays into five major groups from a historical perspective as follows:

1. Diagnostic bioassays: Bioassays of extreme specificity for use in the positive identification of a particular plant growth substance.
2. Bioassays for determination of structure-activity relationships: Bioassays displaying differences in sensitivity of response to structurally different members of the same class of plant growth substances.
3. Detection and screening bioassays: Bioassays of great sensitivity and rapidity of response, coupled with adequate specificity, to allow their use in the detection of low concentrations of plant growth substances in fractions from various chromatographic methods.
4. Total class activity bioassays: Bioassays of equal or nearly equal sensitivity to structurally diverse members of a class of plant growth substances to allow determination of total activity of that particular class in an extract.
5. Simple requirement bioassays: Bioassays that require very little specialized equipment and space and employ low-cost, easily obtained plant material.

In the following sections selected bioassays for auxins, gibberellins, cytokinins, abscisic acid, ethylene, and brassinosteroids will be discussed briefly; for more details on these bioassays and others see Yopp et al. (1986).

Selected Auxin Bioassays

1. *Avena* (oat) coleoptile curvature bioassay: The principle of this bioassay is the ability of auxin to promote curvature and it is based on elongation. This assay requires strict and rapid polar transport in the basipetal (tip to base) direction and little lateral diffusion.
2. *Nicotiana* (tobacco) gene expression bioassay: This bioassay is based on the expression of a chimeric gene in transformed tobacco (*Nicotiana tabacum*) mesophyll protoplasts in response to both auxins and cytokinins. Quantitation of each plant growth substance is based on a color reaction visualized by light (Boerjan et al. 1992).

3. *Avena* (oat) coleoptile segment straight growth bioassay: This bioassay is based on the ability of auxin to promote cell elongation. It is not as specific as the *Avena* curvature bioassay; however, it does not require polar transport of the compound being assayed.

4. *Phaseolus* (bean) internode bioassay: This bioassay is based on the ability of auxin to promote curvature. It is probably the easiest of all curvature bioassays to use because it is not as sensitive to temperature and can be performed under ambient light conditions.

5. *Vigna* (mung bean) adventitious root induction bioassay: This bioassay is based on the ability of auxin to stimulate adventitious roots on stem cuttings.

Selected Gibberellin Bioassays

1. *Hordeum* (barley) endosperm reducing sugar production bioassay: This bioassay is based on the ability of gibberellins to stimulate α-amylase activity which produces reducing sugars.

2. *Rumex* (broad leaf dock) leaf bioassay: Based on the ability of gibberellins to delay senescence (yellowing) in broad-leaf dock. The parameter being assayed for quantification is chlorophyll.

3. *Lactuca* (lettuce) hypocotyl bioassay: Based on the ability of gibberellins to promote elongation of lettuce hypocotyls.

4. Dwarf Maize (corn), *Oryza* (rice), and *Pisum* (pea) bioassays: This bioassay utilizes single-gene dwarf mutants and is based on the ability of gibberellins to promote shoot elongation. Note that these single recessive dwarfing mutants segregate three tall to one dwarf; therefore, it is necessary to sow many seeds in order to have adequate plants for a given experiment.

Selected Cytokinin Bioassays

1. *Nicotiana* (tobacco) stem pith callus bioassay: In this bioassay when all of the requirements for callus production are provided, with the exception of cytokinin, there will be little or no callus growth. However, when cytokinins are added callus will grow rapidly and the resulting increase in weight is then used as an indicator of the amount of cytokinin provided.

2. *Nicotiana* (tobacco) gene expression bioassay: This bioassay is based on the expression of a chimeric gene in transformed tobacco (*Nicotiana tabacum*) mesophyll protoplasts in response to both auxins and cytokinins. Quantitation of each plant growth substance is based on a color reaction visualized by light (Boerjan et al. 1992).

3. *Raphanus* (radish) cotyledon expansion bioassay: This bioassay is based on the ability of cytokinins to promote expansion of radish cotyledons. The amount of expansion is an indicator of the amount of cytokinin provided.

4. *Glycine* (soybean) hypocotyl elongation bioassay: This bioassay is based on the ability of cytokinins to promote elongation.

5. *Amaranthus* (pigweed) dark betacyanin promotion bioassay: The red pigment betacyanin normally requires light in order to be produced. The principle of this bioassay is based on the ability of cytokinins to promote betacyanin production in dark-grown pigweed seedlings.

Selected Abscisic Acid Bioassays

1. *Lactuca* (lettuce) seed germination inhibition bioassay: This bioassay is based on the ability of ABA to inhibit lettuce seed germination by restricting hypocotyl and radicle extension growth. The degree of inhibition in seed germination is an indicator of the amount of ABA provided.

2. *Gossypium* (cotton) petiole abscission bioassay: This bioassay is based on the ability of ABA to promote abscission in cotton. The greater the degree of abscission, the higher the levels of ABA provided.

3. *Oryza* (rice) seedling growth inhibition bioassay: This bioassay is based on the ability of ABA to inhibit growth of the leaf sheath. The reduction in length of the leaf sheath is proportional to the level of ABA provided.

4. *Commelina* (dayflower) stomatal closure bioassay: The ability of ABA to promote stomatal closure is the principle of this bioassay. The amount and degree of stomatal closure is measured in order to evaluate the amount of ABA present in the sample.

Selected Ethylene Bioassays

1. *Pisum* (pea) etiolated stem inhibition, swelling and diageotropism induction (triple response) bioassay: The ability of ethylene to inhibit elongation and epicotyl hook opening and to promote horizontal growth is the principle of this bioassay. Each of the three responses can be used to quantify the level of ethylene present in a sample.

2. *Lycopersicon* (tomato) leaf and stem epinasty induction: Epinasty is defined as downward bending of the petiole. The degree of petiole bending is proportional to the amount of ethylene contained within a sample.

3. Fruit-ripening promotion bioassay: The principle of this bioassay is the ability to promote fruit ripening. The time it takes for a given fruit (tomato, banana, lemon, etc.) to ripen is proportional to the amount of ethylene present in a sample.

4. *Gossypium* (cotton) debladded cotyledonary petiole abscission bioassay: This bioassay is based on the ability of ethylene to promote abscission. The degree of abscission is proportional to the amount of ethylene present in a sample. This is the same bioassay as for ABA.

Selected Brassinosteroid Bioassays

1. *Phaseolus* (bean) first internode bioassay: In this assay auxin is applied to one side of the internode resulting in bending following a lag period. BR reduces the lag period when applied one hour prior to auxin. When higher amounts of BR

are in a plant sample, IAA-induced curvature occurs more rapidly. Therefore, by comparing the degree of curvature between internodes treated with both BR plus IAA and the IAA control a quantitative estimate of the level of BR in a sample can be determined.

2. *Oryza* (rice) lamina inclination bioassay: This bioassay is based on the ability of BR to promote bending of the leaf lamina. The degree of bending of the petiole is proportional to the amount of BR contained within a sample.

3. *Pisum* (pea) inhibition test: At optimum concentrations, BR will induce elongation and curvature. However, at supraoptimal concentrations, BR inhibits growth and promotes splitting of the tissue. The degree of reduction in elongation below control values is proportional to the amount of BR contained within a sample (Kohout et al. 1991).

Physiochemical Assays

The two most commonly used physiochemical assays used today for plant growth substances are high-performance liquid chromatography (HPLC) and capillary gas chromatography (GC). Since the introduction of capillary GC columns their superior resolving power has made gas chromatography the most commonly used method for the detection of plant growth substances. Today HPLC is only used with highly selective detectors or with highly purified samples. This section will provide some general information on physiochemical methods for the detection of plant growth substances, however, for more details in this area see the review by Hedden (1993).

Gas Chromatography–Mass Spectrometry (GC-MS) Detection of Plant Growth Substances.   Isotopic dilution GC-MS is considered by most to be the best method for the analysis of plant growth substances because of its relative simplicity and accuracy. This topic has been the subject of many reviews on its use for IAA and related compounds, ABA and its metabolites, gibberellins, cytokinins, brassinosteroids and jasmonates (papers cited in Hedden (1993)). Unlike the detectors used for HPLC methods, GC detectors are usually destructive and do not easily allow the recovery of a radioactively labeled internal standard. The development of mass spectrometric methods for measuring isotope dilution has overcome these problems and is now commonly used for the identification and quantification of all classes of plant growth substances with the exception of ethylene, since there are better methods available for this.

Quantification of Ethylene.   Ethylene is a gas, which makes it much easier to analyze then other plant growth substances. Since no extraction is necessary and there are generally very few contaminants, researchers are able to use simple gas chromatographic procedures utilizing a flame ionization detector. Recently, a highly sensitive on-line detection system using a photoacoustic laser detector has enabled ethylene to be quantified at levels much lower than those with gas chromatography. However, at the present time this technology is not readily available, but shows promise in future research where continuous, nondestructive monitoring of ethylene is necessary (Hedden 1993).

HPLC Detection of Plant Growth Substances. Although HPLC may not be the best method for the detection and quantification of plant growth substances, there are specialized cases using specific detectors when they are used. Fluorimetric detection has been used to quantify IAA directly (Crozier et al. 1980), whereas ABA and jasmonic acid must be converted to fluorescent hydrazones (Anderson 1986) and brassinosteroids to fluorescent bisboronates (Fujioka et al. 1988) prior to fluorimetric detection. Although this method is very sensitive and specific, it is very important that samples are highly purified prior to analysis. Electrochemical detection, both amperometric and coulometric methods, has been described in the literature for IAA directly (Law and Hamilton 1982; Wright and Doherty 1985) and brassinosteroids following the conversion to bisferroceneboronates (Gamoh et al. 1990). Due to their limited application and the need to carefully maintain detectors to maximize sensitivity, the use of electrochemical detectors has not been common in the detection of plant growth substances.

The potential use of liquid chromatography-mass spectrometry (LC-MS) to identify underivatized conjugates of gibberellins (Moritz et al. 1992) and IAA (Ostin et al. 1992) has been demonstrated. Although it has a great deal of potential the sensitivity of LC-MS is not as good as GC-MS. The technology in this area is developing very rapidly and major improvements in LC-MS sensitivity will probably be apparent in the near future (Hedden 1993). It is possible that with these improvements in technology someday it will be more commonly used than GC-MS because of the many other benefits associated with it.

Immunological Assays. Immunoassays are currently very popular and have a lot of potential value in the plant sciences. The original claims that they would require minimal sample purification, thereby making analysis of large numbers of samples very rapid and inexpensive, have not yet become a reality. Interferences from samples are just as much of a problem as they are for physiochemical methods; in fact, in many cases it has been shown that an HPLC step was required to provide the purity to obtain consistent and comparable results to physiochemical assays. Interferences in immunoassays can be specific leading to overestimates of antigen levels and other problems. Therefore, once a purification scheme has been established it is very important to validate the assay with another technique such as GC-MS.

Immunoassays for plant growth substances which use polyclonal antiserum or monoclonal antibodies are described in the literature. Prior to defining polyclonal antiserum or monoclonal antibodies it is appropriate to define an antibody, which is an immunoglobulin present in the serum of an animal and synthesized by plasma cells in response to invasion by an antigen, conferring immunity against later infection by the same antigen. With that definition in mind, polyclonal antiserum may be defined as a serum sample containing antibodies against a specific antigen. Since most antigens have large numbers of epitopes, an antiserum will contain many different antibodies against a given antigen, each antibody having been produced by a single clone of plasma cells. A monoclonal antibody is an immunoglobulin produced by a single clone of lymphocytes and recognizes only a single epitope on an antigen. A monoclonal antibody is produced in the laboratory by hybridoma cells which are immortalized antibody-secreting cell lines created by fusing an antibody-secreting plasma cell (lymphocyte), producing the desired antibody, with a myeloma cell line (an immortal antibody-secreting tumor cell). The myeloma cell line used is typically a mutant that has lost the ability to produce its own antibodies, so that the resultant hybridoma cell line secretes only the desired antibody (Walker and Cox 1988).

For plant growth substance analysis monoclonal antibodies have many advantages over polyclonal antiserum. These advantages are as follows:

1. They are homogeneous chemical reagents that can be renewed whenever needed. Once a useful hybrid is obtained, it is relatively simple and inexpensive to produce hundreds of milligrams of that antibody.

2. One can select antibodies with respect to their degree of specificity, cross-reactivity, affinity, ability to immunoprecipitate and other physical properties to suit individual needs.

3. Small amounts of antigen can be used to elicit a response.

Therefore, if antibodies need to be made against a novel plant growth substance or a derivative of existing plant growth substances, it is not necessary to get large amounts of material. One of the only cases where it would not be feasible to use monoclonal antibodies is if they are not commercially available or when there is no access to an inexpensive centralized facility for their production. If a researcher does not have experience in the production of monoclonal antibodies it would be time-consuming, sometimes frustrating, and expensive without the proper equipment and expertise. There are two commonly used immunological assays for the determination of plant growth substances: radioimmunoassay (RIA), which utilizes isotopes such as ^{125}I, ^{3}H, and ^{14}C as a tracer and the enzyme-linked immunosorbent assay (ELISA), which utilizes enzymes such as alkaline phosphatase or horseradish peroxidase as a tracer. Initial studies in the use of immunological techniques for the detection of plant growth substances utilized RIAs, however, their requirement for radioactive tracers of high specific activity and scintillation counting was a serious disadvantage in many laboratories. Therefore, since this time the use of ELISA has become more common because it requires less expensive equipment, does not require the use of radioactivity, and generally has a higher sensitivity.

There are several RIAs which have been described in the literature for plant growth substances, such as ammonium sulphate precipitation, immunoprecipitation, and the use of dextran-coated charcoal. For ammonium sulphate or immunoprecipitation the sample plus a constant amount of radiolabeled plant growth substance and antibody are incubated for a designated period of time. After they have reacted, either ammonium sulphate or the appropriate antibody is added to separate the free and bound plant growth substance, and the bound form is precipitated. The supernatant is decanted off and the pellet is counted. Note: With decreasing amounts of radiolabeled plant growth substance in the pellet more of it is contained within the sample (Figure 2.2). The use of dextran-coated charcoal follows the same procedure as that used with the ammonium sulphate or immunoprecipitation procedure; however, the major difference is that this method removes free plant growth substances from the solution (and the amount which is found free in supernatant rather than the pellet is counted). Note: When using the dextran-coated charcoal method the less radiolabeled growth substance in the supernatant, the more is contained in the sample (Figure 2.2).

There are numerous ways in which an ELISA may be run; however, two which are commonly used for plant growth substances will be described. The first is the immobilized antibody assay. With this assay the less color observed, the more plant growth substance is present in the sample. The procedure shown in Figure 2.3 is as follows: Rabbit antimouse IgG antibody is coated to the microtiter plate to saturate polystyrene surfaces. The plate is then washed three times with phosphate-buffered saline and specific mouse antibodies added. These antibodies will bind to the rabbit antibodies (at this stage the plates may be stored in the freezer) and following a designated period of time the plate will be washed. The sample or standard is then added together with the enzyme-labeled plant growth substance and allowed to react. The plate is then washed and substrate added, after a designated period of time the reaction may be stopped by adding KOH and absorbance read at 405 nM (Figure 2.3a). The immobilized antigen assay is another ELISA method which will be described. In this assay a plant growth substance protein conjugate is attached to the polystyrene plate and after a designated period of time the plate is washed (at this point the plate may be stored in the freezer) and the monoclonal antibody for the plant growth substance and the plant sample or standard are added. The plate is then washed and enzyme-labeled second antibody added. After a designated period of time the plate is washed and substrate added, the reaction is then stopped by adding KOH and absorbance read at 405 nM. If high amounts of a plant growth substance are present little or no antibody will attach

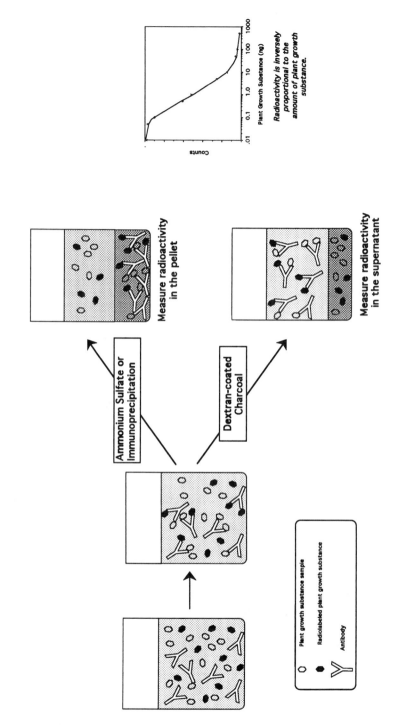

Figure 2.2. Schematic representation of a radioimmunoassay (RIA) using ammonium sulfate, immunoprecipitation, and dextran-coated charcoal.

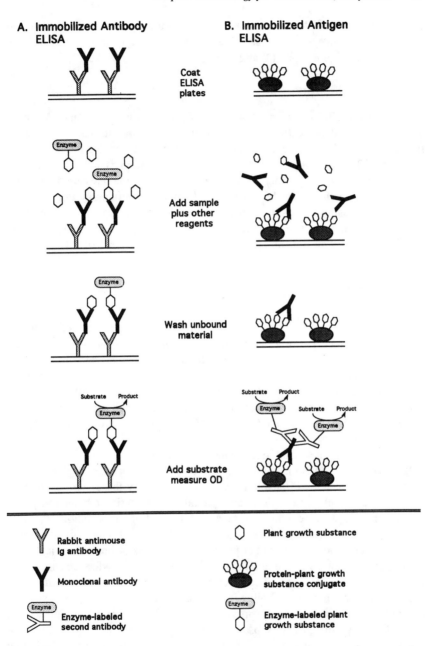

Figure 2.3. Schematic representation of an enzyme-linked immunosorbent assay (ELISA) using an immobilized antibody (a) and an immobilized antigen (b).

to the plate and no color will result after the addition of substrate; however, as the levels of plant growth substance are reduced there is an increase in color development (Figure 2.3b) (Wang 1986).

Final Identification

In order to definitively state that a specific plant growth substance has been assayed, a final identification of this compound is required for chromatographic methods not utilizing mass spectrometry as a detector and for immunological methods. When utilizing bioassays a final identification is not necessary because plant growth substance–like compounds are being measured, not one specific form such as indole-3-acetic acid. Final identification is commonly accomplished using mass spectrometry either alone or in tandem with a GC or LC. This method provides conclusive proof that a specific compound being assayed is present in the sample by providing the equivalent of a fingerprint.

CONCLUSIONS

There are many ways in which plant growth substances can be analyzed depending on the question which the researcher needs to have answered. However, with any of the analytical techniques used it is very important to prepare the tissue properly prior to analysis, to achieve complete extraction, to purify the sample to a stage where there is no interference with the method being used, to maximize recovery through extraction and purification and to monitor this with the use of internal standards, and to provide final identification when necessary.

There are many new analytical techniques which are currently emerging that will show great promise in the future. LC-MS for the analysis of plant growth substance conjugates will enable researchers to quantify these compounds reliably. The use of photoacoustic laser detectors for evaluating ethylene in plants should provide a wide range of new and interesting information. Properly validated immunoassays can be of considerable value because of their high sensitivity and convenience. Other practical advantages of immunological techniques are as follows: immunolocalization of plant growth substances, an alternative to bioassays, and use of anti-idiotypic antibodies to identify plant growth substance receptors. Following further developments in immunoaffinity chromatography will probably provide faster and more efficient purification methods for plant growth substances, thereby accelerating the number of samples which can be processed in a day. New sensitive bioassays utilizing recombinant DNA technology such as the one recently developed for auxins and cytokinins (Boerjan et al. 1992) show a great deal of promise as an analytical technique. The sensitivity of older existing methods such as GC-MS are continually being improved to increase sensitivity and will continue to be used in the future.

REFERENCES

Anderson, J. M. (1986). "Fluorescent hydrazides for the high-performance liquid chromatographic determination of biological carbonyls". *Anal. Biochem.* 152:146–153.

Bieleski, W. J. (1964). "The problem of halting enzyme action when extracting plant tissues". *Anal. Biochem.* 9:431–442.

Boerjan, W., Genetello, C., van Montagu, M., and Inzé, D. (1992). "A new bioassay for auxins and cytokinins". *Plant Physiol.* 99:1090–1109.

Crozier, A., Loferski, K., Zaerr, J. B., and Morris, R. O. (1980). "Analysis of picogram quantities of indole-3-acetic acid by high performance liquid chromatography-fluorescence procedures". *Planta* 150:366–370.

Fujioka, S., Yamane, H., Spray, C. R., Gaskin, P., and MacMillan, J. (1988). "Qualitative and quantitative analysis of gibberellins in vegetative shoots of normal *dwarf*-1, *dwarf*-2, *dwarf*-3, and d*warf*-5 seedlings of *Zea mays* L.". *Plant Physiol.* 88:1367–1372.

Gamoh, K., Sawamoto, H., Takatsuto, S., Watabe, Y., and Arimoto, H. (1990). "Ferroceneboronic acid as a derivatization reagent for the determination of brassinosteroids by high-performance liquid-chromatography with electro-chemical detection". *J. Chromato.* 515:227–231.

Hedden, P. (1993). "Modern methods for the quantitative analysis of plant hormones". *Annu. Rev. Plant Physiol. Plant Mol. Biol.* 44:107–29.

Kohout, L., Zhabinskii, V. N., and Litvinovskaya, R. P. (1991). "Types of brassinosteroids and their bioassays". In *Brassinosteroids. Chemistry, Bioactivity, and Applications*, eds., H. G. Cutler, T. Yokata, and G. Adam, American Chemical Society, Washington, DC, pp. 56–73.

Laloue, M., Terrine, C., and Gawer, M. (1974). "Cytokinins: Formation of the nucleoside-5-triphosphate in tobacco and *Acer* cells". *FEBS Letts.* 46:45–50.

Law, D. M. and Hamilton, R. H. (1982). "A rapid isotope dilution method for analysis of indole-3-acetic acid and indoleacetyl aspartic acid from small amounts of plant tissue". *Biochem. Biophys. Res. Commun.* 106:1035–1041.

Moritz, T., Schneider, G., and Jensen, E. (1992). "Capillary liquid chromatography/fast atom bombardment mass spectrometry of gibberellin glucosyl conjugates". *Biol. Mass Spectrom.* 21:554–559.

Östin, A., Mortiz, T., and Sandberg, G. (1992). "Liquid chromatography-mass spectrometry of conjugates and oxidative metabolites of indole-3-acetic acid". *Biol. Mass Spectrom.* 21:292–298.

Powell, L. and Maybee, C. (1985). "Hormone analysis employing 'Baker'-10 SPE™ disposable columns". *Plant Physiol.* 77:76.

Rivier, L. and Crozier, A. (1987). *Principles and Practice of Plant Hormone Analysis, Volumes 1 and 2*, Academic Press, London.

Walker, J. M. and Cox, M. (1988). *The Language of Biotechnology: A Dictionary of Terms*, American Chemical Society, Washington, DC.

Wang, T. L. (1986). *Immunology in Plant Science*, Cambridge University Press, Cambridge, U.K.

Wright, M. and Doherty, P. (1985). "Auxin levels in single half nodes of *Avena fatua* estimated using high performance liquid chromatography with coulometric detection". *J. Plant Growth Reg.* 4:91–100.

Yopp, J. H., Aung, L. H. and Steffens, G. L. (1986). *Bioassays and Other Special Techniques for Plant Hormones and Plant Growth Regulators*, Plant Growth Regulator Society of America, Lake Alfred, FL.

Chemistry, Biological Effects, and Mechanism of Action of Plant Growth Substances

Dose-response curves for all of the known plant growth substances are bell-shaped, as shown in Figure 3.1. At lower concentrations the effects are typically stimulatory reaching a maximum beyond which they become inhibitory. There are two general classes of hormones found in animal systems, steroid and peptide, both of which probably also occur in plant systems. The steroid class forms a hormone/receptor (defined as those molecules that specifically recognize and bind the hormone and, as a consequence of this recognition, can lead to other changes or series of changes which ultimately result in the biological response) complex in the cytoplasm, which is then transported into the nucleus where mRNA is synthesized, resulting in a given response (Figure 3.2a). The second class are peptide hormones which bind to a receptor at the plasmamembrane, altering the enzyme adenylate cyclase and activating cyclic AMP from ATP, which acts as a secondary messenger for a given response (Figure 3.2b). In order for hormone binding in either class to be specific, the following criteria must be met (Cuatrecasas et al. 1977):

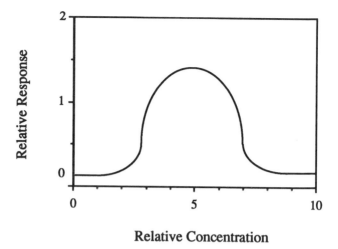

Relative Concentration

Figure 3.1.   A typical dose-response curve for the known plant growth substances.

1. There must be strict structural and steric specificity.
2. The response must be saturable, thereby indicating a finite and limited number of binding sites.
3. The response must be tissue-specific.
4. The hormone must bind with a high affinity in order to show physiological relevance.
5. Hormone binding must be reversible showing kinetics consistent with a physiological response observed and biological activity.

Following the binding of a given hormone, a cascade of events takes place in transducing this signal. Today there is a considerable amount of excitment about how plant growth substances act as signals in the promotion of gene expression resulting in a biochemical and/or physiological response. However, we are still in the infancy stage of development in our understanding on how this occurs for all of the major classes of plant growth substances. In fact, we are still not sure whether it is a plant growth substance alone or various combinations which cause a given response. In addition, we still do not know if a given response occurs in a dose-dependent manner or in response to changes in tissue sensitivity (receptors). With that preface the chemistry, biological effects, and proposed mechanism of action of each of the major classes of plant growth substances will be discussed.

A   *Steroid Hormones*

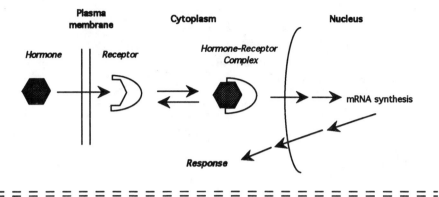

= :

B   *Peptide Hormones*

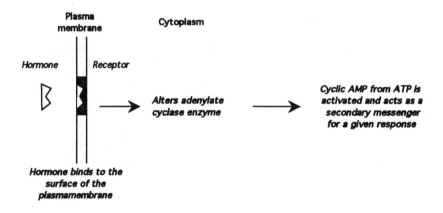

Figure 3.2.   Schematic representation of the mechanism of action of (a) steroid and (b) peptide hormones.

AUXINS

Biosynthesis of Indole-3-Acetic Acid

Following the discovery of IAA, many indole compounds have been reported in plants; however, auxin activity of each of these can probably be attributed to their conversion to IAA. Some of these conversions are shown in Figure 3.3. Tryptophan is now generally accepted as the primary precursor for

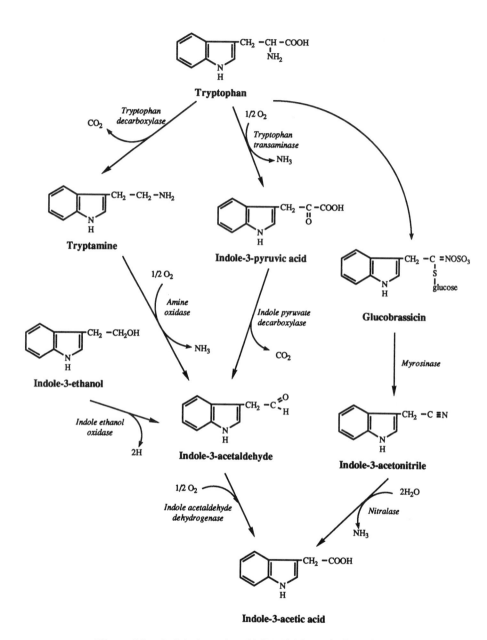

Figure 3.3.   Indole-3-acetic acid (IAA) biosynthetic pathway.

the biosynthesis of IAA. After Thimann (1935) showed that tryptophan could be converted to IAA in mold, Wildman et al. (1947) found an enzyme system capable of converting tryptophan to IAA in spinach leaves. Since this time many other researchers have demonstrated that tryptophan is the primary precursor in higher plants. This pathway came under fire by researchers suggesting that bacterial contamination was responsible for the production of IAA (Libbert et al. 1966). However, it was then shown that tryptophan could be converted to IAA under sterile conditions by Muir and Lantican (1968) and has since been shown by many other researchers. There are two major pathways in which tryptophan may be converted to IAA of which one or both may function in a given plant species. The first pathway is via the conversion of tryptophan to indole-3-pyruvic acid by the enzyme tryptophan transaminase; it is then decarboxylated by indolepyruvate decarboxylase to indole-3-acetaldehyde (IAAld), which is subsequently converted to IAA via indoleacetaldehyde dehydrogenase. The second pathway involves the decarboxylation of tryptophan to tryptamine by tryptophan decarboxylase, which is then converted to IAAld by amine oxidase and finally to IAA by IAAld dehydrogenase. It has been reported that in some plant systems other precursors can be used. For example, in cucumber seedlings indole-3-ethanol has been identified and when exogenously applied can readily be converted to IAAld via indole ethanol oxidase (Rayle and Purves 1967). Members of the Cruciferae or Brassicaceae contain the natural product glucobrassicin which is converted to indole-3-acetonitrile via myrosinase (Gmelin and Virtanen 1961) and finally to IAA by nitralase (Thimann and Mahadevan 1958) (Figure 3.3). Although other systems exist in different plants, thus far most of the plants tested utilize tryptophan and only a limited number utilize alternative systems. For more details on the biosynthesis of IAA in higher plants see Cohen and Bialek (1984).

Free versus Bound Auxins

There are two categories of auxins in plants free and bound. Free auxins can readily diffuse out of the tissue, are easily extracted with various solvents, and can be immediately used to regulate physiological processes in plants. Examples are indole-3-acetaldehyde, indole-3-acetonitrile, indole-3-ethanol, or indole-3-pyruvic acid. Whereas bound auxins are not readily available and are only made available after they are subjected to hydrolysis (chemical splitting of a bond and adding a water, can be by acid or base), enzymolysis (enzymatic breakdown) or autolysis (self-digestion). Bound auxins typically serve as reserve or storage (glucosides) and detoxification (amino acid or protein complexes) forms of auxin. Examples of reserve forms of IAA are IAA glucosides which are abundant in seeds and are inactive until IAA is released naturally by enzymolysis (Zenk 1961). Detoxification forms of IAA include amino acid and

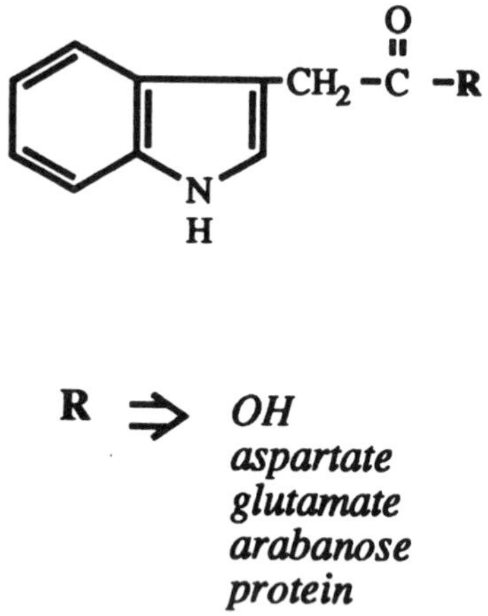

Figure 3.4.   Inactivation of indole-3-acetic acid though conjugation into inactive forms.

protein complexes which can only be liberated to free IAA by acid or base hydrolysis (Figure 3.4) (Andreae and Good 1957; Andreae and van Ysselstein 1960). Synthetic auxins such as 2,4-dichlorophenoxyacetic acid (2,4-D) can also be converted to amino acid or glucoside derivatives but at a much slower rate (Zenk 1962, 1968). For more information on bound versus free auxins see Cohen and Bialek (1984).

Destruction of IAA

The presence or absence of auxin has a profound effect on plant growth and development. In addition to reversible and irreversible bound auxins which occur within the plant, enzymatic oxidation and photooxidation are two other means of destruction or inactivation of IAA. Enzymatic oxidation of IAA was first shown by Tang and Bonner (1947). They isolated IAA oxidase, which is involved in the oxidation of IAA in pea epicotyls. This enzyme system has since been shown to be ubiquitous in the plant kingdom. The main product formed by the oxidative decarboxylation pathway is 3-methylene-oxindole; however, others such as 3-hydroxymethyl oxindole, indole-3-methanol, indole-3-aldehyde, and indole-3-carboxylic acid have also been shown to occur (Bandurski et al. 1986). The nondecarboxylation oxidative pathway for IAA results in oxindole-3-acetic acid as the major product; however, dioxindole-3-acetic acid has also been reported in some plant species (Reinecke and Bandurski 1987). Photooxidation of IAA is a very different reaction from enzymatic oxidation and at the present time does not appear to have physiological significance because very large light dosages are required and the process does not appear to be related to phototropism or other physiological phenomena. For more details on the metabolism of IAA see Bandurski (1984).

Synthetic Auxins

There are many synthetic auxins commercially available today that are used for many purposes. Synthetic auxins are compounds which are similar in activity but not in structure to IAA and may be broken down into six groups (Figure 3.5):

1. Indole derivatives—indole-3-acetic acid (IAA), indole-3-butyric acid (IBA).

2. Benzoic acids—2,3,6-trichlorobenzoic acid; 2-methoxy-3,6-dichlorobenzoic acid (Dicamba).

3. Naphthalene acids—α and β-naphthaleneacetic acid (α and β-NAA).

4. Chlorophenoxyacetic acids—2,4,5-trichlorophenoxy acetic acid (2,4,5-T), 2,4-dichlorophenoxyacetic acid (2,4-D).

5. Naphthoxyacetic acids—α and β-napthoxyacetic acid (α and β-NOA).

6. Picolinic acids—4-amino-3,5,6-trichloropicolinic acid (Tordon or Pichloram).

Both chlorophenoxyacetic acids and picolinic acids are herbicides commonly used in modern agriculture. Zimmerman and Hitchcock (1942) discovered that the substitution of various groups on the ring or side chain had a profound effect on auxin activity of phenoxy acids. Due to their selectivity, the phenoxyacetic acids particularly, 2,4-D and 2,4,5-T have been widely used as broad-leaf herbicides for many years. They were developed because of their potential usefulness for chemical warfare. The phenoxyacetic acid compounds are very stable. In fact, most broad-leaf dicots cannot break down these compounds; therefore, very low concentrations can be used. Many plants have the enzyme IAA oxidase which normally causes the breakdown of IAA; however, it does not have the ability to break down 2,4-D or 2,4,5-T. The most commonly used formulations of 2,4-D and 2,4,5-T are free acids, salts, and amine salts. It should be noted that free acids are very volatile and present a large drift problem. Agent Orange, which was used as a defoliant in Vietnam, is a mixture of free 2,4-D and the n-butyl ester of 2,4,5-T. Since they are both very volatile, maximum drift could be obtained, which is a problem in agriculture but in Vietnam served as an excellent defoliant. During the synthesis of 2,4,5-T and other chlorinated phenols numerous side products were generated such as chlorodioxins which are very harmful to humans, one of which is 2,3,8,9-tetrachlorodibenzo-para-dioxin (Figure 3.6). Since the use of chlorinated hydrocarbons in Vietnam many years ago, procedures to avoid the production of harmful by-products have been developed and are currently being used today.

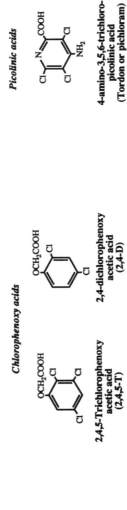

Figure 3.5. Structural formulas for the six groups of synthetic auxins.

52

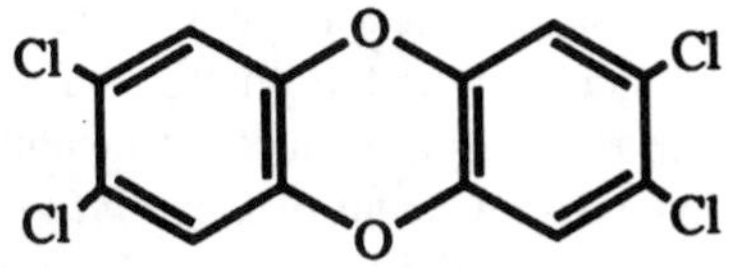

Figure 3.6. Structural formula for dioxin (2,3,8,9-tetrachlorodibenzo-p-dioxin).

Physiological Effects of IAA

Auxins have many physiological effects in plant systems; however, covering all auxin-mediated responses is beyond the scope of this book. Therefore, in this section a brief description of the following auxin responses will be presented: cellular elongation, phototropism, geotropism, apical dominance, root intitiation, ethylene production, and fruit development.

Cellular Elongation. Many studies in this area have been with coleoptiles or excised root tissues. These tissues have little IAA and when placed in contact with exogenous IAA there is a large increase in growth. The steps involved in cell elongation in response to auxin may be summarized as follows:

1. There is first a decrease in resistance of the cell wall to stretching. This occurs due to the breaking of noncovalent bonds between xyloglucans and cellulose in the cell wall.

2. There is a change in the water relations within the cell. Even though the osmotic potential in the cell does not change, the water potential in the auxin-treated cell becomes more negative due to a reduction in pressure potential.

3. The decrease in water potential allows water to move into the cell and exerts pressure outward on a plastic cell wall resulting in elongation.

4. When elongation is complete noncovalant linkages between cellulose and polysaccharides reform. This process is not reversible.

There are many mechanisms proposed on how cell-wall loosening occurs (Ray 1987) many of which have been rejected. The most popular mechanism, which has been around for some time, is the acid-growth hypothesis. This theory states that auxins cause receptive cells in coleoptile or stem sections to secrete hydrogen ions into the surrounding primary walls, which, in turn, lowers the pH and causes cell-wall loosening and fast growth to occur. The low pH presumably activates certain cell-wall-degrading enzymes which are normally inactive at a higher pH (Cleland 1987).

Phototropism. The definition of phototropism is the movement of a plant organ in response to directional fluxes or gradients in light. The Cholodny-Went theory on phototropism suggests that light from one side causes the transport of auxin toward the shaded side, thereby causing a higher concentration of auxin on the shaded side than on the illuminated side (Figure 3.7). The unequal distribution of auxin is thought to cause the bending response observed in stems; however, when lateral movement of auxin was blocked there was no difference observed. This theory continues to be a highly debated topic, therefore, for more details in this area see a review by Briggs and Baskin (1988).

Geotropism. The definition of geotropism is the movement of a plant organ in response to gravity. If a plant is placed horizontally the shoots will bend upward against gravity (negative geotropism), whereas roots will bend downward along with the force of gravity (positive geotropism). It is thought that plant roots and shoots perceive these changes in gravity by statoliths, which are starch-containing plastids such as amyloplasts or chloroplasts (Haberlandt 1902; Nemec 1901). The cells which contain statoliths (Greek *lithos*, meaning stone) are called statocytes. A statolith is a body that changes position in the cell as a direct consequence of the change of an organ with reference to the direction of the force of gravity. Although proposed many years ago, the statolith theory is still debated by researchers today as to whether it is correct or incorrect.

The Cholodny-Went theory on geotropism suggests that stems and roots in response to gravity accumulate IAA on the lower side. In the stems IAA promotes growth on the bottom side of the stem resulting in upward bending. It has been shown that when the root tip is removed the ability of the root to respond to gravity is lost, and when the root tip is replaced geotropism is regained. Initial studies in the 1970s suggested that an inhibitor such as ABA was coming from the root tip causing the inhibition of growth on the bottom side of the root and promoting downward bending. It is now apparent that IAA is that inhibitor because it can inhibit root growth at concentrations 100 to 1000 times lower than ABA. Today, it is generally accepted that since the roots are much more sensitive to IAA than the shoots its accumulation on the lower side of the root in response to gravity will inhibit growth in that region while normal growth continues on the top resulting in downward bending (Figure 3.8) (Evans 1985).

Apical Dominance. Long before regulation of growth by plant growth substances was known, botanists recognized that the apical bud causes lateral bud suppression. Initial experiments showed that when the apical bud was removed lateral buds grew; however, with time one of the lateral buds became dominant, suppressing subsequent growth of lateral buds. Following the discovery of auxin it was shown that relatively high levels were contained in shoot tips. Thimann and Skoog (1934) were the first to present evidence that auxin suppressed lateral bud growth in much the same manner as did the apical bud. Since this time many other researchers have shown this effect in other plants.

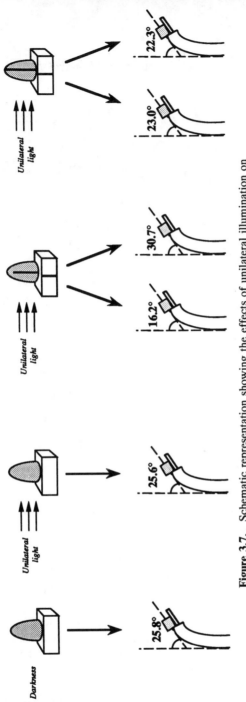

Figure 3.7. Schematic representation showing the effects of unilateral illumination on auxin distribution in coleoptile tips.

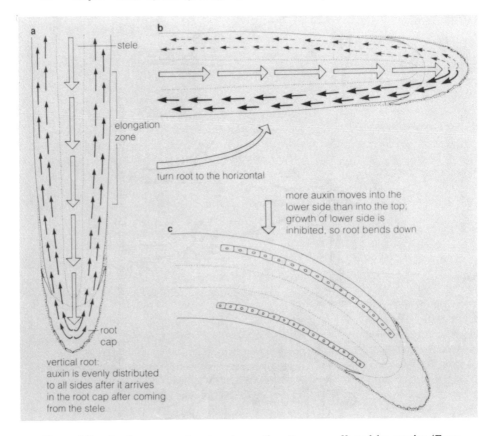

Figure 3.8.   Auxin movement and root growth patterns as affected by gravity (From Salisbury and Ross (1992)).

Apical dominance has since been shown to be widespread in plants and to be controlled by plant growth substances (Hillman 1984; Tamas 1987).

Root Initiation and Elongation.   Julius von Sachs (1880) was the first to suggest that young leaves and active buds had a transmissible plant growth substance which promoted root initiation. Since this time IAA has been shown to stimulate root initiation from stem cuttings, thereby demonstrating a practical use for auxins. Synthetic auxins are generally used instead of the naturally occurring IAA because they are not destroyed by IAA oxidase or other enzymes and will persist in the tissue for longer periods of time. Exogenous applications of auxins can promote root initiation and early development of the root, whereas elongation of the root is generally inhibited unless concentrations are sufficiently low (Wightman et al. 1980). Since auxin has been shown to stimulate ethylene production in a variety of tissues it was thought for many years that inhibition of growth by IAA was due to its ability to stimulate ethylene. It was recently shown that IAA can inhibit elongation of attached roots of pea seedlings without any effect on ethylene production (Eliasson et al. 1989).

Ethylene Production. Zimmerman and Wilcoxon (1935) were the first to show that auxin stimulated ethylene evolution by tomato plants. Since this time there have been many reports on the ability of auxins to stimulate ethylene production in both intact and detached plant part systems (Arteca 1990). Today many of the responses which were once attributed to auxins are now found to be due to ethylene production.

Fruit Growth. Increases in fruit size are mainly due to cell enlargement. Auxins are known to be involved in cell extension and are thought to play a fundamental role in determining growth patterns in fruits. The potential role of auxins in fruit growth are supported by two lines of evidence. First there is a correlation between seed development and final size and shape of the fruit. Second, application of auxin to certain fruits at particular stages of their development induces a growth response (Crane 1949). One of the best examples of auxin involvement in size and shape of fruit is with the strawberry where it has been shown that the endosperm and embryo in the achene produce auxin, which moves out and stimulates growth (Nitsch 1950). The location of the achenes on the fruit has a great effect on its shape. In a strawberry the achenes are located outside the fleshy receptacle and are easy to manipulate. When all of the achenes are removed from the fruit there is no growth; however, if they are all removed and auxin is applied to the surface of the receptacle there is normal growth. When only several achenes are left on the receptacle, growth will occur directly below the achene (Nitsch 1950).

Auxin Transport

Early experiments by Went (1934) and others suggested that auxin movement in the plant was strictly polar, moving basipetally (movement from the tip to the base). Now although basipetal transport appears to predominate in coleoptiles and shoots it is probably not as strict as once thought. Jacobs (1961) tested the transport of auxin in coleus stem sections. He found that the ratio of basipetal to acropetal (movement from base to tip) transport was 3 to 1. Although acropetal transport is 1/3 that of basipetal transport in this test system it is real and significant. Transport of IAA in stems is mainly basipetal at a rate of 6–26 mm/hour. In the roots IAA transport is mainly due to basipetal movement from the root tip to the physiological base of the plant but at a rate

Figure 3.9.   Structural formula for the auxin transport inhibitor TIBA (2,3,5-triiodobenzoic acid).

of 1–2 mm/hour. Danielli (1954) established the following criteria for an active transport system which is as follows:

1. The compound should move at a velocity greater than diffusion. This is true for auxin since it moves 6–26 mm/hour.

2. Movement of the compound should be driven by metabolic forces. This is also true for auxin transport because metabolic inhibitors and anaerobic conditions block auxin transport.

3. The compound should be able to move against a concentration gradient. Auxins can move very rapidly against concentration gradients in a basipetal direction.

4. The transport system should show specificity for a certain compound. IAA and α-NOA are both active and transported at the same velocity, whereas, β-NOA is inactive and not transported. The transport system for auxin can also discriminate between optical isomers of indole-α-propionic acid. The plus form is more active and is transported better than the minus form. Specific inhibitors of auxin transport such as 2,3,5-triiodobenzoic acid (Figure 3.9) mimic auxin and therefore block its transport.

5. The compound being transported should show a saturation effect, when all transport sites are saturated the addition of more plant growth substance should have no effect. Auxin transport has been shown to be saturable in many test systems.

Transport of IAA can be experimentally shown by excising a segment of plant tissue and placing an agar block containing IAA on the top portion and an agar block minus IAA on the bottom of the segment. If the segment is placed in its natural orientation transport will occur, however, when placed in the opposite orientation there will be no transport. Transport can be measured in two ways: One way is to run bioassays on the bottom block over time following the addition of IAA to the top. The second way which is more accurate utilizes ^{14}C- or ^{3}H-labeled IAA in the upper block followed by a chromatographic step and scintillation counting to evaluate how much of the IAA has been transported to the bottom block (Figure 3.10).

The chemiosmotic hypothesis of polar transport is very popular (Figure 3.11) (Rubery and Sheldrake 1974; Raven 1975). This hypothesis states that

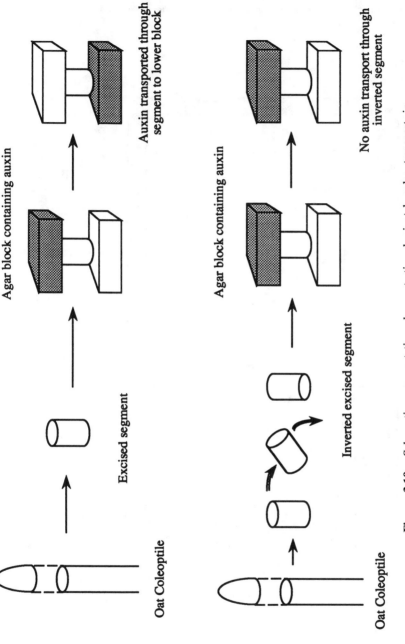

Figure 3.10. Schematic representation demonstrating basipetal polar transport in *Avena* coleoptiles.

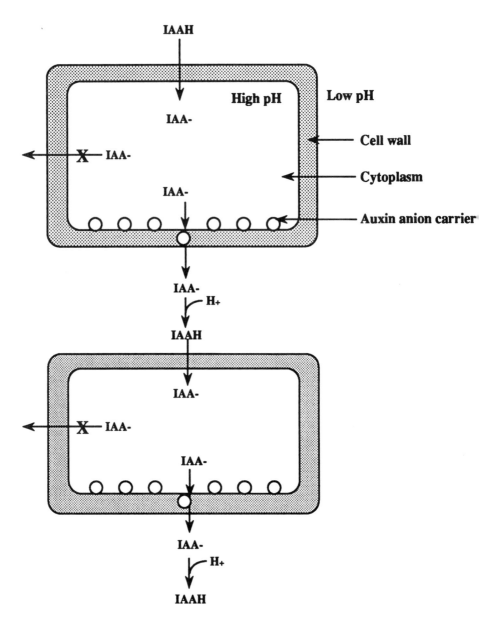

Figure 3.11.   Schematic representation of the chemiosmotic hypothesis of auxin transport.

the cell expends energy to maintain a pH gradient across the plasma membrane with the cell wall being more acidic than the cytoplasm. At an acidic pH native auxin will be undissociated (IAAH) and can readily enter the cell by diffusion. Once in the cytoplasm auxin dissociates to IAA⁻ which cannot diffuse out of the cell. As the IAA⁻ concentration in the cell increases, movement of IAA out of the cell is thermodynamically favored via a specific IAA⁻ carrier located at the base of the cell. Jacobs and Gilbert (1983) using indirect immunofluorescence along with monoclonal antibodies provided direct evidence for the location of the presumptive auxin anion carrier in pea tissue. They showed that the carrier was located in the plasma membrane at the basal ends of the parenchyma cells sheathing the vascular bundles. This research is the first direct evidence for an auxin transport carrier in pea stem cells.

GIBBERELLINS

Biosynthesis of Gibberellins

Gibberellins belong to a large group of naturally occurring compounds called terpenoids (e.g., carotinoids). Terpenoids are built from five carbon isoprene units, the immediate precursor to gibberellins is a diterpene which contains four isoprene units. The mevalonic acid pathway is used for the biosynthesis of gibberellins. Each of the steps in this pathway up to GA_{12} aldehyde are the same in all plants; however, from this point on different species use different pathways to form the more than 90 gibberellins known today (Figure 3.12). Once produced there are a number of interconversions between gibberellins which can take place depending on the plant. There is also evidence that bound gibberellins exist as glucosides; however, it is still not clear whether they are for inactivation or as a storage source. Young leaves are thought to be the major site of gibberellin biosynthesis, which can subsequently be transported throughout the plant in a nonpolar fashion. Roots also have the ability to synthesize gibberellins which are transported to the shoots via the xylem sap. High levels of gibberellins have been found in immature seeds, and cell-free extracts from seeds possess the ability to synthesize gibberellins which are not transported out of the seeds (Takahashi et al. 1991).

There are a number of commercially available growth retardants which are currently used for height control by inhibiting GA biosynthesis (Davis and Curry 1991). Onium compounds (Phosphon D, AMO-1618, Cycocel, mepiquate chloride, piperidium bromide), pyrimidine compounds (ancymidol, flurprimidol), triazole compounds (paclobutrazol, uniconazole, triapenthenol, BAS 111, Lab 150), tetcyclacis, prohexadione calcium, and inabenfide have been used to inhibit gibberellin biosynthesis (Grossmann 1990; Davis and Curry 1991).

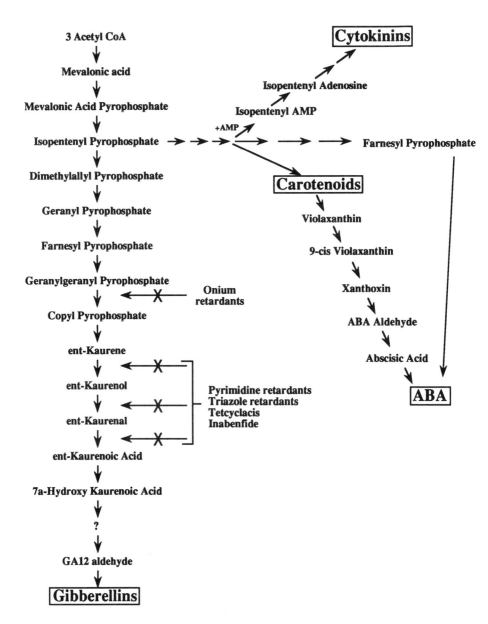

Figure 3.12. Mevalonic acid pathway for the biosynthesis of gibberellins, cytokinins, and abscisic acid.

Physiological Effects of Gibberellins

In this section a general outline of the major physiological processes affected by gibberellins including growth of intact plants, genetic dwarfism, bolting/ flowering, mobilization of storage compounds, effects on seed germination, and dormancy will be discussed; however, for more details see Takahashi et al. (1991).

Effects on Growth of Intact Plants.   There are numerous reports in the literature showing that gibberellins promote growth of intact plants. All of the more than 90 gibberellins known today are able to promote either stem elongation, cell division, or both, but their effectiveness varies greatly. The differences in a plant's responsiveness to a chemical depends on many factors and it is not uncommon for one gibberellin to be more active in one plant system then another. As shown in Figure 3.13, different gibberellins have varying degrees of effectiveness on lettuce hypocotyl elongation (Rai and Laloraya 1967). Another example is the cucumber hypocotyl bioassay which is less

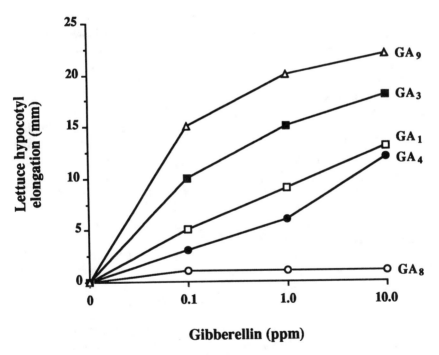

Figure 3.13.   The effectiveness of different gibberellins on lettuce hypocotyl elongation (From Rai and Laloraya (1967)).

responsive to gibberellins having hydroxyl groups in the 7 position such as GA_3 (Yopp et al. 1986). In general, growth is promoted by gibberellins in many plant species, especially dwarfs and biennials in the rosette stage. Typically gibberellins stimulate growth in intact plants more effectively than in excised sections, which is very different from auxins.

Genetic Dwarfism.   There are many gibberellin biosynthesis mutants which have been developed over the years (Reid 1990). They are single gene muta-tions, the plants are generally about one-fifth the size of normal plants and they are characterized by shortened internode length. When gibberellins are exogenously applied to these mutants there is a striking increase in size making them similar in appearence to their normal counterparts. There are also gib-berellin sensitivity mutants which do not respond to exogenously applied gib-berellins and contain levels similar to their normal counterparts while still remaining dwarfs (Reid 1990; Scott 1990). Speculation on why this occurs is that there may be an excess of natural inhibitors contained within the plant or they are receptor mutants which prevents their growth in response to gibberel-lins.

Bolting and Flowering.   Gibberellins have been shown to be involved in the promotion of flowering in a wide range of higher plants and they have been the subject of numerous reviews (Takahashi et al. 1991). In rosette plants their leaf development is profuse and internode elongation is retarded; however, just prior to the reproductive stage of development the stem elongates five to six times the original height of the plant. Plants with a rosette growth habit require long days or cold treatment prior to flowering. If gibberellins are applied to rosette plants under noninductive conditions they will typically promote bolting and flowering. The involvement of gibberellins in the promotion of flowering came under some controversy when it was reported that the use of smaller dosages of GA induced bolting without flowering. This led some to believe that gibberellins had an indirect effect on flowering (Stuart and Cathey 1961). Gibberellins have since been shown to be implicated in flowering of a wide range of plants, not only in the bolting response (Takahashi et al. 1991). The influence of gibberellins on bolting includes a stimulation of cell division and cell elongation. In general GA promotes cell division and cell elongation in the subapical region of the plant. Plant growth retardants which block gibberellin biosynthesis cause an inhibition of cell division in the subapical meristem region and induces lateral expansion of the apex (Sachs and Kofranek 1963).

Mobilization of Storage Compounds, Effects on Seed Germination, and Dormancy.   Independent research by Yomo (1960) and Paleg (1960 a, 1960b) showed that gibberellic acid stimulated α-amylase and other hydrolytic enzymes promoting hydrolysis of storage reserves. Later it was shown that the

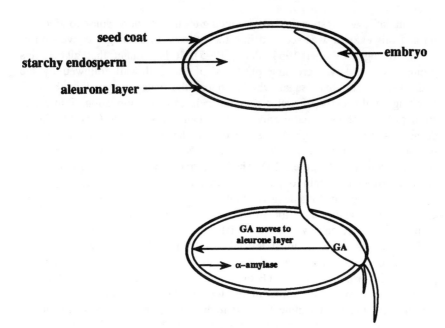

α-Amylase synthesized in aleurone layer.

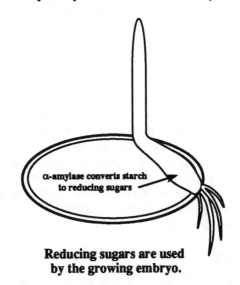

**Reducing sugars are used
by the growing embryo.**

Figure 3.14.   Schematic representation of the production of gibberellins in barley seeds and subsequent transport to the aleurone layer of cells, where it stimulates α-amylase which causes the conversion of starch to sugar for energy to be used for growth.

aleurone layer was responsible for producing α-amylase in response to gibberellin and this enzyme was secreted into the endosperm causing a conversion of starch to sugar (Varner 1964; Paleg 1965). It is thought that gibberellins promote growth by increasing plasticity of the cell wall followed by the hydrolysis of starch to sugar which reduces the water potential in the cell, resulting in the entry of water into the cell causing elongation. Filner and Varner (1967) showed that α-amylase produced in response to GA_3 was due to de novo synthesis through the use of density labeling of proteins. Since this time the mechanism by which GA_3 promotes hydrolysis of starch has been studied extensively (Figure 3.14). The basic steps involved are as follows: GA is produced in the embryo transferred to the aleurone layer of cells where α-amylase is produced via de novo synthesis, which promotes the conversion of starch to sugar and is used for growth of the seedling. Mobilization of endosperm starch in cereals which are initiated by GA-induced α-amylase can be inhibited at the transcriptional level by ABA (Davies and Jones 1991). In addition to α-amylase there are many other hydrolytic enzymes which are produced in response to GA_3 (Brown and Ho 1986). Gibberellins have also been shown to mimic the effect of red light in the stimulation of lettuce seed germination and to substitute for low temperatures or long days in order to break dormancy. It is now generally accepted that gibberellins are potent promoters of seed germination and possess the ability to break dormancy in a variety of crops (Takahashi et al. 1991).

CYTOKININS

Biosynthesis of Cytokinins

It is known that the initial steps of the mevalonic acid pathway up to the isopentenyl pyrophosphate step are involved in the production of cytokinins (Figure 3.12). At this step isopentenyl pyrophosphate together with AMP produces isopentenyl AMP, which is converted to isopentenyl adenosine followed by a series of steps to produce cytokinins. Information on the biosynthesis of cytokinins is incomplete and additional work is in progress to better understand the biosynthesis of cytokinins in higher plants (Kaminek et al. 1992; Mok and Mok 1994).

Bound versus Free and Degradation of Cytokinins

Examples of free cytokinins are zeatin and isopentenyladenine. There are also conjugated forms of cytokinins which can be produced in the following ways. Glucosides can be formed by the attachment of carbon 1 of glucose to the hydroxyl group on the side chain of zeatin or the carbon 1 can attach to the N atom of the C-N bond at either position 7 or 9 on the adenine ring. Another alternative is to form an alanine conjugate at position 9. The glucoside conjugates may represent storage forms or in some cases facilitate transport of certain cytokinins, whereas alanine conjugates are more likely to be irreversibly formed products which can serve as a detoxification mechanism in the plant. Degradation of cytokinins occurs largely by cytokinin oxidase, this enzyme removes the five carbon side chain and releases free adenine from zeatin and free adenosine from zeatin riboside (Kaminek et al. 1992; Mok and Mok 1994).

Physiological Effects of Cytokinins

Cell Division and Organ Formation.   The major function of cytokinins in plants is to promote cell division. Jablonski and Skoog (1954) noted that tobacco stem pith callus grew in response to kinetin or IAA alone, however, for continued growth of the pith tissue in culture both kinetin and IAA had to be present in the culture. The explanation for this is that initially endogenous (occur within the plant) IAA or cytokinin may interact with exogenous (applied externally) applications of IAA or cytokinin, however, with time endogenous levels are depleted and no further growth occurs. It is now known that by manipulating the ratio of IAA to cytokinin one can obtain callus, roots, and/or shoots. The ability to regenerate plants from calli is a biotechnological tool commonly used today for the selection of plants with resistance to drought, salt stress, pathogens, herbicides, and others.

Seed Germination, Cell and Organ Enlargement.   Kinetin has been shown to overcome the inhibitory effect which far-red light has on lettuce seed germination (Miller 1956). In general cytokinins are thought of as cell-division-promoting substances; however, there are specific instances where cytokinins also have effects on cell enlargement. Cytokinins have been shown to promote cell enlargement of excised cotyledons from radish, pumpkin, cocklebur, flax, and other dicotyledonous plants. When the cotyledons are detached from the plant they are removed from their natural cytokinin source (Yopp et al. 1986); however, when exogenous applications of cytokinins are made they promote cell expansion. The enlargement of the cells is due to water uptake caused by a decrease in the osmotic potential of the cell promoted by the conversion of lipids (which serve as the storage material in cotyledons) to reducing sugars (glucose and fructose). Cytokinins have also been shown to be involved in phytochrome-mediated cell expansion in bean leaf discs by substituting for red light (Miller 1956).

Root Initiation and Growth.   There are a limited number of studies on cytokinin effects in the root system. It appears that depending upon the concentration and particular plant, cytokinins can either promote and/or inhibit root initiation and development. It has been shown that kinetin can stimulate dry weight and elongation of roots in *Lupin* seedlings, whereas at higher concentrations both were inhibited (Fries 1960). Lateral pea root development in excised root segments is stimulated by low concentrations of kinetin, whereas higher concentrations inhibited elongation and lateral branching (Torrey 1962). More recently Dong and Arteca (1982) showed that when kinetin was applied to the roots at very low concentration there was a stimulation in photosynthesis and growth; however, if roots were in contact with 0.47 µm kinetin for longer than two days, growth of the roots and the entire plant were severely reduced.

Bud and Shoot Development. Cytokinins were shown by Wickson and Thimann (1958) to promote lateral bud break in pea stem sections and to partially overcome apical dominance. More recently, using genetic engineering techniques, cytokinin levels were increased in tobacco and *Arabidopsis* plants by Medford et al. (1989). In this study a bacterial gene encoding isopentenyl AMP synthase (an enzyme responsible for cytokinin production) together with a heat shock promoter was inserted into the genome of tobacco and *Arabidopsis* plants. This new gene was turned on by subjecting the transgenic plants to 40°–45°C for short periods, which resulted in an increase in zeatin riboside monophosphate, zeatin riboside, and zeatin of 23, 46, and 80 times, respectively. The transgenic plants which were overproducing cytokinins exhibited multiple lateral bud breaks and an overall lack of apical dominance. Another example of genetic engineering to overcome apical dominance was to insert a gene encoding for an enzyme which is responsible for converting free IAA into an inactive amino acid conjugate. By incorporating this gene the levels of free IAA were reduced, thereby promoting lateral bud break (Klee et al. 1991). Apical dominance appears to be controlled by a balance between endogenous cytokinin and auxin levels. Two hypotheses on how cytokinins are involved in apical dominance have been suggested. The first suggests that cytokinins may inhibit IAA oxidase found in the lateral buds, thereby allowing auxin levels to build and causing lateral bud elongation. A second hypothesis is that cytokinins may initiate a sink mechanism at the lateral bud promoting the transport of nutrients, vitamins, minerals, and other growth substances, all of which may be limiting their growth. Before either hypothesis can be proven or disproven more research is necessary in this area.

Delay of Senescence and Promotion of Translocation of Nutrients and Organic Substances. When mature leaves are detached from the plant there is a rapid breakdown of protein, chloroplasts degrade resulting in a loss of chlorophyll, and there is an outflow of nonprotein nitrogen, lipids, and nucleic acid components through the leaky membranes. It has been shown that the formation of roots on detached leaves retarded the onset of senescence symptoms (Chibnall 1954). Cytokinins have been shown to partially replace the need for roots in delaying senescence (van Staden et al. 1988). In the darkness senescence is greatly accelerated; however, cytokinins have the ability to replace the light effect by delaying the senescence process probably by maintaining the integrity of the tonoplast membrane (Thimann 1987). When cytokinins are applied to etiolated leaves or cotyledons several hours before exposure to light the etioplasts are converted to chloroplasts resulting in an increase in the rate of chlorophyll production (Parthier 1979; Kaminek et al. 1992; Mok and Mok 1994). Cytokinins can also delay senescence of cut flowers (van Staden et al. 1990) and fresh vegetables (Weaver 1972; Ludford 1987). Mothes and Engelbrecht (1961) showed that kinetin had the ability to promote the transport of organic substances in excised leaves kept in the dark. Leopold and Kawase (1964) showed that when one bean leaf was sprayed with kinetin the leaf adjacent to it began to senesce. The ability of cytokinins to promote the translocation of nutrients and the creation of sinks has since been shown in many plants (Thimann 1987; Mok and Mok 1994).

ABSCISIC ACID

Biosynthesis of Abscisic Acid

ABA is a sesquiterpene composed of three isoprene units. It is synthesized via the early steps of the mevalonic acid pathway with two potential routes from isopentenyl pyrophosphate as shown in Figure 3.12. The two possible routes are first via farnesyl pyrophosphate to ABA and second via carotinoids through a series of steps to ABA.

Inactivation of Abscisic Acid

ABA can be metabolized in two different ways: It can be converted to abscisyl-β-D-glucopuranoside, which is a reversible reaction, or it can be irreversibly converted to 6'-hydroxymethyl ABA, phaseic acid, or 4'-dihydrophaseic acid (Figure 3.15). ABA can also be inactivated by the attachment of a glucose to the carboxyl group of ABA to form an ABA-glucose ester. Inactivation due conjugation of ABA is similar to what occurs with IAA, gibberellins, or cytokinins.

Physiological Effects of ABA

Stomatal Closure. The importance of ABA as a stress-induced plant growth substance has been known for many years (Davies and Jones 1991). Exogenous applications of ABA have since been shown to close stomates in the light and will remain closed until ABA is metabolized (Raschke 1987;

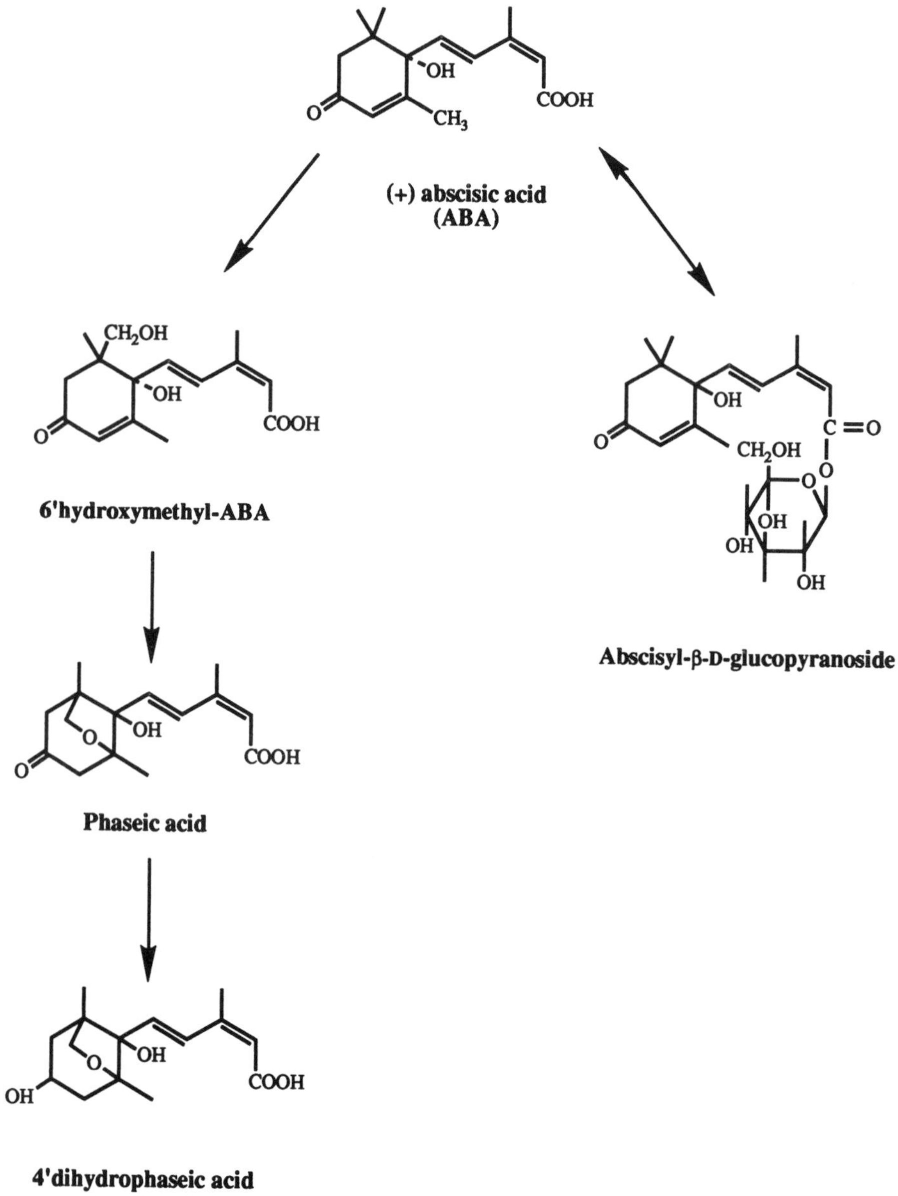

**(+) abscisic acid
(ABA)**

6'hydroxymethyl-ABA

Abscisyl-β-D-glucopyranoside

Phaseic acid

4'dihydrophaseic acid

Figure 3.15.   Metabolism of abscisic acid.

Davies and Jones 1991). Harris and Outlaw (1990) using an immunoassay for ABA were able to quantify ABA in a single guard cell showing the ability to detect ABA in the femtogram range (1×10^{-15}). Using this technique they were able to show that following water stress there was a 20-fold increase in ABA levels. Water-stressed roots also produce ABA which can be transported throughout the plant. It is thought that when water is limited the root tips perceive this stress and produce ABA, which is transported to the leaves causing stomatal closure. The net result is a reduction in water loss and photosynthesis, thereby acting as a survival mechanism (Davies and Jones 1991).

When plants perceive light under unstressed conditions there is an influx of potassium into the guard cells which is facilitated by an ATP-dependent pump for K^+ which is located on the plasma membrane of the guard cells while H^+ and organic acids such as malic acid are transported out. When this occurs, the osmotic potential becomes negative within the cell reducing the water potential and causing water to rush into the cell resulting in stomatal opening (Figure 3.16). When plants are stressed ABA will cause K^+ to leak out of the guard cells while H^+ and organic acids enter causing the guard cells to close. Once closed ABA will prevent opening in response to light by blocking the previously mentioned process until it is inactivated by metabolism. The signal transduction mechanism by which ABA produces these changes has not yet been fully established. However, there is evidence suggesting that Ca^{2+} and GTP-binding proteins have a role in the signal transduction chain (Davies and Jones 1991; Li and Assmann 1993).

Defense Against Salt and Temperature Stress.   ABA levels increase in response to salt, cold, and high-temperature stress, each of which are known to cause a deficiency of water. It has been suggested that ABA synthesis is regulated at the transcriptional level and following an increase in ABA there is a modification of gene expression in the stressed plants (Davies and Jones 1991). It has also been shown that exogenous applications of ABA have the ability to harden plants against frost damage (Guy 1990; Tanino et al. 1990) and excess salt (Skriver and Mundy 1990). Research in this area has a number of practical implications which should be realized in the near future.

Dormancy.   Since the discovery of ABA researchers have been trying to provide evidence that it is the dormancy hormone but they have been unsuccessful. Early work by Eagles and Wareing (1963) showed that under short days ABA levels were increased in leaves and buds resulting in dormancy and that exogenous applications of ABA to nondormant buds promoted dormancy. It has also been shown that short-day treatments which induce dormancy in some species do not promote increases in ABA levels correlating with the induction of dormancy. There are also numerous studies which show that

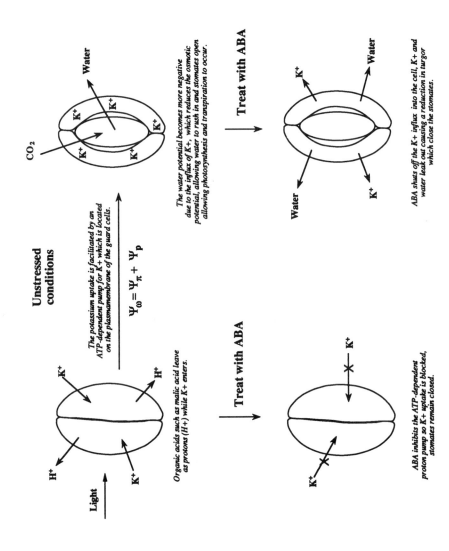

Figure 3.16. Schematic representation showing stomatal opening/closing and the effects of abscisic acid on this process.

exogenous application of ABA promotes seed dormancy in many species. One conclusion which might be drawn from this information is that in some species ABA is important by itself, whereas interactions with other known or unknown plant growth substances may be involved in other plant systems (Davies and Jones 1991).

Abscission, Seed Germination, and Growth. It was initially suggested that ABA was responsible for causing abscission of leaves, fruits, and flowers; however, it is now generally accepted that it is not. Osborne (1989) reviewed the effects of ethylene and ABA on abscission and concluded that ABA had no direct role in the abscission process. ABA probably acts indirectly by causing premature senescence, which causes a rise in ethylene production, which turns on numerous genes involved in abscission.

It is known that ABA has a wide variety of physiological, biochemical, and molecular effects in seeds and that ABA is universally present in developing seeds; however, at the present time the direct role for ABA in these processes remains unknown (Davies and Jones 1991).

ETHYLENE

Biosynthesis of Ethylene

Ethylene is a gaseous plant growth substance which has been shown to be involved in numerous aspects of plant growth and development from seed germination through senescence and death of the plant. The biosynthesis of ethylene is one of the best characterized pathways for any of the known plant growth substances. Ethylene is synthesized from methionine by the reactions outlined in Figure 3.17. Methionine is converted to S-adenosylmethionine (AdoMet) via the enzyme AdoMet synthetase. A portion of AdoMet is recycled in the following sequence: methyl thioadenosine to methylthioribose to methyl-thioribose-1-phosphate to 2-keto-4-methylbutyrate and back to methionine (Abeles et al. 1992). AdoMet also proceeds in two other ways, one is to S-adenosylmethyl thiopropylamine by the enzyme AdoMet decarboxylase and on to polyamine biosynthesis. The other is to 1-aminocyclopropane-1-carboxylic acid (ACC) via the enzyme ACC synthase (rate limiting step in this pathway) which in turn has two possible fates, it may be converted to ethylene by ACC oxidase or to malonyl-ACC (an inactive end product) via the enzyme ACC N-malonyltransferase. The HCN given off in the conversion of ACC to ethylene can combine with cysteine to form cyanoalanine plus H_2S via the enzyme β-cyanoalanine synthase. Through labeling experiments it has been shown that methionine is utilized in the following way: C-5 along with sulphur is recycled, C-3,4 are used for ethylene, C-2 in HCN, and C-1 is found in CO_2 (Figure 3.18). However, many of the studies on ethylene biosynthesis have

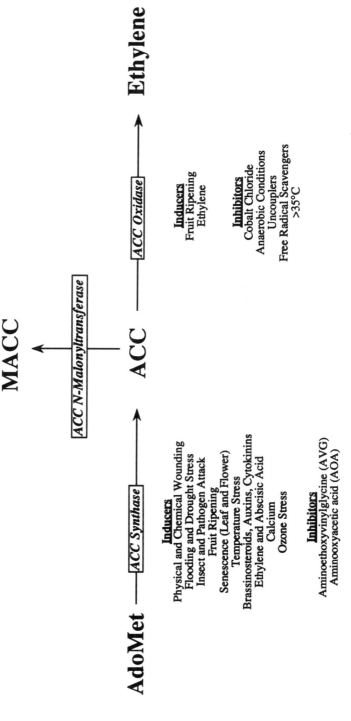

Figure 3.17. Biosynthetic reactions from S-adenosylmethionine (AdoMet) to ethylene plus inducers and inhibitors.

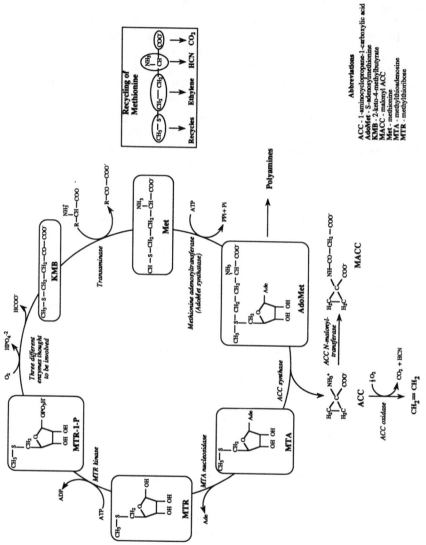

Figure 3.18. The ethylene biosynthetic pathway.

focused on the steps between AdoMet and ethylene. As shown in Figure 3.17, there many factors which can stimulate ethylene biosynthesis, one of which is ethylene itself. This process is called autocatalytic ethylene production. Ethylene first stimulates ACC oxidase followed by a large increase in ACC synthase activity. Within recent years ACC synthase and ACC oxidase proteins have been purified and their genes cloned in many tissues, whereas ACC N-malonyltransferase has recently been purified and characterized, and work is in progress to clone the gene for this enzyme (Guo et al. 1992, 1993). Since the genes for ACC synthase and ACC oxidase were cloned they have been utilized to modify fruit ripening and are currently being investigated for their ability to regulate other plant physiological processes (Kende 1993; Abeles et al. 1992).

Properties of Ethylene and Structure/Activity Relationships

Ethylene is an unsaturated hydrocarbon with a molecular weight of 28.05. It has a boiling point of $-103°C$. It is colorless, flammable, lighter than air (the relative density for ethylene is 0.979 while air is 1.0), and it is soluble in water (at 0°C 315 µl/L and at 25°C 140 µl/l). The structure/activity studies outlined in Table 3.1 show that ethylene is more effective than other hydrocarbons in regulating a given response although others are still active.

Table 3.1.   Structure-activity studies.

Compound	Formula	Relative Activity
Ethylene	$CH_2 = CH_2$	1
Propylene	$CH_3CH = CH_2$	100
Vinyl chloride	$CH_2 = CHCl$	1,400
Carbon monoxide	CO	2,700
Acetylene	$CH \equiv CH$	2,800
Vinyl fluoride	$CH_2 = CHF$	4,300
Propyne	$CH_3C \equiv CH$	8,000
Vinyl methyl ether	$CH_2 = CH\text{-}O\text{-}CH_3$	100,000
1-Butene	$CH_3CH_2CH = CH_2$	270,000
Carbon dioxide	CO_2	300,000

Induction of Ethylene by Auxin

The initial discovery that auxin regulates ethylene production was made by Zimmerman and Wilcoxon (1935). They observed that the application of heteroauxin (IAA) to plant shoots promoted epinasty of other untreated plants enclosed in the same container. Their observation plus the ablility of auxins and ethylene to cause a number of similiar effects suggested that some of those responses attributed to auxin might be due to ethylene. This idea went unnoticed for the next 29 years when Morgan and Hall (1964) showed a parallel relationship between auxin and ethylene responses and independently discovered the capacity of auxin to promote ethylene biosynthesis. Today it is generally accepted that auxin stimulates ethylene production and it is thought that in vegetative tissues ethylene production is regulated by endogenous levels of auxin (Abeles et al. 1992; Arteca 1990).

Stress Ethylene Production

The tendency of stress to promote ethylene production has given rise to the term *stress ethylene*. Numerous stresses such as chemicals, drought, flooding, radiation, insect damage, disease, mechanical wounding, and others have been shown to stimulate ethylene production. Stress ethylene is metabolic in origin and is produced by living cells. Once cells die there is no further ethylene produced. Ethylene produced in response to stress appears to be a secondary messenger allowing a plant to respond (Abeles et al. 1992).

Physiological Effects of Ethylene

Fruit Ripening.   Ancient Egyptians unknowingly took advantage of increased ethylene production resulting from injury by cutting immature sycamore figs to stimulate ripening. In more recent times, fruit ripening was shown to be regulated by ethylene. It is now generally accepted that ethylene plays an essential role in the ripening of climacteric fruits and the expression "One rotten apple spoils the whole bushel" now has a scientific basis (Theologis 1992). The term *climacteric* refers to fruits which will ripen in response to ethylene and *nonclimacteric* refers to those which do not. Ethylene goes from undetectable to 0.1 to 1 µl/l, which stimulates ripening of fruits which exhibit a climacteric rise in respiration (e.g., apples and pears), whereas, nonclimacteric fruits synthesize very little ethylene and are not induced to ripen (e.g., grapes and cherries). The term *climacteric* was initially used to designate an increase in fruit respiration; however, it now includes a rise in ethylene production (Figure 3.19). Synthesis of ethylene has been greatly reduced in transgenic tomato plants by expression of antisense gene constructs of ACC oxidase (Hamilton et al. 1990) or ACC synthase (Oeller et al. 1991) and by expression of sense gene construct of ACC deaminase (Klee et al. 1991) and in all cases significant delays in fruit ripening were achieved. All evidence to date with transgenic plants indicates that the degree to which ethylene is reduced is correlated with the delay in ripening (Klee

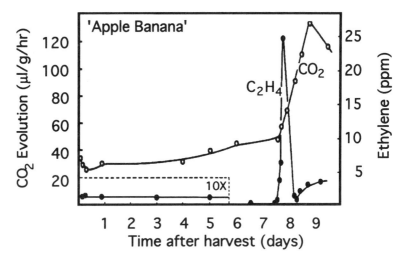

Figure 3.19.   The relationship between ethylene production and respiration during the climacteric rise in banana (From Burg and Burg (1965)).

1993). Recently, ACC N-MTase, which is the enzyme responsible for the conversion of ACC to MACC, which is an inactive end product was purified and characterized by Guo et al. (1992, 1993). This discovery will now allow these researchers to isolate the gene for ACC N-MTase and to use this gene to genetically transform plants to delay fruit ripening in the same manner as with the deaminase gene (Klee 1993). However, this time a plant gene will be used instead of a bacterial gene in order to inactivate ACC. For more details on the involvement of ethylene in fruit ripening see Chapter 10 of this text and Abeles et al. (1992).

Seedling Growth.   The first effect of ethylene was shown in etiolated pea seedlings by Neljubow (1901), who demonstrated the triple response showing that ethylene inhibited elongation and promoted lateral expansion and horizontal growth (Figure 3.20). It is known today that ethylene can inhibit or promote elongation of growing stems, roots, or other organs. The inhibition of elongation has been shown to be rapid and reversible. Ethylene has also been shown to promote elongation in stems and roots; however, this occurs at a slower rate then the inhibitory response. The longer lag period which has been reported in many studies is consistent with the hypothesis that growth promotion by ethylene may be an indirect effect (Abeles et al. 1992).

The triple response promoted by ethylene can act as a survival mechanism in seedlings. For example, when a pea seedling encounters a barrier such as soil crusting, a rock, or something which restricts its emergence it responds

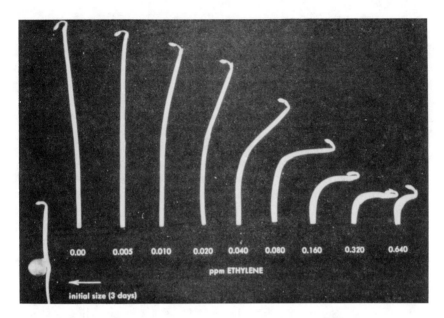

Figure 3.20.   Promotion of the triple response in etiolated pea seedlings by ethylene.

with the triple response, allowing it to either break through or go around the obstruction, thereby permitting it to reach the soil surface and grow. When ethylene production in the hook region of a seedling is very high it remains tightly hooked (pea or bean) enabling it to make its way through physical forces encountered in the soil without being damaged. Once the seedling perceives red light there is a decrease in ethylene production resulting in hook opening. This acts as a safety mechanism which prevents premature hook opening and damage to the seedling prior to emergence (Abeles et al. 1992).

Abscission.   Abscission may be defined as the separation of an organ or plant part from the parent plant. The process of abscission is very important in agriculture since abscission or nonabscission of flowers, fruits, and leaves influences yields and efficiency of harvesting operations, both of which will be discussed later in this text. Ethylene is thought to have a natural role in the regulation of the rate of abscission. Three lines of evidence supporting the role of ethylene in abscission have been proposed by Reid (1985) which still apply today. First, ethylene production increases prior to abscission in many abscising plant organs. Second, treatment of a wide range of plant species with ethylene or ethylene-releasing compounds stimulates abscission. Third, inhibitors of ethylene biosynthesis or action will inhibit abscission.

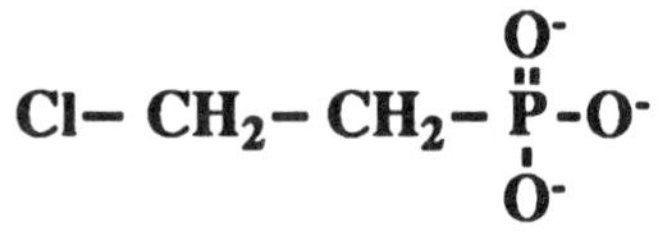

$$Cl-\ CH_2-\ CH_2-\ \overset{\overset{O^-}{\|}}{\underset{\underset{O^-}{|}}{P}}-O^-$$

Figure 3.21. Structural formula for the ethylene-releasing compound ethrel (2-chloro-ethylphosphoric acid).

Flowering. Many years ago it was observed that smoke from wood would hasten flowering in pineapples and mangos (Traub et al. 1940). It is now known that ethylene is the key component in smoke which speeds the flowering process. Ethylene in most cases inhibits flowering; however, it has a stimulatory effect in pineapples (De Greef et al. 1989), mangos (Chacko et al. 1974), lychee (Chen and Ku 1988), and *Plumbago indica* (Nitsch and Nitsch 1969). Plants can be treated with ethylene either directly or through the use of ethylene-releasing substances such as Ethrel (trade name) 2-chloroethylphosphoric acid (Figure 3.21) or indirectly with auxins which stimulate ethylene production naturally (Abeles et al. 1992).

Senescence. Senescence may be defined as a general failure of many synthetic reactions that precede cell death and is the phase of plant growth which extends from full maturity to death and is characterized by chlorophyll, protein, or RNA degradation as well as other factors. Workers have shown with both leaves and flowers that exogenous ethylene treatments enhance the senescence process and that increased ethylene production occurs during aging. It must be noted that there are exceptions where ethylene may not be solely involved in the senescence process. Bleecker et al. (1988) has shown that the leaves of the ethylene-insensitive *Arabidopsis* mutant (etr) senesce at a slower rate then the wild type; however, they still senesce, indicating that ethylene is not the only factor involved. In addition, there are some plants which exhibit no increase in ethylene production prior to senescence, in these cases it may be that there is an increase in sensitivity to ethylene rather than by having its effect as a result of increased production (Abeles et al. 1992).

Other Physiological Effects. In addition to the previously described physiological processes, ethylene has been shown to be involved in photosynthesis, respiration, transpiration, dormancy, seed germination, bud break, apical dominance, cell growth, tissue culture, embryogenesis, epinasty, root initiation, storage organs, xylem formation, floral inhibition, sexual development, gravitropism, thigmomorphogenesis, exudation, resin ducts, latex formation, gummosis, hypertrophy, and others (Abeles et al. 1992).

BRASSINOSTEROIDS

Biosynthesis of Brassinosteroids

At the present time the biosynthesis of brassinolide has not yet been thoroughly investigated; however, a proposed pathway for the biosynthesis of brassinolide has been outlined in two review articles (Mandava 1988; Yokota et al. 1991).

Structure/Activity Relationships

All of the naturally occurring brassinosteroids are known to be derivatives of 5 α-cholestane. Variations of kinds and orientation on the skeleton have been shown to have an effect on activity. In order to have brassinosteroid activity the following structural requirements must be met:

1. A trans A/B ring system (5 α-hydrogen)
2. A 6-ketone or a 7-oxa-6-ketone system in ring B
3. *Cis* α-oriented hydoxyl groups at C-2 and C-3 positions
4. *Cis* hydroxy groups at C-22 and C-23 as well as a methyl or ethyl group at C-24
5. The α-orientation at C-22, C-23, and C-24 are more active then β-oriented compounds.

These requirements were shown in several different test systems each showing the same result (Thompson et al. 1981, 1982; Arteca et al. 1985; Takasuto et al. 1983).

Transport and Metabolism of Brassinosteroids

At the present time it is not known how brassinosteroids are transported. When brassinosteroids are applied to the roots of tomato plants there is a stimulation in ethylene biosynthesis resulting in epinasty (Figure 3.22) (Schlagnhaufer and Arteca 1985). Prior to this work indirect evidence was presented by several workers indicating that BR could be transported from the roots to the shoots of plants. It was shown that when BR was applied to the roots little or no ACC was found in the xylem sap indicating that there was a signal (presumably BR) from the roots which stimulated ACC synthesis in leaf tissue. Others have shown that when BR is applied to the roots of tomato and radish plants there was an increase in petiole and hypocotyl elongation, and when applied to the base of mung bean cuttings it promoted elongation of the epicotyls (Sasse 1991).

In a subsequent study by Schlagnhaufer and Arteca (1991) the transport and

Figure 3.22.   Brassinosteroid-induced epinasty in *Lycopersicon esculentum*.

metabolism of [³H]BR was studied in tomato plants. The application of [³H]BR to the roots of tomato plants for 12 hours led to the production of unknown metabolites. When the plants were returned to a solution minus BR after 24 hours, the ACC content in these tissues decreased and there was a large increase in the two BR metabolites suggesting that the plant metabolizes BR to inactive forms resulting in a decrease in ethylene production. Additional evidence that plants have the ability to metabolize BR was presented in a study by Yokota et al. (1991). In this study they showed that when radiolabeled castasterone or brassinolide were fed to mung bean or rice seedlings there was an increase in polar metabolites.

Physiological Effects of Brassinosteroids

Comparisons with Other Plant Growth Substances in Different Bioassays.
Since the discovery of brassinolide its biological activity in bioassay systems designated for auxins, gibberellins and cytokinins has been investigated (Table 3.2).

One of the main effects of brassinolide appears to be its close relationship with IAA, typically acting synergistically. Although in most cases brassinolide acts in a similar manner to auxins, gibberellins, or cytokinins, in auxin bioassays based on root formation in mung bean, pea shoot lateral decapitated bud

Table 3.2. Activity of brassinolide (BR) in selected auxin (IAA), gibberellin (GA₃), and cytokinin (KIN) bioassays.

Auxin bioassays	Activity[1]	
	BR	IAA
Azuki bean epicotyl section elongation	P (10 μm)[2]	P (100 μm)
Maize etiolated mesocotyl section elongation	P (10 μm)	P (1 μm)
Pea etiolated epicotyl section growth	P (10 μm)	P (3 μm)
Mung bean hypocotyl root formation	I (1 μm)	P (10 μm)
Pea decapitated shoot lateral bud growth	NE (500 ppm)	I (500 ppm)
Cress seedling root elongation	NE (10 μm)	I (10 μm)
Bean etiolated hypocotyl hook opening	I (10 μm)	I (1 μm)
Jerusalem artichoke tuber slice weight change	IN (10 μm)	IN (1 μm)
Ethylene production in mung bean hypocotyls	P (1 μm)	P (1 μm)
Bending of dwarf rice leaf lemina	P (1 μm)	P (1 μm)
Epinasty in tomato plants	P (1 μm)	P (100 μm)

Gibberellin bioassays	Activity[1]	
	BR	GA3
Bean etiolated hypocotyl elongation	P (10 μm)	P (1 μm)
Mung bean epicotyl elongation	P (1 μm)	P (10 μm)
Cucumber hypocotyl elongation	P (1 μm)	P (10 μm)
Pea (dwarf) excised apical section elongation	P (10 μm)	P (1 μm)
Pigweed betacyanin formation in light	I (10 μm)	I (1 μm)
Dock leaf disc senescence	P (50 μm)	R (50 μm)

Cytokinin bioassays	Activity[1]	
	BR	KIN
Cucumber cotyledon expansion	P (100 μm)	P (10 μm)
Pea (dwarf) apical hook and tip section expansion	NE (10 μm)	P (10 μm)
Pigweed betacyanin formation in dark	NE (100 μm)	P (100 μm)
Cocklebur leaf disc senescence	P (50 μm)	R (50 μm)

[1] The following code is used in the table: P = promotes; I = inhibits; IN = increases; R = retards; and NE = no effect.

[2] Concentrations given are for optimal responses, not threshold detection limits.

growth, and cress seedling root elongation, BR and IAA act differently. In the dock leaf disc senescence bioassay for gibberellins BR promotes senescence, whereas gibberellins delay senescence. In cytokinin bioassays using the dwarf pea apical hook and tip expansion, pigweed betacyanin formation, and cockelbur leaf disc senescence bioassays BR and cytokinin also act differently (Yopp et al. 1981).

Promotion of Ethylene Biosynthesis and Epinasty.   In etiolated mung bean hypocotyl segments BR increases ethylene biosynthesis at the step between AdoMet and ACC by stimulating ACC synthase activity. BR-induced ethylene can be inhibited by aminooxyacetic acid (AOA), Co^{2+}, fusicoccin (a fungal toxin), and the auxin transport inhibitors 2,3,4-triiodobenzoic acid and 2-(p-chlorophenoxy)-2-methylpropionic acid. BR acts synergistically with active auxins and calcium, whereas it has an additive effect when used in combination with cytokinins in the stimulation of ethylene production. Light has been shown to inhibit BR-induced ethylene production while having little effect on ethylene produced in response to IAA (Arteca 1990). BR applications to the roots of hydroponically grown tomato plants have also been shown to promote a dramatic increase in the step between AdoMet and ACC resulting in an increase in ACC, ethylene, and petiole bending (Schlagnhaufer and Arteca 1985). BR shows similar effects in the promotion of ethylene production in plant parts as well as in a whole plant system unlike auxin which is typically more effective in eliciting a response in a detached plant system.

Shoot Elongation.   Brassinosteroids have been shown to promote elongation of vegetative tissue in a wide variety of plants at very low concentrations. Wang et al. (1993) has shown that brassinosteroid can stimulate hypocotyl elongation by increasing wall relaxation without a concomitant change in wall mechanical properties in Pakchoi. In a subsequent paper by Zurek et al. (1994) they showed BR stimulates wall loosening in soybean epicotyl segments. However, they found that the loosening in soybean appears to alter mechanical properties of the wall, since they observed an increase in plastic extensibility as measured by Instron analysis. The promotive effect of BR on elongation has clearly been shown under white, green, or weak red light. However, little or no effects have been found in complete darkness suggesting that brassinolide action may result by overcoming the inhibitory effects of light (Mandava 1988; Cutler et al. 1991; Kamuro and Inada 1991).

Root Growth and Development.   Brassinosteroids are powerful inhibitors of root growth and development. BR and IAA effects are generally similar and a synergism between the two is typically reported. However, in the case of root initiation they act quite differently, IAA stimulating and BR having an inhibitory effect. The possible reasons for these differences may be that BR acts independently of IAA in roots, or it acts as an antagonist of IAA. It has been documented that ethylene has an inhibitory effect on root growth (Roddick and Guan 1991) and BR stimulates ethylene production; therefore, it is possible that the inhibition of root growth is due to BR-induced ethylene production. However, further research is necessary before definitive statements can be made.

Plant Tissue Culture.   24-Epibrassinolide has been shown to mimic culture conditioning factors and to synergize with these factors in promoting carrot cell growth (Bellincampi and Morpurgo 1991). However, in transformed tobacco cells brassinosteroids have been shown to significantly inhibit cell growth at concentrations as low as 10^{-8} (Bach et al. 1991).

Antiecdysteroid Effects in Insects.   Structurally brassinosteroids are very similar to ecdysteroids, which are moulting hormones of insects and other arthropods. Brassinosteroids have been shown to interfere with ecdysteroids at their site of action and are the first true antiecdysteroids observed thus far. Since brassinosteroids are natural products they are good candidates for safer insect pest control. However, their cost must be reduced before becoming economically feasible (Richter and Koolman 1991).

Other Biological Effects.   Preliminary results indicate that brassinosteroids have a number of other biological effects such as the promotion of changes in plasmalemma energization and transport, assimilate uptake; enhancement of xylem differentiation, enhanced resistance to chilling, disease, herbicide, and salt stress; promotion of germination, and decreased fruit abortion and drop (Cutler et al. 1991; Iwahori et al. 1990).

Effects of Brassinosteroids on Nucleic Acid and Protein Synthesis

When bean plants were treated with BR there was a significant increase in RNA and DNA polymerase activities and synthesis of RNA, DNA, and protein (Kalinich et al. 1985). Putative inhibitors of RNA and protein synthesis have since been shown to interfere with BR-induced epicotyl elongation indicating that the growth effects induced by BR depend on the synthesis of nucleic acids and proteins (Mandava 1988). BR has the ability to stimulate elongation of soybean epicotyls (Clouse et al. 1992) and *Arabidopsis* peduncles (Clouse et al. 1993) at low concentrations. It was found that when BR stimulated elongation in soybean, gene expression patterns were altered by BR either plus or minus IAA indicating that BR was having its effect alone. However, the possibility still exists that it could be acting with endogenous auxin. In order to take this work one step further the effects of BR on several known auxin-regulated genes were evaluated. This work indicated that the molecular mechanism of BR-induced elongation is different than auxin-induced elongation in this system (Clouse et al. 1992). More recently a brassinosteroid-regulated gene from elongating soybean epicotyls has been identified and characterized (Zurek and Clouse 1994). There is a synergistic relationship in the stimulation of ACC synthase when BR and IAA are used in combination in etiolated mung bean hypocotyl sections. Recently, a full-length cDNA (pAIM-1) for IAA-induced

ACC synthase was identified and characterized in this tissue (Botella et al. 1992). Using this cDNA as a probe it was shown that BR could turn on the gene for ACC synthase (Arteca et al. 1993).

Kulaeva et al. (1991) presented evidence that BR can protect cereal leaf cells from heat shock and salt stress. They showed that when wheat leaves were subjected to temperatures of 40°C both 22S, 23S-homobrassinolide and 24-epibrassinolide activated total protein synthesis and de novo synthesis of different polypeptides at these high temperatures as well as at normal temperatures. The 22S, 23S-homobrassinolide also stimulated heat shock granules in the cytoplasm and increased thermotolerance of total protein synthesis under heat shock. 24-epibrassinolide protected the leaf cell ultrastructure in leaves under salt stress and also prevented nuclei and chloroplast degradation.

Practical Applications of Brassinosteroids

In the early 1980s USDA scientists showed that BR could increase yields of radishes, lettuce, beans, peppers and potatoes. However, subsequent results under field conditions were disappointing because inconsistent results were obtained. For this reason testing was phased out in the United States. More recently large-scale field trials in China and Japan over a six-year period have shown that 24-epibrassinolide, an alternative to brassinolide, increased the production of agronomic and horticultural crops (including wheat, corn, tobacco, watermelon, and cucumber). However, once again depending on cultural conditions, method of application, and other factors the results sometimes were striking while other times they were marginal. Further improvements in the formulation, application method, timing, effects of environmental conditions, and other factors need to be investigated further in order to identify the reason for these variable results (Cutler et al. 1991).

Are Brassinosteroids a New Class
of Plant Growth Substances?

Brassinosteroids are thought to be a new class of plant growth substances (Sasse 1990). However, there are gaps in our knowledge in some areas which allow for a degree of skepticism. The first line of evidence supporting the idea that brassinosteroids are a new class of plant growth substances is that they are widely distributed within the plant kingdom. Second, they have an effect at extremely low concentrations, both in bioassays and whole plants. Third, they have a range of effects which are different from the other classes of plant substances and there are strict structural requirements for a brassinosteroid to be active in promoting a physiological response. Fourth, they can be applied to one part of the plant and transported to another location where, in very low amounts, they elicit a biological response. At the present time the actual mechanism of BR action remains unclear. Recent studies using molecular technology suggest that BR has the ability to regulate gene expression resulting in elongation (Clouse et al. 1992; Zurek and Clouse 1994) and ethylene production (Arteca et al. 1993). In addition, a BR-insensitive mutant has been identified which should facilitate identification and cloning of genes for the BR receptor and possibly genes regulating other aspects of the diverse effects of BR in plants (Clouse et al. 1993). However, before definitive conclusions on the mechanism of action can be made more work is necessary.

SALICYLATES

Biosynthesis of Salicylic Acid

In order to manipulate levels of salicylic acid (SA) in plants utilizing molecular tools available today there needs to be a better understanding of the salicylic acid biosynthetic pathway. A proposed pathway in plants has been outlined in a review by Raskin (1992). This shows that the shikimic acid pathway is used to produce cinnamic acid which can be converted in two ways to salicylic acid, one via o-coumaric acid and the other via benzoic acid (Figure 3.23).

Transport and Inactivation of Salicylic Acid

At the present time there is no direct evidence that SA is transported in plants. However, the physical properties indicate that it could be readily transported throughout the plant (Raskin 1992). SA can be inactivated in two different ways, by conjugation and metabolic inactivation. It has been shown that plants have the capacity to produce o-glucosides or glucose esters of SA. Plants can also metabolically inactivate SA by additional hydroxylation of its aromatic ring (Raskin 1992).

Physiological Effects of Salicylic Acid

Effects on Flowering. There are a number of reports in the literature showing that SA can stimulate flowering. The first report showing that SA had a promotive effect on flowering was organogenic tobacco callus where it was shown that SA in combination with kinetin and IAA promoted bud formation. While this finding was interesting it did not provide solid evidence that SA was involved in flowering because there are many compounds that can induce flower buds in tobacco cell cultures. The effects of SA on flowering has been reviewed by Raskin (1992) and although there are many cases where SA has a promotive effect there are also conflicting reports.

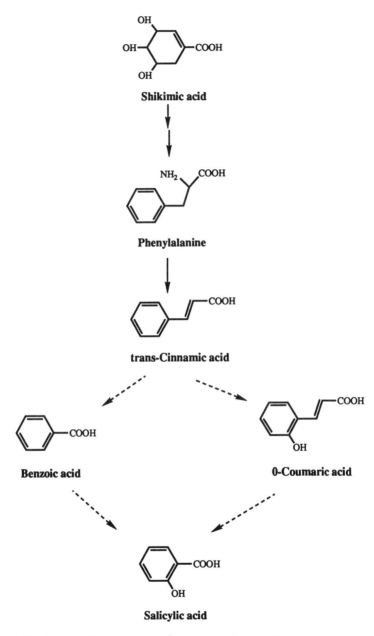

Figure 3.23. Proposed biosynthetic pathway for salicylic acid in plants (from Raskin (1992)).

Relationship Between Salicylic Acid and Heat Production in Plants.
Thermogenicity (heat production) in plants was first described by Lamarck
(1778) in the genus *Arum*. Today it is known that heat production occurs in
male reproductive structures of cycads and in flowers or inflorescences of some
angiosperm species belonging to several families. The heating is thought to be
associated with a large increase in the cyanide-resistant nonphosphorylating
pathway which is unique to the mitochondria. *Arum* lilies (voodoo lily) produce an inflorescence which can be up to 80 cm long. It was suggested by van
Herk (1937) that a burst of metabolic activity in the appendix of the voodoo
lily was triggered by *Calorigen* a water-soluble substance produced in the male
flower primordia just below the appendix. In an attempt to identify calorigen
it was found that SA was the actual compound inducing the heat production in
voodoo lilies. In this system one day before blooming, SA moves from the male
flowers of the voodoo lily to the appendix (Figure 3.24a). It then induces heat,
which is a product of cyanide-insensitive respiration resulting in the volatilization of foul-smelling amines and indoles which are attractive to insect pol-
linators. Temperature increases as much as 14°C above ambient have been
reported to occur in the appendix (Raskin 1992).

Relationship Between Salicylic Acid and Disease Resistance in Plants.
Some disease-resistant plants restrict the spread of infection by pathogens to a
small area around the point of initial penetration where a necrotic lesion
appears. This protective cell suicide is called a hypersensitive reaction (HR).
An HR can lead to a systemic acquired resistance. Commonly associated with
HR and systemic aquired resistance is the production of low-molecular-weight
pathogenesis-related (PR) proteins. Resistance to pathogens and the production
of some PR proteins in plants can be induced by SA or acetylsalicylic acid
even in the absence of pathogenic organisms (Raskin 1992). During the development of the hypersensitive response to pathogens, large amounts of SA are
produced from cinnamic acid in the vicinity of the neucrotic lesions (Yalpani et
al. 1993). A large portion of SA is immobilized as β-O-D-glucosylsalicylic acid
and free SA enters the phloem where it is subsequently detected in the upper
leaves. Increases in SA are sufficient for the systemic induction of PR proteins
and resistance to subsequent infection (Figure 3.24b). It is not clear at the
present time whether the export of SA from the infected leaf can account for
all the SA present during systemic acquired resistance (Raskin 1992).

Is Salicylic Acid a New Class
of Plant Growth Substances?

SA is thought by some to be a new class of plant growth substances (Raskin
1992). It is a chemically characterized compound, ubiquitously found in the
plant kingdom and has an effect on many physiological processes in plants at

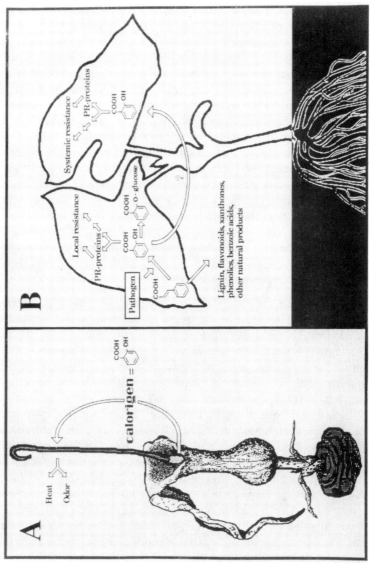

Figure 3.24. Schematic representation showing: (a) Movement of salicylic acid from the male flowers of the voodoo lily to the appendix; and (b) Increases in salicylic acid leading to the systemic induction of pathogenesis proteins and resistance to subsequent infection (From Raskin (1992)).

low concentrations. At the present time the mechanism of action of SA remains unclear. SA regulates some aspects of disease resistance and thermogenesis in plants. The biochemical link between the action of SA in plant disease resistance and its thermogenic and odor-producing effects in *Arum* lilies remains to be elucidated. Further research on the SA biosynthetic pathway and metabolism along with molecular studies on SA signal transduction should be conducted to better understand the mechanism of action of this important regulatory compound.

JASMONATES

Biosynthesis, Metabolism, and Transport of Jasmonates

Jasmonates are formed biosynthetically from linolenic acid by a series of reactions shown in Figure 3.25 and are metabolized as outlined in Figure 3.26 (Sembdner and Parthier 1993). Transport, intracellular location, and regulation of jasmonic acid biosynthesis is poorly understood. At the present time there is no direct evidence that jasmonates are transported from a site of synthesis to another site where they elicit a response (Staswick 1992; Sembdner and Parthier 1993).

Physiological Effects of Jasmonates in Plants

Jasmonates have been shown to elicit both inhibitory and promotive effects on the plant's morphology and physiology, some of which are similar to abscisic acid and ethylene. Exogenous applications of JA have inhibitory effects on seedling longitudinal growth, root length growth, mycorrhizial fungi growth, tissue culture growth, embryogenesis, seed germination, pollen germination, flower bud formation, pulvinule opening, carotenoid biosynthesis, chlorophyll formation, rubisco biosynthesis, and photosynthetic activities. In addition to its inhibitory effects it has also been shown to have promotive or inductive effects on sugarcane-cutting elongation, differentiation in plant tissue culture, adventitious root formation, breaking seed dormancy, pollen germination, turion germination, fruit ripening, pericarp senescence, leaf senescence, leaf abscission, tuber formation, tendril coiling, stomatal closure, microtubule disruption, chlorophyll degradation, respiration, ethylene biosynthesis, and protein synthesis. Providing details on all of the effects of jasmonates previously mentioned is beyond the scope of this book and therefore will be limited to a few effects and their relationship with other plant growth substances (Sembdner and Parthier 1993).

When exogenously applied jasmonates promote leaf senescence there is chlorophyll degradation and many other effects associated with the senescence

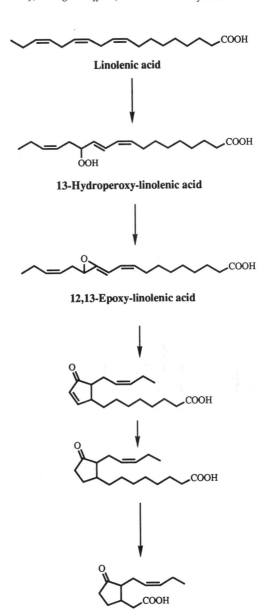

Figure 3.25.   Biosynthesis of jasmonic acid from linolenic acid (From Sembdner and Parthier (1993)).

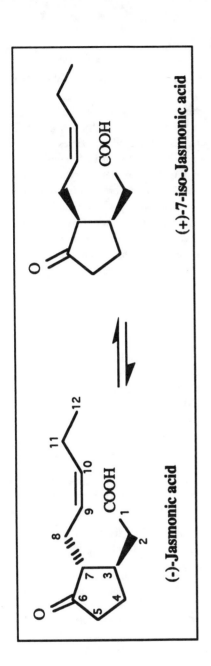

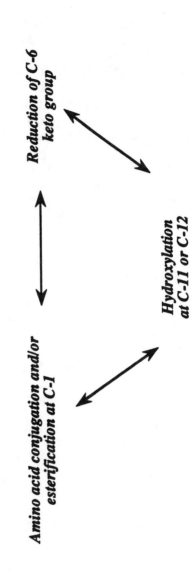

(−)-Jasmonic acid

(+)-7-iso-Jasmonic acid

Amino acid conjugation and/or esterification at C-1

Hydroxylation at C-11 or C-12

Reduction of C-6 keto group

Figure 3.26. Potential routes for metabolism of jasmonic acid in plants.

93

process. At the present time there is only indirect evidence on the role of JA in leaf senescence. It has been shown that cytokinins delay senescence and also counteract JA effects on chlorophyll and rubisco degradation. JA-Me has been shown to stimulate ethylene biosynthesis by increasing the activities of ACC synthase or oxidase, however, its effect depends on the plant species and the stage of plant development. It is possible that JAs effects on leaf senescence, fruit ripening, and other processes may be due to its stimulatory effect on ethylene biosynthesis. There are similarities as well as differences in structure, physical properties, and activity of abscisic acid and jasmonic acid. Both promote stomatal closure and induce proteinase inhibitors and *Brassica* seed storage proteins, but only JA has been shown to induce soybean vegetative storage proteins. The relationship between JA and other plant growth substances requires further elucidation before definitive statements can be made on its role in different plant processes (Sembdner and Parthier 1993).

Jasmonate-Induced Proteins and Regulation of Gene Expression

JA has been shown to induce proteins (Table 3.3) in all of the plant species tested. At the present time the function of many of these proteins remains unclear. Two functional types of proteins appear to be regulated by JA: storage proteins, and proteins which are involved in the plant's defense against attack from pathogens, herbivores, and physical and chemical stress. At the molecular level there is a large amount of evidence showing that JA affects gene expression in plants in a number of test systems (Sembdner and Parthier 1993).

Table 3.3. Proteins induced by jasmonic acid (From Sembdner and Parthier (1993)).

Protein	Source	Other inducers
Chalcone synthase	Soybean cell culture	Wounding
Napin	Rapeseed embryos	Wounding
Trypsin inhibitor	Alfalfa leaves	Wounding
Lipoxygenase	Soybean, pea	Wounding, desiccation
Vegetative storage proteins	Soybean	Wounding, desiccation, soluble sugars
Proteinase inhibitors	Tomato leaves	Wounding, ABA
Proteinase inhibitor II	Potato leaves	ABA, sugars
Phenylalanine ammonia lyase	Soybean cell culture	Oligosaccharides
Thionin	Barley leaves	—
Unknown function	Barley leaves	ABA, sorbitol, sugars, desiccation
Unknown function	Cotton cotyledons	ABA
Unknown function	Tomato leaves	Wounding

Potential of Jasmonates as a Systemic Signal and in the Stress Signal Transduction Chain

Methyl jasmonate (MeJA) can elicit a physiological response in a gaseous state. In fact, proteinase inhibitors and vegetative storage proteins can be induced by MeJA at nanomolar concentrations. Sagebrush (*Artemesia tridentata*) can release MeJA into the air in sufficient quantity to induce proteinase inhibitors in plants incubated in the same chamber (Farmer and Ryan 1992). This is similar to the response found by Zimmerman and Wilcoxon (1935) who showed that when IAA was applied to one tomato plant, ethylene was given off, promoting epinasty in other plants enclosed in the same chamber. It remains to be determined whether volatile MeJA is an effective natural signal between plants.

It is thought that the wound signal is perceived triggering lipases which attack the membrane releasing polyunsaturated fatty acids (linolenic acid) which are metabolized to jasmonates via the jasmonic acid biosynthetic pathway, thereby triggering a response. Support for this model is that the synthesis of inducible proteinase inhibitors and tendril coiling are triggered by jasmonates and also their precursors (Staswick 1992; Sembdner and Parthier 1993).

Is Jasmonic Acid a New Class of Plant Growth Substances?

Jasmonic acid is considered by some to be a new class of plant growth substances (Staswick 1992). It is a chemically characterized compound and has been identified in most organs of many plant species. It has physiological effects at very low concentrations and indirect evidence suggests that it is transported throughout the plant. There is a considerable amount of interest on the mechanism of jasmonic acid in plants and although a lot of progress has been made there are still major gaps in our understanding in this area of research (Sembdner and Parthier 1993).

REFERENCES

Abeles, F. B., Morgan, P. W., and Saltveit Jr., M. E. (1992). *Ethylene in Plant Biology. Second Edition*, Academic Press, Inc., San Diego, CA.

Andreae, W. A. and Good, N. E. (1957). "Studies in indoleacetic acid metabolism. IV: Conjugation with aspartic acid and ammonia as processes in the metabolism of carboxylic acid". *Plant Physiol.* 32:566–572.

Andreae, W. A. and van Ysselstein, M. W. (1960). "Studies of indoleacetic acid metabolism. VI: Indoleacetic acid uptake and metabolism by pea roots and epicotyls". *Plant Physiol.* 35:225–232.

Arteca, J. M., Botella, J. R., and Arteca, R. N. (1993). "Effects of plant hormones on ACC synthase gene expression in etiolated mung beans". *Plant Physiol.* 102S:131.

Arteca, R. N. (1990). "Hormonal stimulation of ethylene biosynthesis". In *Polyamines and Ethylene: Biochemistry, Physiology, and Interactions*, eds., H. E. Flores, R. N. Arteca, and J. C. Shannon, American Society of Plant Physiologists, Rockville, MD, pp. 216–223.

Arteca, R. N., Bachman, J. M., Yopp, J. H., and Mandava, N. B. (1985). "Relationship of steroidal structure to ethylene production by etiolated mung bean segments". *Physiol. Plant.* 64:13–16.

Bach, T. J., Roth, P. S., and Thompson, M. J. (1991). "Brassinosteroids specifically inhibit growth of tobacco tumor cells". In *Brassinosteroids. Chemistry, Bioactivity, and Applications*, eds., H. G. Cutler, T. Yokota, and G. Adam, American Chemical Society, Washington, DC, pp. 176–188.

Bandurski, R. S. (1984). "Metabolism of indole-3-acetic acid". In *The Biosynthesis and Metabolism of Plant Hormones*, eds., A. Crozier and J. R. Hillman, Cambridge University Press, Cambridge, U.K., pp. 183–200.

Bandurski, R. S., Schulze, A., and Reinecke, D. M. (1986). "Biosynthetic and metabolic aspects of auxins". In *Plant Growth Substances*, eds., M. Bopp, Springer-Verlag, Berlin, pp. 83–91.

Bellincampi, D. and Morpurgo, G. (1991). "Stimulation of growth induced by brass-inosteroid and conditioning factors in plant cell cultures". In *Brassinosteroids. Chemistry, Bioactivity, and Applications*, eds., H. G. Cutler, T. Yokota, and G. Adam, American Chemical Society, Washington, DC, pp. 189–199.

Bleecker, A. B., Estelle, M. A., Somerville, C., and Kende, H. (1988). "Insensitivity to ethylene conferred by a dominant mutation in *Arabidopsis thaliana*". *Science* 231:1086–1089.

Botella, J. R., Arteca, J. M., Schlagnhaufer, C. D., Arteca, R. N., and Phillips, A. T. (1992). "Identification and characterization of a full-length cDNA encoding for an auxin-induced 1-aminocyclopropane 1-carboxylate synthase from etiolated mung bean hypocotyl segments and expression of its mRNA in response to indole-3-acetic acid". *Plant Mol. Biol.* 20:425–436.

Briggs, W. R. and Baskin, T. I. (1988). "Phototropism in higher plants – controversies and caveats". *Botanica Acta* 101:133–139.

Brown, P. H. and Ho, T-H. D. (1986). "Barley aleurone layers secrete a nuclease in response to gibberellic acid. Purification and partial characterization of the associated ribonuclease, deoxyribonuclease, and 3′-nucleotidase activities". *Plant Physiol.* 82:801–806.

Chacko, E. K., Kohli, R. R., and Randhawa, G. S. (1974). "Investigations on the use of (2-chloroethyl)phosphonic acid (ethephon, CEPA) for the control of biennial bearing in mango". *Sci. Hort.* 2:389–398.

Chen, W.-S. and Ku, M.-L. (1988). "Ethephon and kinetin reduce shoot length and increase flower bud formation in lychee". *HortSci.* 23:1078.

Chibnall, A. C. (1954). "Protein metabolism in rooted runner-bean leaves". *New Phytol.* 53:31.

Cleland, R. E. (1987). "Auxin and cell elongation". In *Plant Hormones and Their Role in Plant Growth and Development*, ed., P. J. Davies, Martinus Nijhoff Publishers, Boston, pp. 132–148.

Clouse, S. D., Hall, A. F., Langford, M., McMorris, T. C., and Baker, M. E. (1993). "Physiological and molecular effects of brassinosteroids on *Arabidopsis thaliana*". *J. Plant Growth Reg.* 12:61–66.

Clouse, S. D., Zurek, D. M., McMorris, T. C. and Baker, M. E. (1992). "Effect of brassinolide on gene expression in elongating soybean epicotyls". *Plant Physiol.* 100:1377–1383.

Cohen, J. D. and Bialek, K. (1984). "The biosynthesis of indole-3-acetic acid in higher plants". In *The Biosynthesis and Metabolism of Plant Hormones*, eds., A. Crozier and J. R. Hillman, Cambridge University Press, Cambridge, pp. 165–181.

Crane, J. C. (1949). "Controlled growth of fig fruits by synthetic hormone applications". *Proc. Amer. Soc. Hort. Sci.* 54:102–108.

Cuatrecasas, P., Hollenberg, M. D., Chang, K., and Bennett, V. (1977). "Hormone receptor complexes and their modulation of membrane function". *Recent Progress in Hormone Research* 31:37–52.

Cutler, H. G., Yokota, T., and Adam, G. (1991). *Brassinosteroids: Chemistry, Bioactivity and Applications*, American Chemical Society, Washington, DC.

Danielli, J. F. (1954). "Morphological and molecular aspects of active transport". *Soc. Exp. Biol.* 8:502–515.

Davies, W. J. and Jones, H. G. (1991). *Abscisic Acid: Physiology and Biochemistry*, Bios Scientific Publishers, Oxford.

Davis, T. D. and Curry, E. A. (1991). "Chemical regulation of vegetative growth". *Critical Reviews in Plant Science* 10:151–188.

De Greef, J. A., De Proft, M. P., Mekers, O., Van Dijck, R., Jacobs, L., and Philippe, L. (1989). "Floral induction of bromeliads by ethylene". In *Biochemical and Physiological Aspects of Ethylene Production in Lower and Higher Plants*, eds., H. Clijster, M. De Proft, R. Marcelle and M. Van Poucke, Kluwer Academic Publishers, Dordrecht, pp. 313–322.

Dong, C.-N. and Arteca, R. N. (1982). "Changes in photosynthetic rates and growth following root treatments of tomato plants with phytohormones". *Photosynthesis Research* 3:45–52.

Eagles, C. F. and Wareing, P. F. (1963). "Experimental induction of dormancy in *Betula pubescens*". *Nature* 199:874.

Eliasson, L., Bertell, G., and Bolander, E. (1989). "Inhibitory action of auxin on root elongation not mediated by ethylene". *Plant Physiol.* 91:310–314.

Evans, M. L. (1985). "The action of auxin on plant cell elongation". *CRC Critical Reviews in Plant Sciences* 2:317–365.

Farmer, E. E. and Ryan, C. A. (1992). "Octadecanoid prescursors of jasmonic acid activate the synthesis of wound-inducible proteinase inhibitors". *Plant Cell* 4:129–134.

Filner, P. and Varner, J. E. (1967). "A test for de novo synthesis of enzymes: Density labeling with $H_2^{18}O$ of barley α-amylase induced by gibberellic acid". *Proc. Natl. Acad. Sci. USA* 58:1520–1526.

Fries, N. (1960). "The effect of adenine and kinetin on growth and differentiation of *Lupinus*". *Physiol. Plant.* 13:468.

Gmelin, R. and Virtanen, A. I. (1961). "Glucobrassicin, the precursor of indolylacetyl-nitrile, ascorbigen and SCN in *Brassica oleracea*". *Suomen Kem.* 34:15–18.

Grossmann, K. (1990). "Plant growth retardants as tools in physiological research". *Physiol. Plant.* 78:640–648.

Guo, L., Arteca, R. N., Phillips, A. T., and Liu, Y. (1992). "Purification and characterization of 1-aminocyclopropane-1-carboxylate N-malonyltransferase from etiolated mung bean hypocotyls". *Plant Physiol.* 100:2041–2045.

Guo, L. G., Phillips, A. T., and Arteca, R. N. (1993). "Amino acid N-malonyltransferases in mung beans: Action on 1-aminocyclopropane-1-carboxylic acid and D-phenylalanine". *J. Biol. Chem.* 268:25,389–25,394.

Guy, C. L. (1990). "Cold acclimation and freezing stress tolerance: Role of protein metabolism". *Annu. Rev. Plant Physiol. Plant Mol. Biol.* 41:187–233.

Haberlandt, G. (1902). "Über die Statolithefunktion der Stärkekörner (about the statolith function of starch grains)". *Berichte der Deutschen Botanisches Gesellschaft* 20:189–195.

Hamilton, A. J., Lycett, G. W., and Grierson, D. (1990). "Antisense gene that inhibits synthesis of the hormone ethylene in transgenic plants". *Nature* 346:284–287.

Harris, M. J. and Outlaw Jr., W. H. (1990). "Histochemical technique: A low-volume, enzyme-amplified immunoassay with sub-fmol sensitivity. Application to measurement of abscisic acid in stomatal guard cells". *Physiol. Plant.* 78:495–500.

Hillman, J. R. (1984). "Apical dominance". In *Advanced Plant Physiology*, eds., M. B. Wilkins, Pitman, London, pp. 127–148.

Iwahori, S., Tominaga, S., and Higuchi, S. (1990). "Retardation of abscission of citrus leak and fruitlet explants by brassinolide". *Plant Growth Reg.* 9:119–125.

Jablonski, J. R. and Skoog, F. (1954). "Cell enlargement and cell division in excised tobacco pith tissue". *Physiol. Plant.* 7:16–24.

Jacobs, M. and Gilbert, S. F. (1983). "Basal localization of the presumptive auxin transport carrier in pea stem cells". *Science* 220:1297–1300.

Jacobs, W. P. (1961). "The polar movement of auxin in the shoots of higher plants: Its occurrence and physiological significance". In *Plant Growth Regulation*, ed., R. M. Klein, Iowa State University Press, Ames, IA, pp. 397–409.

Kalinich, J. F., Mandava, N. B., and Todhunter, J. A. (1985). "Relationship of nucleic acid metabolism to brassinolide-induced responses in beans". *J. Plant Physiol.* 120:207–214.

Kaminek, M., Mok, D. W. S., and Zazimalova, E. (1992). *Physiology and Biochemistry of Cytokinins in Plants*, SPB Academic Publishing, Hague, The Netherlands.

Kamuro, Y. and Inada, K. (1991). "The effect of brassinolide on the light-induced growth inhibition in mung bean epicotyl". *Plant Growth Reg.* 10:37–43.

Kende, H. (1993). "Ethylene biosynthesis". *Annu. Rev. Plant Physiol. Plant Mol. Biol.* 44:283–307.

Klee, H. J. (1993). "Ripening physiology of fruit from transgenic tomato (*Lycopersicon esculentum*) plants with reduced ethylene synthesis". *Plant Physiol.* 102:911–916.

Klee, H. J., Hayford, M. B., Kretzmer, K. A., Barry, G. F., and Kishore, G. M. (1991). "Control of ethylene synthesis by expression of a bacterial enzyme in transgenic tomato plants". *Plant Cell* 3:1187–1193.

Kulaeva, O. N., Burkhanove, E. A., Fedina, A. B., Khokhlova, V. A., Bokebayeva, G. A., Vorbrodt, H. M., and Adam, G. (1991). "Effect of brassinosteroids on protein synthesis and plant cell ultrastructure under stress conditions". In *Brassinosteriods. Chemistry, Bioactivity and Applications*, eds., H. G. Cutler, T. Yokota, and G. Adam, American Chemical Society, Washington, DC, pp. 141–157.

Lamarck J. B. (1778). *Flore Francaise 3*, L'Imprimerie Royale, Paris.

Leopold, A. C. and Kawase, M. (1964). "Benzyladenine effects on bean leaf growth and senescence". *Am. J. Botany* 51:294–298.

Li, W. and Assmann, S. M. (1993). "Characterization of a G-protein regulated outward K^+ current in mesophyll cells of *Vicia faba* L.". *Proc. Natl. Acad. Sci. USA* 90:262–266.

Libbert, E., Wichner, S., Schiewer, U., Risch, H., and Kaiser, W. (1966). "The influence of epiphytic bacteria on auxin metabolism". *Planta* 68:327–334.

Ludford, P. M. (1987). "Postharvest hormone changes in vegetables and fruit". In *Plant Hormones and Their Role in Plant Growth and Development*, ed., P. J. Davies, Martinus Nijhoff, Boston, pp. 574–592.

Mandava, N. B. (1988). "Plant growth-promoting brassinosteroids". *Annu. Rev. Plant Physiol. Plant Mol. Biol.* 39:23–52.

Medford, J. I., Horgan, R., El-Sawi, Z., and Klee, H. J. (1989). "Alterations of endogenous cytokinins in transgenic plants using a chimeric isopentenyl transferase gene". *The Plant Cell* 1:403–413.

Miller, C. O. (1956). "Similarity of some kinetin and red light effects". *Plant Physiol.* 31:318.

Mok, D. W. S. and Mok, M. C. (1994). *Cytokinins. Chemistry, Activity, and Function*, CRC Press, Boca Raton, FL.

Morgan, P. W. and Hall, W. C. (1964). "Effect of 2,4-D on the production of ethylene by cotton and grain sorghum". *Physiol. Plant.* 15:308–311.

Mothes, K. and Engelbrecht, L. (1961). "Kinetin-induced directed transport of substances in excised leaves in the dark". *Phytochem.* 1:58–62.

Muir, R. M. and Lantican, B. P. (1968). "Purification and properties of the enzyme system forming indoleacetic acid". In *Biochemistry and Physiology of Plant Growth Substances*, eds., F. Wightman and G. Setterfield, Runge Press, Ottawa, pp. 259–272.

Neljubow, D. N. (1901). "Uber die horizontale nutation der stengel von *Pisum sativum* und einiger anderen". *Pflanzen Beitrage Botanik Zentralblatt* 10:128–139.

Nemec, B. (1901). "Über die Wahrnehmung des Schwerkraftreizes bei den Pflanzen. (About the perception of gravity by plants)". *Jahrb. Wiss. Bot.* 36:80–178.

Nitsch, C. and Nitsch, J. P. (1969), Floral induction in a short-day plant. "*Plumbago indica* L., by 2-chloroethanephosphonic acid". *Plant Physiol.* 44:1747–1748.

Nitsch, J. P. (1950). "Growth and morphogenesis of the strawberry as related to auxin". *Am. J. Bot.* 37:211–215.

Oeller, P. W., Min-Wong, L., Taylor, L. P., Pike, D. A. and Theologis, A. (1991). "Reversible inhibition of tomato fruit senescence by antisense RNA". *Science* 254:437–439.

Osborne, D. J. (1989). "Abscission". *CRC Critical Reviews in Plant Sciences* 8:103–129.

Paleg, L. G. (1960a). "Physiological effects of gibberellic acid: I. On carbohydrate metabolism and amylase activity of barley endosperm". *Plant Physiol.* 35:293.

Paleg, L. G. (1960b). "Physiological effects of gibberellic acid: II. On starch hydrolyzing enzymes of barley endosperm". *Plant Physiol.* 35:902.

Paleg, L. G. (1965). "Physiological effects of gibberellins". *Annu. Rev. Plant Physiol.* 16:291–322.

Parthier, B. (1979). "The role of phytohormones (cytokinins) in chloroplast development (a review)". *Biochemie Physiologie Pflanzen* 174:173–214.

Rai, V. K. and Laloraya, M. M. (1967). "Correlative studies on plant growth and metabolism II. Effect of light and of gibberellic acid on the changes in protein and soluble nitrogen in lettuce seedlings". *Plant Physiol.* 42:440–444.

Raschke, K. (1987). "Action of abscisic acid on guard cells". In *Stomatal Function*, eds., E. Zeiger, G. D. Farquhar and I. R. Cowan, Stanford University Press, Stanford, CA, pp. 253–270.

Raskin, I. (1992). "Role of salicylic acid in plants". *Annu. Rev. Plant Physiol. Plant Mol. Biol.* 43:439–463.

Raven, J. A. (1975). "Transport of indoleacetic acid in plant cells in relation to pH and electrical gradients, and its significance for polar IAA transport". *New Phytol* 74:163.

Ray, P. M. (1987). "Principles of plant cell growth". In *Physiology of Cell Expansion During Plant Growth*, eds., D. J. Cosgrove and D. P. Knievel, American Society of Plant Physiology, Rockville, MD, pp. 1–17.

Rayle, D. L. and Purves, W. K. (1967). "Conversion of indole-ethanol to indoleacetic acid in cucumber seedling shoots". *Plant Physiol.* 42:1091–1093.

Reid, J. B. (1990). "Phytohormone mutants in plant research". *J. Plant Growth Reg.* 9:97–111.

Reid, M. S. (1985). "Ethylene and abscission". *HortSci.* 20:45–50.

Reinecke, D. M. and Bandurski, R. S. (1987). "Auxin biosynthesis and metabolism". In *Plant Hormones and Their Role in Plant Growth and Development*, ed., P. J. Davies, Martinus Nijhoff Publishers, Boston, pp. 24–42.

Richter, K. and Koolman, J. (1991). "Antiecdysteroid effects of brassinosteriods in insects". In *Brassinosteroids. Chemistry, Bioactivity, and Applications*, eds., H. G. Cutler, T. Yokota and G. Adam, American Chemical Society, Washington, DC, pp. 265–279.

Roddick, J. G. and Guan, M. (1991). "Brassinosteroids and root development". In *Brassinosteroids. Chemistry, Bioactivity, and Applications*, eds., H. G. Cutler, T. Yokota and G. Adam, American Chemical Society, Washington, DC, pp. 231–245.

Rubery, P. H. and Sheldrake, A. R. (1974). "Carrier mediated auxin transport". *Planta* 118:101–121.

Sachs, J. (1880). "Stoff und Form der Pflanzenorgane. I". *Arb. Bot. Inst. Wurzburg* 2:452–488.

Salisbury, F. B. and Ross, C. W. (1992). *Plant Physiology, Fourth Edition*, Wadworth Publishing Co., Belmont, CA.

Sasse, J. M. (1991). "Brassinosteroids—Are they endogenous plant hormones?" *PGRSA Quarterly* 19:1–18.

Schlagnhaufer, C. and Arteca, R. N. (1985). "Brassinosteroid-induced epinasty in tomato plants". *Plant Physiol.* 78:300–303.

Schlagnhaufer, C. D. and Arteca, R. N. (1991). "The uptake and metabolism of brassinosteroid by tomato *(Lycopersicon esculentum)* plants". *J. Plant Physiol.* 138:191–194.

Scott, I. M. (1990). "Plant hormone response mutants". *Physiol. Plant.* 78:147–152.

Sembdner, G. and Parthier, B. (1993). "The biochemistry and the physiological and molecular actions of jasmonates". *Annu. Rev. Plant Physiol. Plant Mol. Biol.* 44:569–589.

Skriver, K. and Mundy, J. (1990). "Gene expression in response to abscisic acid and osmotic stress". *The Plant Cell* 2:503–512.

Staswick, P. E. (1992). "Jasmonate, genes, and fragrant signals". *Plant Physiol.* 99:804–807.

Stuart, N. W. and Cathey, H. M. (1961). "Applied aspects of the gibberellins". *Ann. Rev. Plant Physiol.* 12:369.

Takahashi, N., Phinney, B. O. and MacMillan, J. (1991). *Gibberellins*, Springer-Verlag, Berlin.

Takasuto, S., Yazawa, N., Ikekawa, N., Takematsu, T., Takeuchi, Y., and Koguchi, M. (1983). "Structure-activity relationship of brassinosteroids". *Phytochem.* 22:2437–41.

Tamas, I. A. (1987). "Hormonal regulation of apical dominance". In *Plant Hormones and Their Role in Plant Growth and Development*, ed., P. J. Davies, Martinus Nijhoff Publishers, Boston, pp. 393–410.

Tang, Y. W. and Bonner, J. (1947). "The enzymatic inactivation of indoleacetic acid". *Arch. Biochem. Biophys.* 3:11–25.

Tanino, K., Weiser, C. J., Fuchigami, L. H. and Chen, T. T. H. (1990). "Water content during abscisic acid induced freezing tolerance in bromegrass cells". *Plant Physiol.* 93:460–464.

Theologis, A. (1992). "One rotten apple spoils the whole bushel: The role of ethylene in fruit ripening". *Cell* 70:181–184.

Thimann, K. V. (1935). "On the plant growth hormone produced by *Rhizopus sinuis*". *J. Biol. Chem.* 109:279–291.

Thimann, K. V. (1987). "Plant senescence: A proposed integration of the constituent processes". In *Plant Senescence: Its Biochemistry and Physiology*, eds., W. W. Thomson and E. A. Nothnagel, American Society of Plant Physiologists, Rockville, MD, pp. 1–19.

Thimann, K. V. and Mahadevan, S. (1958). "Enzymatic hydrolysis of indoleacetonitrile". *Nature* 181:1466–1467.

Thimann, K. V. and Skoog, F. (1934). "On the inhibition of bud development and other functions of growth substance in *Vicia faba*". *Proc. Royal Soc.* B114:317–339.

Thompson, M. J., Mandava, N. B., Meudt, W. J., Lusby, W. R., and Spaulding, D. W. (1981). "Synthesis and biological activity of brassinolide and its 22,23–isomer. Novel plant growth promoting steroids". *Steroids* 38:567–580.

Thompson, M. J., Meudt, W. J., Mandava, N. B., Dutky, S. R., Lusby, W. R. and Spaulding D. W. (1982). "Synthesis of brassinosteroids and relationship of structure to plant growth-promoting effects". *Steroids* 39:89–105.

Torrey, J. G. (1962). "Auxin and purine interactions in lateral root initiation in isolated pea root segments". *Physiol. Plant.* 15:177.

Traub, H. P., Cooper, W. C., and Reece, P. C. (1940). "Inducing flowering in the pineapple, *Ananas sativus*". *Proc. Amer. Soc. Hort. Sci.* 37:521–525.

van Herk, A. W. H. (1937). "Die chemischen Vorgange im Sauromatum Kolben III. Mitteilung". *Proc. K. Ned. Akad. Wet.* 40:709–719.

van Staden, J., Bayley, A. D., Upfold, S. J., and Drewes, F. E. (1990). "Cytokinins in cut carnation flowers. VIII. Uptake, transport and metabolism of benzyladenine and the effect of benzyladenine derivatives on flower longevity". *J. Plant Physiol.* 703–707.

van Staden, J., Cook, E. L. and Noodén, L. D. (1988). "Cytokinins and senescence". In *Senescence and Aging in Plants*, eds., L. D. Noodén and A. C. Leopold, Academic Press, New York, pp. 281–328.

Varner, J. E. (1964). "Gibberellic acid-controlled synthesis of α-amylase in barley endosperm". *Plant Physiol.* 39:413–415.

von Sachs, R. M. and Kofranek, A. M. (1963). "Comparative cytohistological studies on inhibition and promotion of stem growth in *Chrysanthemum morifolium*". *Am. J. Bot.* 50:772.

Wang, T.-W., Cosgrove, D. J., and Arteca, R. N. (1993). "Brassinosteroids stimulation of hypocotyl elongation and wall relaxation in pakchoi (*Brassica chinensis* cv Lei-Choi)". *Plant Physiol.* 101:965–968.

Weaver, R. J. (1972). *Plant Growth Substances in Agriculture*, W. H. Freeman and Company, San Francisco.

Went, F. W. (1934). "On the pea test method for auxin, the plant growth hormone". *K. Akad. Wetenschap. Amsterdam Proc. Sect. Sci.* 37:547.

Wickson, M. and Thimann, K. V. (1958). "The antagonism of auxin and kinetin in apical dominance". *Physiol. Plant.* 11:62.

Wightman, F., Schneider, E. A. and Thimann, K. V. (1980). "Hormonal factors controlling the initiation and development of lateral roots. II. Effects of exogenous growth factors on lateral root formation in pea roots". *Physiol. Plant.* 49:304–314.

Wildman, S. G., Ferri, M. G., and Bonner, J. (1947). "The enzymatic conversion of tryptophan to auxin by spinach leaves". *Arch. Biochem. Biophys.* 13:131.

Yalpani, N., León, J., Lawton, M. A., and Raskin, I. (1993). "Pathway of salicylic acid biosynthesis in healthy and virus-inoculated tobacco". *Plant Physiol.* 103:315–321.

Yokota, T., Ogino, Y., Suzuki, H., Takahashi, N., Saimoto, H., Fujioka, S., and Sakurai, A. (1991). "Metabolism and biosynthesis of brassinosteroids". In *Brassinosteroids. Chemistry, Bioactivity, and Applications*, eds., H. G. Cutler, T. Yokota and G. Adam). American Chemical Society, Washington, DC, pp. 86–96.

Yomo, H. (1960). "Studies on the α-amylase activity substance. IV. On the amylase activating action of gibberellin". *Hakko Kyokaishi* 18:600–602.

Yopp, J. H., Aung, L. H. and Steffens, G. L. (1986). *Bioassays and Other Special Techniques for Plant Hormones and Plant Growth Regulators*, Plant Growth Regulator Society of America, Lake Alfred, FL.

Yopp, J. H., Mandava, N. B., Thompson, M. J., and Sasse, J. M. (1981). *Brassinosteroids in Selected Bioassays, Proc. Plant Growth Reg. Soc. Am.*, Plant Growth Regular Society of America, St. Petersburg, FL.

Zenk, M. H. (1968). "The action of light on the metabolism of auxin in relation to phototropism". In *Biochemistry and Physiology of Plant Growth Substances*, eds., F. Wightman and G. Setterfield, Runge Press, Ottawa, pp. 1109–1128.

Zenk, M. H. (1962). "Aufnahme and Stoffwechsel von Naphthylessigsäure durch Erbsenepicotyle". *Planta* 58:75–94.

Zenk, M. H. (1961). "Indoleacetyl glucose, a new compound in the metabolism in indoleacetic acid in plants". *Nature* 191:493–494.

Zimmerman, P. W. and Hitchcock, A. E. (1942). "Substituted phenoxy and benzoic acid growth substances and the relation of structure to physiological activity". *Contrib. Boyce Thompson Inst.* 12:321–343.

Zimmerman, P. W. and Wilcoxon, F. (1935). "Several chemical growth substances which cause initiation of roots and other responses in plants". *Contrib. Boyce Thompson Inst.* 7:209–229.

Zurek, D. M. and Clouse, S. D. (1994). "Molecular cloning and characterization of a brassinosteroid-regulated gene from elongating soybean (*Glycine max* L.) epicotyls". *Plant Physiol.* 104:161–170.

Zurek, D. M., Rayle, D. L., McMorris, T. C., and Clouse, S. D. (1994). "Investigation of gene expression, growth kinetics, and wall extensibility during brassinosteroid-regulated stem elongation". *Plant Physiol.* 104:505–513.

CHAPTER 4

Seed Germination and Seedling Growth

In order to understand factors regulating germination and subsequent growth of the seedling we should first be knowledgeable as to the processes involved during this period. Propagation by seeds is the major method of reproduction in nature and the most widely used method in agriculture due to its high efficiency. A seed is a ripened ovule which when shed from the parent plant consists of an embryo and a stored food supply both of which are enclosed in a seed coat or covering. Seed germination may be defined as a series of events which take place when dry quiescent seeds imbibe water resulting in an increase in metabolic activity and the initiation of a seedling from the embryo. In order for germination to be initiated the following criteria must be meet:

1. The seed must first be viable (the embryo is alive and capable of germination).

2. Appropriate environmental conditions such as available water, proper temperature, oxygen, and, in some cases, light must be supplied.

3. Primary dormancy in the seed must be overcome.

In most cases the first visible sign of seed germination is emergence of the radicle from the seed coat. There are also specific cases where the shoot is the first visible sign, an example of this is *Salsola* seeds (Mayer and Poljakoff-Mayber 1989). Following emergence of the radicle the seedling grows as a subterranean organism not yet relying on photosynthesis for growth. When the seedling emerges from the soil, photosynthesis and active growth begins.

The four stages involved in seed germination and seedling growth which will be discussed in this chapter are:

1. Imbibition of water.
2. Formation or activation enzyme systems.
3. Metabolism of storage products, subsequent transport and synthesis of new materials.
4. Emergence of the radicle and growth of the seedling.

THE GERMINATION PROCESS

Imbibition of Water

During the imbibition stage two factors occur simultaneously, water uptake and changes in respiration (CO_2 evolution or O_2 uptake).

Water Uptake. In germinating seeds water uptake follows a triphasic pattern as shown in Figure 4.1. The water potential of a mature dry seed can exceed -100 MPa due mainly to matric potential (based on the ability of the matrices, e.g., cell walls, starch, etc. to be hydrated and bind water) which promotes rapid water uptake during the first phase of imbibition. This uptake occurs equally well in dead and living seeds and therefore is independent of metabolic activity. The rate of uptake depends on the texture of the germination medium, degree of soil packing, and the closeness of the soil to the seed. When the seed removes moisture from the soil, the area closest to the seed becomes dry and must be replenished by water from adjacent pores. Therefore, it is very important that a firm, fine-textured soil is in close contact with the seed in order to maintain a uniform water supply during the germination process. During phase 2 there is a plateau in water uptake because matric potential no longer plays a significant role and the water potential of the seed is based largely on the negative osmotic potential (solute concentration) of the living cells in the seeds and pressure potential (turgor) of the cell wall, which is positive. Although there is little water taken up during this time it is still a period of active metabolism (enzyme synthesis) in preparation for germination in nondormant seeds, or active metabolism in dormant seeds or inertia in dead seeds. Phase 3 is characterized by a second burst in water uptake in nondormant

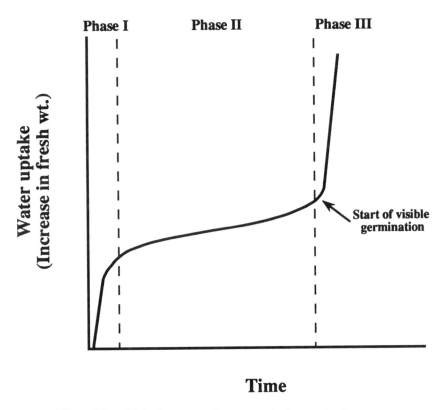

Figure 4.1.   Triphasic pattern of water uptake in germinating seeds.

seeds together with radicle emergence and elongation resulting from enlargement of cells rather than cell division (Bewley and Black 1985). This is promoted by a decrease in water potential in the seed due to hydrolysis of storage reserves which results in decreased osmotic potential. During this phase fats, proteins, carbohydrates, phosphorus-containing compounds, and other materials stored in the endosperm, cotyledons, perisperm, or female gametophyte (conifers) are broken down into simpler chemical substances, which are translocated to the growing points of the embryo axis, resulting in seedling growth. Dormant seeds may have an active phase 2; however, do not undergo a phase 3 (Figure 4.1). The duration of each of these phases depends on seed properties such as content of hydratable substances, seed coat permeability, seed size, and oxygen uptake. In addition, the conditions during hydration of the seed such as temperature, moisture levels, substrate availability and composition will also determine the length of each phase.

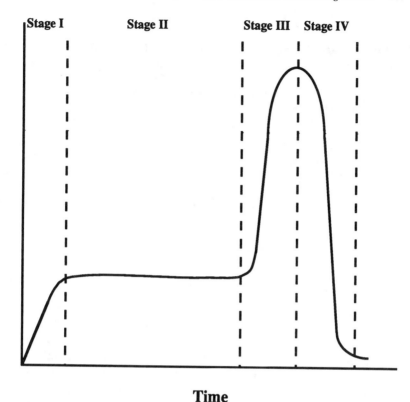

Time

Figure 4.2.   Four phases of respiration occurring during imbibition of water by seeds.

Changes in Respiration.   The first easily observable metabolic change is a rapid increase in respiration which occurs shortly after the seed is put in water. During imbibition there are four phases of respiration as shown in Figure 4.2. An initial sharp rise in respiration occurs in phase 1, which is attributed in part to activation and hydration of preexisting mitochondrial enzymes associated with the citric acid cycle and electron transport. There is a plateau in respiration during phase 2 due to low O_2 present in the seed which is caused by the seed coat restricting air movement from outside, thereby promoting anaerobic respiration. Not all seeds go through the phase 2 lag phase; examples of these are *Avena fatua* and *Phaseolus lunatus.* During phase 3 there is a secondary respiratory burst which is attributed to increased O_2 supply and newly synthesized mitochondrial and respiratory enzymes in dividing cells resulting in aerobic respiration of carbohydrates. During phase 4 there is a dramatic decline in respiration due to emergence of the seedling through the soil surface resulting in photosynthesis (Figure 4.2).

Formation or Activation of Enzyme Systems

Evidence for activation or de novo (newly formed) synthesis of enzymes during germination has been shown in several ways:

1. Appearance of enzyme activity prior to germination or by an increase in activity during germination.
2. Use of protein synthesis inhibitors.
3. Incorporation of radioactive precursors into proteins.
4. Immunological techniques.
5. Molecular techniques.

During germination the formation of enzyme systems can occur in several ways:

1. From preexisting enzymes which are active upon hydration.
2. Activation of preexisting enzymes which are inactive.
3. De novo synthesis of enzymes from preexisting or de novo produced mRNA.

Shortly following the absorption of water by the seed, enzyme activity begins to appear. During the germination process numerous enzymes such as lipases, proteinases, phosphatases, hydrolyases, calmodulin, carboxypeptidases, and others (Mayer and Poljakoff-Mayber 1989; Bernier and Ballance 1993; Bewley and Black 1985; Bernhardt et al. 1993; Cocucci and Negrini 1991; Washio and Ishikawa 1992) are either activated or de novo synthesized by one of the processes previously mentioned. Each of these enzymes are involved in the germination process and evidence now suggests that the synthesis or activation of some of these enzymes are regulated by plant growth substances.

Metabolism of Storage Products and Subsequent Transport

The second change which is easily observed during germination is the breakdown of reserve materials by enzymes in the seed. This is caused by the formation or activation of enzyme systems discussed in the previous section. Three types of chemical changes which occur during the germination process are:

1. Breakdown of reserve materials in the seed.
2. Transport of breakdown products from one part of the seed to another.
3. Synthesis of new materials from breakdown products.

Prior to seedling emergence the only substances taken up by the seed are water and oxygen. During the initial stages of seed germination there is a loss of dry weight due to oxidation and leakage from the seed. An increase in dry weight begins when the root has emerged, enabling the seedling to take up minerals and make its way to the soil surface where it is exposed to light resulting in photosynthesis.

The major storage materials in seeds are lipids and carbohydrates. There are also other materials present in dried seeds but they are generally of lesser importance (Mayer and Poljakoff-Mayber 1989). Metabolic changes which occur in the early stages of germination are due to various enzymes which break down starch, proteins, hemicellulose, polyphosphates, lipids and other storage materials.

Carbohydrates. The major storage carbohydrates are starch, oligo- and polysaccharides of the cell wall, and soluble sugars. Starch is typically broken down by α and β-amylases. Dry seeds contain mainly β-amylases and after the seed imbibes water there is a rise in α-amylase activity which accounts for about 90% of the total amylolytic activity in the endosperm. An exception to this rule occurs during germination of alfalfa seeds. During this process α-amylase activity has been shown to decrease while, in contrast, β-amylase activity increases in the cotyledons of germinating seeds. These changes appear to be specific for germinating alfalfa seeds but it is possible that this phenomenon may occur in other plants as well (Kohno and Nanmori 1991). α-Amylase converts starch to a variety of sugars, maltose, and glucose, whereas β-amylase converts soluble oligosaccharrides produced via α-amylase action to maltose (Swain and Dekka 1966). Other starch-degrading enzymes produced during germination are debranching enzymes and phosphorylases. Additional information on the breakdown and interconversions of carbohydrates during germination can be found in Mayer and Poljakoff-Mayber (1989).

Lipids. Lipids are stored in lipid bodies (spherosomes) which either have preexisting or de novo produced enzymes required for their breakdown into fatty acids and glycerol. Several lipases which differ in pH optimum for activity have been found in seeds (Lin et al. 1982). They break down lipids to fatty acids which are converted via β-oxidation to acetyl-CoA, which can be used in the tricarboxylic acid cycle. Fatty acids may also be broken down via α-oxidation, which peroxidatively decarboxylates fatty acids and forms CO_2. The long-chain aldehyde formed is oxidized to the corresponding acid by a reaction linked to NAD. Lipoxidase also plays a part in fatty acid oxidation by breaking down the unsaturated fatty acid chain into two smaller parts (Mayer and Poljakoff-Mayber 1989).

Proteins. Storage proteins in most seeds are found in protein bodies. During germination storage proteins are broken down, resulting in leakage of the breakdown products from protein bodies. Seeds contain a variety of proteolytic enzymes which are present in dry seeds while other enzymes appear during germination. Proteolytic enzymes can be broken down into proteinases and peptidases depending on the site of the molecule attacked. Peptidases are divided into endopeptidases and exopeptidases, carboxypeptidases and aminopeptidases, depending on the point at which they attack on the molecule. The types of proteolytic enzymes and peptidases in seeds do not typically differ from one another; however, there are a few cases where these enzymes have been shown to exhibit differences (Mayer and Poljakoff-Mayber 1989; Washio and Ishikawa 1992).

Phosphorus-Containing Compounds. The main forms in which phosphorus is stored are nucleic acids, phospholipids, phosphate esters of sugars, nucleotides, and phytin. Since most of the phosphate contained within the seed is in the bound form, orthophosphate may be a limiting factor in many reactions in which phosphate participates, such as phospholipid, nucleic acid and protein synthesis and other energy-generating processes. During germination there is a large decrease in phytin which can make up to 80% of the total phosphate found in seeds (Reddy et al. 1982) and a concomitant increase in inorganic phosphate suggesting that phytin acts as a storage pool of phosphate during germination. The enzyme involved in the release of phosphate from phytin is a phosphatase called phytase. At the present time little is currently known about the metabolism of nucleotides and sugar phosphates. The major enzymes studied thus far are phosphatases and phosphokinases which are responsible for phosphate transfer (Mayer and Poljakoff-Mayber 1989).

Transport of Digested Storage Compounds and Synthesis of New Materials

After storage compounds are broken down into simpler chemical substances, they are transported from the endosperm to the embryo, or from the cotyledons to growing points of the seed. Once these compounds reach their destination they are utilized for many purposes including the production of new enzymes, structural materials, regulatory compounds, plant growth substances, and nucleic acids which carry out cell functions and synthesize new materials (Mayer and Poljakoff-Mayber 1989; Davies and Slack 1981).

Emergence of the Radicle and Seedling Growth

The second burst of water uptake during imbibition is caused by a decrease in osmotic potential resulting from the hydrolysis of storage compounds as previously mentioned. Along with this increase in water uptake is the concurrent emergence of the radicle, which from this point on provides a continual supply of water and nutrients for growth of the seedling. Development of the seedling begins with cell division at the two ends of the embryonic axis, followed by expansion of the seedling structures. Cell division initiated at the growing points appears to be independent of the initiation of cell elongation (Berlyn 1972; Haber and Luippold 1960). The embryo has an axis at the top bearing one or more seed leaves, or cotyledons at the base where the radicle emerges. The plumule is the growing point of the shoot which occurs at the upper end of the embryonic axis, above the cotyledons. The seedling stem is divided into the hypocotyl section which is below the cotyledons, and the epicotyl section, which is above the cotyledons. Once seedling growth begins there are increases in both fresh and dry weight with a decrease in weight of the storage tissues. As the seedling makes its way to the soil surface, respiration increases at a steady rate (Figure 4.2). Storage tissues soon become exhausted and cease to be involved in metabolic activities except in plants where the cotyledons emerge from the ground and become active in photosynthesis. Once the seedling breaks through the soil surface there is a sharp decrease in respiration which is concurrent with the beginning of photosynthesis (Figure 4.2). Water absorption increases steadily as new roots continue to grow in order to provide water, nutrients, and support for the new plant. Plants can store reserves in two different ways, in dicots cotyledons act as the storage tissue source, whereas monocots utilize the endosperm. The initial growth of the dicot seedling follows two growth patterns, epigeous and hypogeous. With epigeous germination the hypocotyl elongates and brings the cotyledons above the ground (Figure 4.3), whereas during hypogeous germination the epicotyl

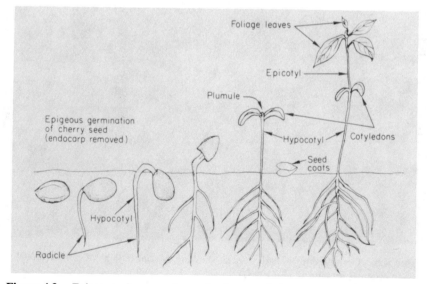

Figure 4.3. Epigeous cherry seed germination; cotyledons are above ground (from Hartmann et al. (1990)).

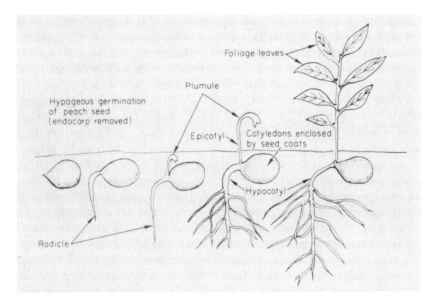

Figure 4.4.   Hypogeous peach seed germination; cotyledons remain below ground (from Hartmann et al. (1990)).

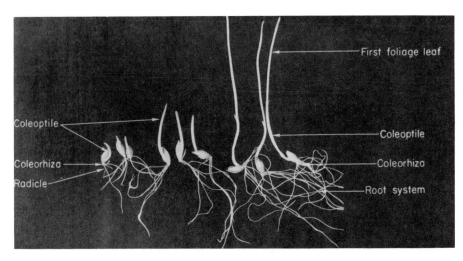

Figure 4.5.   Germination of barley seeds showing special pattern of germination by grasses (from Hartmann et al. (1990)).

emerges and the cotyledons remain below the soil surface (Figure 4.4). In monocotyledenous plants the coleoptile and coleorhiza enclose the plumule and radicle, respectively. The first structures to emerge from the seed coat are the coleoptile and coleorhiza which act as a protective sheath. As growth of the seedling progresses, the first foliage leaf and radicle emerge through these structures to produce the stems and roots (Figure 4.5).

CONTROL OF GERMINATION AND SEEDLING GROWTH

There is a great deal of experimental evidence in the literature showing that endogenous growth-promoting and growth-inhibiting compounds are directly involved in seed germination (Bewley and Black 1985; Lewak 1985; Lalonde and Saini 1992; Sharma et al. 1992; Whitehead and Nelson 1992; Cocucci and Negrini 1991; Kovac et al. 1993; Ni and Bradford 1993; Gallego et al. 1991; Kermode 1990). Evidence for involvement of the different plant growth substances alone and in combination (Bewley and Black 1985; Khan 1975) is based on correlations of their endogenous concentration with specific developmental stages, effects of exogenous applications, and the relationship of plant growth substances to metabolic activity. Plant growth substances have also been implicated in postgermination events such as reserve mobilization (Halmer 1985; Munoz et al. 1990), root growth (Torrey 1962; Wightman et al. 1980; MacIsaac et al. 1989), hypocotyl growth (Sawhney and Srivastava 1974), cotyledon size and weight (Gallego et al. 1991) and chlorophyll synthesis in cotyledons (Parthier 1979). Although there are numerous reports showing that plant growth substances are involved in seed germination and subsequent growth their specific roles are still obscure. In order to have a better understanding of the functions of plant growth substances during these processes, it is important to identify and quantify them from imbibition through growth of the seedling. Methods for the identification and quantification of plant growth substances during recent years have made considerable strides. However, there are still a limited number of studies on changes in plant growth substance levels during germination and subsequent growth of the seedling. Before definitive statements can be made on the role of plant growth substances during these processes more comprehensive studies on their quantification is required.

Specific Plant Growth Substances

Gibberellins and Abscisic Acid.   Gibberellins are a class of plant growth substances which are most directly implicated in the control and promotion of germination (Takahashi et al. 1992; Karssen et al. 1983; Groot and Karssen 1987; Groot et al. 1988; Mayer and Poljakoff-Mayber 1989). Abscisic acid (ABA), however, has been documented to inhibit seed germination (Addicott 1983; Creelman 1989; Mayer and Poljakoff-Mayber 1989) acting as a natural gibberellin antagonist. Recent research involving ABA-deficient mutants of *Arabidopsis thaliana* and *Lycopersicon esculentum* has provided definitive evidence that endogenous levels of ABA are involved in the regulation of seed germination (Karssen et al. 1983, 1987; Reid 1990). Under the proper conditions ABA-deficient mutants have been shown to exhibit vivipary, which is a phenomenon of precocious germination before maturity, indicating that ABA has a role in preventing precocious germination of developing embryos. ABA-deficient mutants produce nondormant seeds that germinate rapidly; whereas GA-deficient seeds will not germinate without exogenous application of GA, and germination percentage increases along with the GA concentration (Karssen et al. 1983, 1987; Groot and Karssen 1987, 1992; Groot et al. 1988).

Seed germination is sensitive to both endogenous plant growth substances and the environment. Both ABA and reduced water potential together with their interactions have been evaluated in several plant species. The inhibitory effect of exogenous applications of ABA to seeds is similar and additive with reduced water potential (Schopfer and Plachy 1985; Welbaum et al. 1990; Ni and Bradford 1992). High levels of ABA in seeds have been shown to promote increased sensitivity of seeds to reduced water potential, thereby reducing their ability to germinate. More recently, it was shown by the same group that ABA- and GA-dependent changes in seed dormancy and germination rates which are due to endogenous or exogenous plant growth substances are based primarily on corresponding shifts in the threshold water potential for radicle emergence (Ni and Bradford 1993).

Cytokinins. There have been many reports showing that cytokinins have a regulatory role in seed germination (Mayer and Poljakoff-Mayber 1989). It has been demonstrated in *Cicer arietinum* seeds that the embryonic axis supplies cytokinins to the cotyledons for the first 12 hours after the start of imbibition. During this time there is a mobilization of reserve materials required for the germination process (Gepstein and Ilan 1980; Munoz et al. 1990; Martin et al. 1987; Pino et al. 1990, 1991). Exogenous applications of cytokinins have also been shown to be involved in germination of Scotch pine seeds (Kovac et al. 1993). In another study it was shown that Scotch pine seeds require red light for maximum germination (Nygren 1987) and that a short pulse of red light increased levels of 2-isopentyladenine, further implicating cytokinins (Qamaruddin and Tillburg 1989). Zeatin riboside-type cytokinins have also been implicated in the regulation of growth which takes place after germination (Qamaruddin 1991).

Ethylene. Vacha and Harvey (1927) were the first to show that ethylene promoted seed germination. Since this time there have been a number of reports showing that exogenous applications of ethylene or ethylene-releasing agents such as ethephon can stimulate this process in a number of species (Ketring 1977; Taylorson 1979; Bewley and Black 1982; Abeles et al. 1992; Adkins and Ross 1981; Corbineau et al. 1989; Hoffman et al. 1983; Saini et al. 1989; Satoh and Esashi 1982; Walker et al. 1989; Lalonde and Saini 1992). Additional evidence supporting the involvement of ethylene in seed germination is found in several reports showing an increase in ethylene production prior to seed germination or a lower level of ethylene evolution from dormant rather than nondormant seeds (Ketring and Morgan 1972; Katoh and Esashi 1975; Kepczvnski and Rudnicki 1975; Ketring 1977; Adkins and Ross 1981; Cardoso and Felippe 1987; Corbineau et al. 1989; Dunlap and Morgan 1977). More recently the use of ethylene biosynthesis and action inhibitors have provided further evidence for the involvement of endogenous ethylene in seed germination (Satoh et al. 1984; Abeles 1986; Saini et al. 1986, 1989; Khan and Huang 1988; Khan and Prusinski 1989). *Arabidopsis thaliana* mutants (etr) are insensitive to ethylene and seeds from these plants exhibit lower germination rates than seeds from the wild type, further supporting the involvement of ethylene in seed germination (Bleecker et al. 1988).

Although carbon dioxide is thought to be an inhibitor of ethylene action it has been shown to stimulate germination. When CO_2 is used in combination with ethylene there is an additive effect in the promotion of ethylene production, indicating that both of these compounds have different modes of action (Abeles et al. 1992). There are also a number of other factors which have been shown to stimulate germination such as aging, phytochrome, temperature, cytokinins, gibberellins, fusicoccin (a fungal toxin), and nitrate. Ethylene has not been shown to be involved in the stimulation of lettuce seed germination by cytokinin (Khan and Prusinski 1989), fusicoccin (Abeles 1986), or gibberellin (Dunlap and Morgan 1977). Although there is a great deal of evidence showing that ethylene is involved in seed germination, it is still not clear if the effect is widespread throughout the plant kingdom and what its mechanism of action is (Abeles et al. 1992; Lalonde and Saini 1992; Mayer and Poljakoff-Mayber 1989).

Others. Potassium nitrate, thiourea, fusicoccin, cotylenin, brassinolide, and strigol have been shown to stimulate germination in specific test systems; however, their role remains unclear. Potassium nitrate has been used in seed testing laboratories for many years but its action is unknown (Hartmann et al. 1990). Thiourea has the ability to maximize germination and overcome certain types of dormancy such as seed coat and high-temperature inhibition (Thomas 1977; Mayer and Poljakoff-Mayber 1989); it is thought that its promotive effects may be due to its cytokinin activity. Both fusicoccin and cotylenin (Khan 1977) have been reported to mimic the stimulatory effect promoted by the combination of GA plus cytokinin. Brassinolide has been shown to speed germination of witchweed (*Striga asiatica*) seeds by reducing the conditioning period required. In a subsequent study, brassinolide was shown to eliminate the inhibitory effects of IAA and light on seed germination in witchweed seed (Takeuchi et al. 1991). Strigol, a natural stimulant from sorghum has also been shown to induce germination of witchweed seeds (Katsumi et al. 1987). At the present time the mechanism of action of each of the previously mentioned compounds remains unknown until further research is done in this area.

ENVIRONMENTAL FACTORS AFFECTING SEED GERMINATION

Water

Water is the most important factor in the initiation of seed germination and survival of the seedling once it emerges. The osmotic potential in the soil solution depends on the presence of solutes (salts). Under high salinity conditions salts accumulate to toxic levels, making the osmotic potential of the soil very negative, resulting in a reduction of water uptake which inhibits germination and causes a reduction in seedling stands (Ayers 1952). These salts originate in the soil or other germination medium used, they can be added in the irrigation water or through fertilization. A lack of water in the soil during the germination process can reduce the germination percentage due to water stress (Doneen and MacGillivray 1943; Hanks and Thorp 1956). Seeds will germinate over a wide range of soil moisture conditions from field capacity to permanent wilting. However, there is generally a decline in the emergence rate as the moisture level drops to halfway between field capacity and the permanent wilting point. There are cases where some seeds such as beet, endive, lettuce, or celery are inhibited as moisture levels are decreased because they require leaching to remove inhibitors. However, in some species (e.g., spinach) exposure to excess water results in the production of a substance which reduces oxygen supply to the embryo and elevates inhibitory substances in the seed which reduce germination (Atwater 1980; Heydeker 1977).

Temperature

Temperature regulates the rate of germination, germination percentage, and subsequent seedling growth. In general the germination rate is low at reduced temperatures but increases as the temperature rises to an optimum level beyond which there is a reduction due to seed injury. On the other hand the germination percentage may remain constant over the middle part of this temperature range if enough time is allowed for germination to occur. All seeds have a minimum, optimum and maximum temperature for germination, but these temperatures differ between species. The minimum temperature is the lowest temperature which will permit effective germination, whereas the maximum temperature is the highest temperature at which germination will occur. The optimum temperature for seed germination is when the largest percentage of seedlings are produced at the highest rate. It should be noted that the optimum temperature for seed germination and seedling growth is not always the same. Seeds from different species can be broken down into several temperature-requiring groups:

1. Cool-temperature-tolerant will germinate at temperatures as low as 4.5°C but will also germinate at higher temperatures, e.g., broccoli, carrot, cabbage, and alyssum.

2. Cool-temperature-requiring seeds such as coleus, primula, and delphinium will not germinate at temperatures higher than 25°C.

3. Warm-temperature requiring seeds will fail to germinate and/or are susceptible to chilling injury at temperatures between 10° and 15°C, e.g., tomato, beans, sorghum, eggplant, and cucumbers.

4. Alternating temperatures are required for some seeds. Fluctuating day-night temperatures (generally a 10°C difference) are required for optimal seed germination and growth.

Aeration

In order to obtain rapid and uniform germination, gas exchange in the germination medium is essential. Oxygen is required for normal respiratory processes to occur in the germinating seed and should be maintained as close to 21% as possible. Carbon dioxide is a product of respiration and when gas exchange is poor can accumulate in the soil, resulting in an inhibition of germination. Seeds from different plant species vary in their oxygen requirements for germination with some requiring 21% and others being able to withstand very low oxygen tensions and still grow.

Light

It has been recognized since around the mid-nineteenth century that light is a critical factor controlling germination (Crocker 1930). The photochemically reactive pigment phytochrome which is widely distributed in plants has been shown to be involved in the mechanism of light sensitivity in seeds (Bewley and Black 1985; Taylorson and Hendricks 1977). When an imbibed seed is exposed to red light (660–760 nm) phytochrome (P) changes to P_{fr} which promotes germination, whereas exposure to far-red light (760–800 nm) promo-

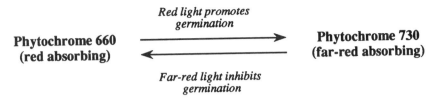

Figure 4.6.   Effects of red and far-red light on the photochemically reactive pigment phytochrome and lettuce seed germination.

tes a shift to the alternate form P_r, which inhibits germination (Figure 4.6). These changes occur very rapidly and can be repeated many times with the last treatment being the one which is effective (Borthwick et al. 1954) (Table 4.1).

The membranes of the seed coats and/or the endosperm act as light sensors; once removed light control disappears. Cytokinins have been shown to be involved in the phytochrome response, having the ability to overcome the inhibitory effect of far-red light in lettuce seeds (Miller 1956) (Table 4.2). Abscisic acid will inhibit germination in both the light or dark and GA has no effect on overcoming the dark inhibition but will speed germination in the light.

Table 4.1.   Effects of red and far-red light on Grand Rapids lettuce seed germination (Borthwick et al. 1954).

Light	% Germination
R	70
R – FR	6
R – FR – R	74
R – FR – R – FR	6
R – FR – R – FR – R	76
R – FR – R – FR – R – FR	6

Table 4.2.   Effects of kinetin on the phytochrome response in lettuce seeds (Miller 1956).

Kinetin (M)	Light treatment	% Germination
0	dark	8
5×10^{-5}	dark	84
0	8 minutes red	96
0	5 minutes red—8 minutes far-red	5
5×10^{-5}	8 minutes far-red	86

PRIMING AND SEED SOAKING

Procedures described in this section are used to initiate germination prior to planting in order to shorten the time to emergence, improve stand uniformity, and overcome potentially adverse conditions in the seedbed.

Soaking

In order to speed up the germination process seeds are sometimes soaked in water prior to planting. Most herbaceous species will show a beneficial effect of short-term soaking. It should be cautioned that longer soaking periods will lead to adverse effects due to injury of the seed (Norton 1986; Heydeker 1977; Mayer and Poljakoff-Mayber 1989). If it is necessary to soak the seeds for long periods of time the water which the seeds are being soaked in should be vigorously aerated.

Osmotic Priming

Osmotic priming involves the imbibition of dry seed with a solution that has a high osmotic potential. This process allows water to enter the seed while still maintaining a low osmotic potential, thereby initiating metabolic activities leading to germination, but prevents or delays emergence of the radicle (Khan 1977; Cantliffe et al. 1981; Mayer and Poljakoff-Mayber 1989; Frett et al. 1991). Inert compounds such as polyethylene glycol are usually used while in some cases salt solutions have been used, even though they may sometimes be cytotoxic (Akers and Holley 1986; Bradford 1986).

Infusion

This procedure utilizes organic solvents for the incorporation of chemicals such as growth regulators, fungicides, insecticides, antibiotics, and herbicidal antidotes into the seed (Khan 1978). The compound to be infused into the seed is first solubilized in an organic solvent (acetone or dichloromethane). The seeds are then immersed for one to four hours (the length of time varies with species) in the organic solvent containing one or more of the previously mentioned compounds. The solvent is removed by evaporation and once dry the seeds are stored until needed. Upon imbibition of water the incorporated chemical is directly absorbed into the embryo and utilized.

Fluid Drilling

This system utilizes seeds which have been treated, pregerminated, suspended in a gel, and placed in special machines which deposit a designated number of seeds per hole in the soil by an automated process (Salter 1978; Gray 1981). Seeds are pregerminated under conditions of aeration, light, and optimum temperatures for the species. They are typically placed in glass jars or plastic columns containing water which is continuously aerated and fresh water continuously supplied to prevent any buildup of inhibitors or pathogens which may accumulate and inhibit the germination process. Although fluid drilling is considered by most to be more beneficial then direct seeding for a number of reasons it has one restriction which is a requirement for high-percentage germination and uniform radicle length. This is important because if there are seedlings which are too large they will clog the fluid drilling apparatus. High-percentage uniform germination has been obtained by pretreating seeds with GA prior to adding them to the gel, thereby overcoming one of the main restrictions for fluid drilling (Coronel and Motes 1982). Growth regulators, fungicides, and other chemicals have also been shown to have beneficial effects when added during imbibition (Ghate et al. 1984; Coronel and Motes 1982). It has also been shown that chilling thermodormant celery seeds has the ability to promote uniform radicle emergence without injury and can be used for fluid drilling (Furatani et al. 1985). Another possible way to overcome this problem with seedlings which are too large is to separate germinated seeds from those which have not germinated by density separation (Taylor and Kenny 1985). There are a number of gels commercially available today. Among the materials used are sodium alginate, hydrolyzed starch-polyacrylonitrile, guar gum, and synthetic clay. At the present time, which gel provides the best results is still open to question.

REFERENCES

Abeles, F. B. (1986). "Role of ethylene in *Lactuca sativa* cv Grand Rapids seed germination". *Plant Physiol.* 81:780–787.

Abeles, F. B., Moran, P. W., and Saltveit Jr., M. E. (1992). *Ethylene in Plant Biology. Second Edition*, Academic Press, San Diego.

Addicott, F. T. (1983). *Abscisic Acid*, Praeger, New York.

Adkins, S. W. and Ross, J. D. (1981). "Studies in wild oat dormancy. I. The role of ethylene in dormancy breakage and germination of wild oat seeds (*Avena fatua* L.)". *Plant Physiol.* 67:358–362.

Akers, S. W. and Holley, K. E. (1986). "SPS: A system for priming seeds using aerated polyethylene glycol or salt solutions". *HortScience* 21:529–531.

Atwater, B. R. (1980). "Germination, dormancy and morphology of the seeds of herbaceous ornamental plants". *Seed Sci. Tech.* 8:523–573.

Ayers, A. D. (1952). "Seed germination as affected by soil moisture and salinity". *Agron. J.* 44:82–84.

Berlyn, G. P. (1972). "Seed germination and morphogenesis". In *Seed biology, Volume 3*, ed., T. T. Kozlowski, Academic Press, New York, pp. 223–312.

Bernhardt, D., Trutwig, A., and Barkhold, A. (1993). "Synthesis of DNA and the development of amylase and phosphatase activities in cotyledons of germinating seeds of *Vaccaria pyramidata*". *J. Exp. Bot.* 44:695–699.

Bernier, A. M. and Ballance, G. M. (1993). "Induction and secretion of alpha-amylase, $(1 \rightarrow 3)$, $(1 \rightarrow 4)$-beta-glucanase, and $(1 \rightarrow 3)$-beta-glucanase activities in gibberellic acid and $CaCl_2$-treated half seeds and aleurones of wheat". *Cereal Chemistry* 70:127–132.

Bewley, J. D. and Black, M. (1982). *Physiology and Biochemistry of Seeds in Relation to Germination. Vol. 2. Viability, Dormancy, and Environmental Control*, Springer-Verlag, Berlin.

Bewley, J. D. and Black, M. (1985). *Seeds: Physiology of Development and Germination*, Plenum Press, New York.

Bleecker, A. B., Estelle, M. A., Somerville, C., and Kende, H. (1988). "Insensitivity to ethylene conferred by a dominant mutation in *Arabidopsis thaliana*". *Science* 241:1086–1089.

Borthwick, H. A., Hendricks, S. B., Toole, E. H. and Toole, V. K. (1954). "Action of light on lettuce seed germination". *Bot. Gazette* 115:205–225

Bradford, K. J. (1986). "Manipulation of seed water relations via osmotic priming to improve germination under stress conditions". *HortScience* 21:1105–1112.

Cantliffe, D. J., Shuler, K. D. and Guedes, A. C. (1981). "Overcoming seed thermodormancy in a heat sensitive romaine lettuce by seed priming". *HortScience* 16:196–198.

Cardoso, V. J. M. and Felippe, G. M. (1987). "Endogenous ethylene and the germination of *Cucumis anguria* L. seeds". *Revista Brasileira de Botânica* 10:29–32.

Cocucci, M. and Negrini, N. (1991). "Calcium-calmodulin in germination of *Phacelia tanacetifolia* seeds: Effects of light, temperature, fusicoccin and calcium-calmodulin antagonists". *Physiol. Plant.* 82:143–149.

Corbineau, F., Rudnicki, R. M. and Côme, D. (1989). "ACC conversion to ethylene by sunflower seeds in relation to maturation, germination and thermodormancy". *Plant Growth Reg.* 8:105–115.

Coronel, J. and Motes, J. E. (1982). "Effect of gibberellic acid and seed rates on pepper seed germination in aerated water columns". *J. Amer. Soc. Hort. Sci.* 107:290–295.

Creelman, R. A. (1989). "Abscisic acid physiology and biosynthesis in higher plants". *Physiol. Plant.* 75:131–136.

Crocker, W. (1930). "Effect of the visible spectrum upon the germination of seeds and fruits". In *Biological Effects of Radiation*, McGraw-Hill, New York, pp. 791–828.

Davies, H. V. and Slack, P. T. (1981). "The control of food mobilization in seeds of dicotyledonous plants". *New Phytol.* 88:41–51.

Doneen, L. D. and MacGillivray, J. H. (1943). "Germination (emergence) of vegetable seed as affected by different soil conditions". *Plant Physiol.* 18:524–529.

Dunlap, J. R. and Morgan, P. W. (1977). "Characterization of ethylene-gibberellic acid control of germination in *Lactuca sativa* L.". *Plant Cell Physiol.* 18:561–568.

Frett, J. J., Pill, W. G., and Morneau, D. C. (1991). "A comparison of priming agents for tomato and asparagus seeds". *HortScience* 26:1158–1159.

Furatani, S. C., Zandstra, B. H., and Price, H. C. (1985). "Low temperature germination of celery seeds for fluid drilling". *J. Amer Soc. Hort. Sci.* 110:149–153.

Gallego, P., Hernandez-Nistal, J., Martin, L., Nicolas, G., and Villalobos, N. (1991). "Cytokinin levels during the germination and seedling growth of *Cicer arietinum* L.: Effect of exogenous application of calcium and cytokinins". *Plant Science* 77:207–221.

Gepstein, S. and Ilan, I. (1980). "Evidence for the involvement of cytokinins in the regulation of proteolytic activity in cotyledons of germinating beans". *Plant Cell Physiol.* 21:57–63.

Ghate, S. R., Phatak, S. C., and Batal, K. M. (1984). "Pepper yields from fluid drilling with additives and transplanting". *HortScience* 19:281–283.

Gray, D. (1981). "Fluid drilling of vegetable seeds". *Hort. Rev.* 3:1–27.

Groot, S. P. C. and Karssen, C. M. (1987). "Gibberellins regulate seed germination in tomato by endosperm weakening: a study with gibberellin-deficient mutants". *Planta* 171:525–531.

Groot, S. P. C. and Karssen, C. M. (1992). "Dormancy and germination of abscisic acid-deficient tomato seeds. Studies with the *sitiens* mutant". *Plant Physiol.* 99:952–958.

Groot, S. P. C., Kieliszewska-Rockika, B., Vermeer, E. and Karssen, C. M. (1988). "Gibberellin-induced hydrolysis of endosperm cell walls in gibberellin-deficient tomato seeds prior to radicle protrusion". *Planta* 174:500–504.

Haber, A. H. and Luippold, H. J. (1960). "Separation of mechanisms initiating cell division and cell expansion in lettuce seed germination". *Plant Physiol.* 35:168–173.

Halmer, P. (1985). "The mobilization of storage carbohydrates in germinated seeds". *Physiologie Végétale* 23:107–125.

Hanks, R. S. and Thorp, F. C. (1956). "Seedling emergence of wheat as related to soil moisture content, bulk density, oxygen diffusion rate, and crust strength". *Proc. Soil Sci. Soc. Am.* 20:307–310.

Hartmann, H. T., Kester, D. E. and Davies Jr., F. T. (1990). *Plant Propagation: Principles and Practices. Fifth Edition*, Prentice Hall, Englewood Cliffs, NJ.

Heydeker, W. (1977). "Stress and seed germination: An agronomic view". In *The Physiology and Biochemistry of Seed Dormancy and Germination*, ed., A. A. Khan. North-Holland Publishing Co., Amsterdam, pp. 237–282.

Hoffman, N. E., Fu, J. R. and Yang, S. F. (1983). "Identification and metabolism of 1-(malonylamino)-cyclopropane-1-carboxylic acid in germinating peanut seeds". *Plant Physiol.* 71:197–199.

Karssen, C. M., Brinkhorst-van der Swan, D. L. C., Breekland, A. E., and Koornneef, M. (1983). "Induction of dormancy during seed development by endogenous abscisic acid: Studies on abscisic acid deficient genotypes of Arabidopsis thaliana". Planta 157:158–165.

Karssen, C. M., Groot, S. P. C., and Koornneef, M. (1987). "Hormone mutants and seed dormancy in Arabidopsis and tomato". In Developmental Mutants in Higher Plants, eds., H. Thomas and D. Grierson. Cambridge University Press, Cambridge, pp. 119–133.

Katoh, H. and Esashi, Y. (1975). "Dormancy and impotency of cocklebur seeds. I. CO_2, C_2H_4, O_2 and high temperature". Plant Cell Physiol. 16:687–696.

Katsumi, M., Tsuda, A., and Sakurai, H. (1987). "Brassinolide-induced stimulation of membrane permeability and ATPase activity in light grown cucumber hypocotyls". Proc. Plant Growth Reg. Soc. Am. 14:215–220.

Kepczynski, J. and Rudnicki, R. M. (1975). "Studies on ethylene in dormancy of seeds. I. Effect of exogenous ethylene on the afterripening and germination of apple seeds". Fruit Science Reports (Poland) 2:25–41.

Kermode, A. R. (1990). "Regulatory mechanisms involved in the transition from seed development to germination". Critical Reviews in Plant Sciences 9:155–195.

Ketring, D. L. (1977). "Ethylene and seed germination". In The Physiology and Biochemistry of Seed Dormancy and Germination, ed., A. A. Khan. North-Holland, Amsterdam, pp. 157–178.

Ketring, D. L. and Morgan, P. W. (1972). "Physiology of oil seeds. IV. Role of endogenous ethylene and inhibitory regulators during natural and induced after-ripening of dormant Virginia-type peanut seeds". Plant Physiol. 50:382–387.

Khan, A. A. (1975). "Primary, preventive and permissive roles of hormones in plant systems". The Botanical Review 41:391–420.

Khan, A. A. (1977). "Preconditioning, germination and performance of seeds". In The Physiology and Biochemistry of Seed Dormancy and Germination, ed., A. A. Khan. North-Holland Publishing Co., Amsterdam, pp. 283–316.

Khan, A. A. (1978). "Incorporation of bioactive chemicals into seeds to alleviate environmental stress". Acta Hort. 83:225–34.

Khan, A. A. and Huang, X. L. (1988). "Syngergistic promotion of ethylene production and germination with kinetin and 1-aminocyclopropane-1-carboxylic acid in lettuce seeds exposed to salinity stress". Plant Physiol. 87:847–852.

Khan, A. A. and Prusinski, J. (1989). "Kinetin enhanced 1-aminocyclopropane-1-carboxylic acid utilization during alleviation of high temperatures stress in lettuce seeds". Plant Physiol. 91:733–737.

Kohno, A. and Nanmori, T. (1991). "Changes in α- and β-amylase activities during germination of seeds of alfalfa (Medicago sativa L.)". Plant Cell Physiol. 32:459–466.

Kovac, M., Horgan, R., and Meilan, R. (1993). "Cytokinins in Scotch pine seedling root exudates and their influence on seed germination". Plant Physiol. Biochem. 31:35–40.

Lalonde, S. and Saini, H. S. (1992). "Comparative requirement for endogenous ethylene during seed germination". Ann. Botany 69:423–428.

Lewak, S. (1985). "Hormones in seed dormancy and germination". In *Hormonal Regulation of Plant Growth and Development*, ed., S. S. Purohit. Martinus Nishoff, Dordrecht, pp. 95–144.

Lin, Y. H., Moreau, R. A. and Huang, A. H. C. (1982). "Involvement of glyoxysomal lipase in the hydrolysis of storage triacylglycerols in the cotyledons of soybean seedlings". *Plant Physiol.* 70:100.

MacIsaac, S. A., Sawhney, V. K., and Poherecky, Y. (1989). "Regulation of lateral root formation in lettuce (*Lactuca sativa*) seedling roots: Interacting effects of α-naphthaleneacetic acid and kinetin". *Physiol. Plant.* 77:287–293.

Martin, L., Diez, A., Nicolas, G., and Villalobos, N. (1987). "Variation of the levels and transport of cytokinins during germination of chick-pea seeds". *J. Plant Physiol.* 128:141–151.

Mayer, A. M. and Poljakoff-Mayber, A. (1989). *The Germination of Seeds. Fourth Edition*, Pergamon Press, London.

Miller, C. O. (1956). "Similarity of some kinetin and red light effects". *Plant Physiol.* 31:318.

Munoz, J. L., Martin, L., Nicolas, G., and Villalabos, N. (1990). "Influence of endogenous cytokinins on reserve mobilization in cotyledons of *Cicer arietinum* L.: Reproduction of endogenous levels of total cytokinins, zeatin, zeatin riboside and their corresponding glucosides". *Plant Physiol.* 93:1011–1016.

Ni, B. R. and Bradford, K. J. (1992). "Quantitative models describing the sensitivity of tomato seed germination to abscisic acid and osmoticum". *Plant Physiol.* 98:1057–1068.

Ni, B-R. and Bradford, K. J. (1993). "Germination and dormancy of abscisic acid-and gibberellin-deficient mutant tomato (*Lycopersicon esculentum*) seeds". *Plant Physiol.* 101:607–617.

Norton, C. R. (1986). "Germination under flooding: Metabolic implications and alleviation of injury". *HortScience* 21:1123–1125.

Nygren, M. (1987). "Germination characteristics of autumn collected *Pinus sylvestris* seeds". *Acta Forestalia Fennica* 1–42.

Parthier, P. (1979). "The role of phytohormones (cytokinins) in chloroplast development". *Biochemie Physiologie Pflanzen* 174:173–214.

Pino, E., Martin, L., Guerra, H., Nicolas, G., and Villalobos, N. (1990). "The effect of dihydrozeatin on the mobilization of protein reserves in the cotyledons of chick-pea seeds". *J. Plant Physiol.* 135:698–702.

Pino, E., Martin, L., Guerra, H., Nicolas, G., and Villalobos, N. (1991). "Effect of dihydrozeatin on the mobilization of protein reserves in protein bodies during the germination of chick-pea seeds". *J. Plant Physiol.* 137:425–432.

Qamaruddin, M. (1991). "Appearance of the zeatin riboside type of cytokinin in *Pinus sylvestris* seeds after red light treatment". *Scand. J. For. Res.* 6:41–46.

Qamaruddin, M. and Tillberg, E. (1989). "Rapid effects of red light on the isopentenyladenosine content in Scotch pine seeds". *Plant Physiol.* 91:5–8.

Reddy, N. R., Sathe, S. K. and Salunkle, D. P. (1982). "Phytates in legumes and cereals". *Adv. Food Res.* 28:1.

Reid, J. B. (1990). "Phytohormone mutants in plant research". *J. Plant Growth Reg.* 9:97–111.

Saini, H. S., Consolacion, E. D., Bassi, P. K. and Spencer, M. S. (1986). "Requirement for ethylene synthesis and action during relief of thermoinhibition of lettuce seed germination by combinators of gibberellic acid, kinetin and carbon dioxide". *Plant Physiol.* 81:950–953.

Saini, S., Consolacion, E. D., Bassi, P. K. and Spencer, M. S. (1989). "Control processes in the induction and relief of thermoinhibition of lettuce seed germination: Action of phytochrome and endogenous ethylene". *Plant Physiol.* 90:311–315.

Salter, P. J. (1978). "Techniques and prospects for 'fluid' drilling of vegetable crops". *Acta Hort.* 72:101–108.

Satoh, S. and Esashi, Y. (1982). "Effects of α-aminoisobutyric acid and D- and L-amino acids on ethylene production and content of 1-aminocyclopropane-1-carboxylic acid in cotyledonary segments of cocklebur seeds". *Physiol. Plant.* 54:147–152.

Satoh, S., Takeda, Y., and Esashi, Y. (1984). "Dormancy and impotency of cocklebur seeds. IX. Changes in ACC-ethylene conversion activity and ACC content of dormant and non-dormant seeds during soaking". *J. Exp. Botany* 35:1515–1524.

Sawhney, V. K. and Srivastava, L. M. (1974). "Gibberellic acid induced elongation of lettuce hypocotyl and its inhibition by colchicine". *Canadian J. Botany* 52:259–264.

Schopfer, P. and Plachy, C. (1985). "Control of seed germination by acsisic acid. II. Effect on embryo growth potential (minimum turgor pressure) and growth coefficient (cell wall extensibility) in *Brassica napus* L.". *Plant Physiol.* 77:676–686.

Sharma, S. S., Sharma, S. and Rai, V. K. (1992). "The effect of EGTA, calcium channel blockers (lanthanum chloride and nifedipine) and their interaction with abscisic acid on seed germination of *Brassica juncea* cv. RLM-198". *Ann. Botany* 70:295–299.

Swain, R. R. and Dekker, E. E. (1966). "Seed germination studies. II. Pathways for starch degradation in germinating pea seedlings". *Biochim. Biophys. Acta* 122:87.

Takahashi, N., Phinney, B. O. and MacMillan, J. (1992). *Gibberellins*, Springer-Verlag, Berlin.

Takeuchi, Y., Worsham, A. D., and Awad, A. E. (1991). "Effects of brassinolide on conditioning and germination of witch*weed (Striga a*siatica) seeds". In *Brassinosteroids: Chemistry, Bioactivity and Applications*, eds., H. G. Cutler, T. Yokota and G. Adam. American Chemical Society , Washington, DC, pp. 298–305.

Taylor, A. G. and Kenny, T. J. (1985). "Improvement of germinated seed quality by density separation". *J. Amer. Soc. Hort. Sci.* 110:347–349.

Taylorson, R. B. (1979). "Response of weed seeds to ethylene and related hydrocarbons". *Weed Sci.* 27:7–10.

Taylorson, R. B. and Hendricks, S. B. (1977). "Dormancy in seeds". *Ann. Rev. Plant Physiol.* 28:331–354.

Thomas, T. H. (1977). "Cytokinins, cytokinin-active compounds and seed germination". In *The Physiology and Biochemistry of Seed Dormancy and Germination*, eds., A. A. Khan. North-Holland Publishing Co., Amsterdam, pp. 111–124.

Torrey, J. G. (1962). "Auxin and purine interactions in lateral root initiation in isolated root segments". *Physiol. Plant.* 15:177–185.

Vacha, G. A. and Harvey, R. B. (1927). "The use of ethylene, propylene and similar compounds in breaking the rest period of tubers, bulb, cuttings and seeds". *Plant Physiol.* 2:187–193.

Walker, M. A., Roberts, D. A., Waite, J. L. and Dumbroff, E. B. (1989). "Relationships among cytokinin, ethylene and polyamines during the stratification-germination process in seeds of *Acer saccharum*". *Physiol. Plant.* 76:326–332.

Washio, K. and Ishikawa, K. (1992). "Structure and expression during the germination of rice seeds of the gene for a carboxypeptidase". *Plant Mol. Biol.* 19:631–640.

Welbaum, G. E., Tissaoui, T., and Bradford, K. J. (1990). "Water relations of seed germination in muskmelon (*Cucumis melo* L.). III. Sensitivity of germination to water potential and abscisic acid during development". *Plant Physiol.* 92:1029–1037.

Whitehead, C. S. and Nelson, R. M. (1992). "Ethylene sensitivity in germinating peanut seeds: The effect of short-chain saturated fatty acids". *Plant Physiol.* 139:479–483.

Wightman, F., Schneider, E. A., and Thimann, K. V. (1980). "Hormonal factors controlling the initiation and development of lateral roots. II. Effects of exogenous growth regulators on lateral root formation in pea roots". *Physiol. Plant.* 49:304–314.

Rooting

The ability of many plants and plant parts to form roots from cuttings under the proper conditions is important in the propagation of many species. Plant parts such as stems, roots, or leaves can serve as cutting sources for propagation when combined with the proper chemical, mechanical, and/or environmental conditions. One of the main benefits of using this type of asexual propagation is that the new plant produced is identical to the parent plant. There is a wide range of adventitious root-forming ability in plants which can be from very easy to root to those which will not root at all (Davis and Haissig 1994). In this chapter the adventitious root formation process in plants and how plant growth substances and/or other factors are involved will be presented.

ANATOMICAL BASIS
FOR ADVENTITIOUS ROOT FORMATION

Stems

Adventitious rooting may be defined as the formation of roots at locations other than where roots occur under natural conditions. Their formation can be broken down into two types: wound-induced and preformed. Wound-induced adventitious roots develop only following wounding (e.g., detachment from the parent plant). It is generally accepted that wound-induced roots are de novo produced (Davies et al. 1982). After the cutting is made the wound healing process begins very rapidly (Cline and Neely 1983). During this process the outer cells die forming a protective surface together with xylem plugging, thereby preventing desiccation and pathogen entry. The cells behind this protective covering divide and within a few days a layer of parenchyma cells (callus) form a wound periderm. Callus typically proliferates at the base of the cutting, however, it is not essential for adventitious rooting. The formation of callus and roots are independent of each other even though they generally occur simultaneously under the proper conditions. Following periderm formation cells near the vascular cambium and phloem will then divide followed by the initiation of adventitious roots. The developmental anatomical changes which occur during wound-induced de novo adventitious root formation can be broken down into four stages:

1. Dedifferentiation of specific differentiated cells
2. Formation of root initials from cells near vascular bundles or vascular tissue which have become meristematic by dedifferentiation
3. Development of root initials into organized root primordia
4. Growth and emergence of root primordia both outward and toward vascular tissues where conducting connections are made (Hartmann et al. 1990).

Preformed, also called *latent*, root initials, as the name implies, develop naturally on plant parts which are still attached to the parent plant. Although emergence prior to detachment can occur (Lovell and White 1986), in general such root initials lie dormant until cuttings are made and placed under environmental conditions favorable for further development and emergence. Preformed root initials typically occur in easy-to-root genera such as *Salix*. The origin of preformed root initials is similar to wound-induced adventitious root formation (Lovell and White 1986). Adventitious roots in herbaceous plants usually originate just outside the vascular bundles (Preistley and Swingle 1929), but the tissues involved at the site of origin can vary depending upon the kind of plant (Hartmann et al. 1990). In woody perennial plants, adventitious roots in stem cuttings usually originate from living parenchyma cells, primarily in the young secondary phloem, but they sometimes arise from vascular rays, cambium, phloem, lenticels, or pith (Lovell and White 1986; Ginzburg 1967) (Figure 5.1).

Leaf Cuttings

There are plant species which can be propagated from leaf cuttings although the origin of new shoots and roots from them is highly variable. In leaf cuttings,

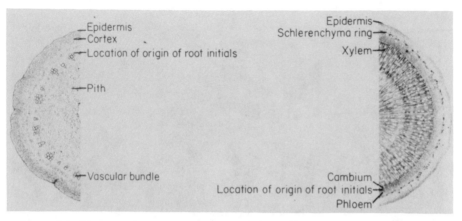

Figure 5.1. Cross section of a stem showing where adventitious roots typically originate. Young, herbaceous, dicotyledonous plant (left) and young, woody plant (right) (from Hartmann et al. (1990)).

roots and shoots can develop from both primary and secondary meristems. An example of primary meristems is with either attached or detached leaves from *Bryophyllum* plants. These plants develop from latent primary meristems (groups of cells directly descended from embryonic cells) that have never ceased to be involved in meristematic activity. Small plants develop from notches around the leaf margin following the detachment of the leaf and placement on moist rooting medium under the proper environmental conditions (Figure 5.2).

More commonly leaf cuttings develop new plants from secondary meristems (e.g., *Saintpaulia*), also called wound meristems, which have differentiated and functioned in some previously differentiated tissue system, and then dedifferentiate into new meristematic zones, resulting in the regeneration of new plant organs (Hartmann et al. 1990).

Root Cuttings

The development of adventitious roots and shoots must take place in order to form plants from root cuttings (Robinson 1975). The production of a new plant from root cuttings occurs in different ways, depending upon the species used. The most common way is for the root cutting to first produce an adventitious shoot followed by root formation which typically occurs at the base of the new shoot. Root cuttings taken from young seedlings have been shown to be more successful than from mature plants, the exact reason for this is unknown. In young roots, buds typically arise endogenously in the pericycle near the vascular cambium, whereas in older roots, buds arise exogenously

Figure 5.2. Development of plants from primary meristems found in notches in the leaf margin of *Bryophyllum* plants.

from callus-like growth. One disadvantage of using root cuttings is that the new plant produced can be different from the parent, thereby removing one of the benefits of asexual reproduction (Hartmann et al. 1990).

POLARITY OF ROOT AND SHOOT FORMATION

Roots and shoots typically show a great deal of polarity with respect to their formation. Stem cuttings form shoots close to the shoot tip (distal) and roots form nearest the junction between the shoot and root (proximal). Root cuttings generally form roots at the distal end and shoots at the proximal end. In leaf cuttings both roots and shoots generally arise from the base of the cutting. Polarity observed with respect to adventitious root initiation has been correlated with auxin movement supporting the role of auxin in root initiation (Hartmann et al. 1990). 2, 3, 5 – triiodobenzoic acid (TIBA), an auxin antagonist, inhibits rooting by inhibiting basipetal transport of IAA and subsequent rooting. Although this compound has no commercial value, it has been used by researchers as a basic tool to study the role of IAA in adventitious rooting.

PHYSIOLOGICAL BASIS OF ADVENTITIOUS ROOT FORMATION IN CUTTINGS

Effects of Plant Growth Substances

Von Sachs (1882) postulated that there was a substance or substances in leaves, buds, and/or cotyledons of plants that moved to the roots and stimulated rooting. This hormone substance was termed *rhizocaline* by Bouillene and Went (1933) and today still remains a hypothetical compound. Different classes of plant-growth-promoting substances, including auxins, cytokinins, gibberellins, ethylene, brassinolide, as well as inhibitory substances such as abscisic acid, growth retardants, and phenolics, influence root initiation. To date, auxins have been shown to have the greatest effect on rooting and are commercially used today. Much of the research on plant growth substance involvement in rooting has dealt with the relationship between the known compounds and the search for new promotive or inhibitory compounds; however, a common problem associated with evaluating the effects of plant growth substances on the rooting process is a high degree of variability. Some factors which may be responsible for this variability are:

1. Environmental conditions and nutrition of the parent plant and the cutting following its removal.
2. Handling of the cutting following removal from the parent plant.
3. Concentration range selected, method of application, and specific substance used, e.g., GA_1 or GA_4, zeatin or isopentyladenine.
4. Stage of the rooting process applied.
5. Interactions between known and unknown plant growth substances.

When viewing the current results in the literature and when planning future experiments in this area variability encountered during the rooting process must be taken into consideration. In order to better understand the rooting process future studies on changes in endogenous levels of plant growth substances during all stages of adventitious root formation and elongation are necessary.

Auxins. Thimann and Koepfli (1935) identified growth-promoting and root-forming substances in plants. They also demonstrated the practical use of indole-3-acetic acid (IAA) in the stimulation of root formation on cuttings. In the same year two synthetic auxins, indole-3-butyric acid (IBA) and naphthaleneacetic acid (NAA), were shown to be even more effective than the naturally occurring IAA for rooting (Thimann 1935; Zimmerman and Wilcoxon 1935). Since this time there have been numerous reports indicating that auxin is involved in the initiation of adventitious roots (Figure 5.3) and that division

Figure 5.3.   Effects of auxin treatment on adventitious root initiation and growth in mung beans, auxin-treated (left) and untreated (right) (courtesy of C. W. Heuser).

of root initials is dependent either upon exogenous or endogenous auxin. Even though there are a number of studies indicating that auxins have a promotive effect on rooting there are exceptions which show that it either has no effect on root elongation or is inhibitory at higher concentrations (Hartmann et al. 1990; Blazich 1989; Flygh et al. 1993; Blakesley and Chaldecott 1993; Blakesley et al. 1991). In addition to either exogenous or endogenous auxin effects it has been suggested that the sensitivity (responsiveness) of the cuttings to auxin may also be very important (Trewavas 1981, 1991; Blakesley and Chaldecott 1993). The concept of plant tissue sensitivity to plant growth substances was presented by Trewavas (1981) and since that time has been the subject of much debate (Trewavas 1991). The theory stated is that in most cases, sensitivity to plant growth substances is more important than their actual concentration. Experimental evidence from studies with transgenic plant tissues have shown that transfer of the Ri (root-inducing) plasmid into plant tissues increased their sensitivity to auxin (Shen et al. 1988; Maurel et al. 1991; Blakesley and Chaldecott 1993). The controlled expression of these genes may some-day offer some insight on the role of plant growth substances in the process of root initiation (Blakesley and Chaldecott 1993). Although the search for new auxins which have a stimulatory effect on rooting continues, IBA and NAA are still the most commonly used auxins on a commercial basis (Blazich 1989) (Table 5.1).

Cytokinins. Since the classical work of Skoog and Miller (1957) which showed that high cytokinin-to-auxin ratios promoted shoot growth and inhibited root development it has been thought that cytokinins always have an inhibitory effect on root growth. There have been reports in the literature supporting that cytokinins inhibit root formation, one of which shows that cuttings from species with high endogenous levels of cytokinins are more difficult to root than those with low cytokinin levels (Okoro and Grace 1978). Experiments where exogenous applications of synthetic cytokinins are applied indicate that root initiation is inhibited in stem cuttings (Mok and Mok 1994; Hartmann et al. 1990). To date none of the commercially available formulations for the induc-tion of roots contains cytokinins (Blazich 1989; Berry 1984; Dirr 1981), further supporting the claim that exogenous applications of cytokinins are not benefi-cial to adventitious root formation (van Staden and Harty 1989). Even though there are reports that cytokinins are inhibitors or have no promotive effect on root formation this may be misleading (van Staden and Harty 1989). There are a limited number of examples in the literature which show that exogenous applications promote adventitious root formation (Meredith et al. 1970; Nemeth 1979; Eriksen 1974; Heide 1965), indicating that the influence of cytokinins in root initiation may depend upon a number of factors outlined earlier in order to have a promotive or inhibitory effect (Hartmann et al. 1990; Blazich 1989; Bollmark and Eliasson 1986; van Staden and Harty 1989). At the present time, results on changes in endogenous levels of the more then 30 known cytokinins are limiting and more research is necessary to better understand changes in cytokinin levels from root initiation through elongation of the root.

Gibberellins. Exogenous applications of gibberellins have been shown to inhibit adventitious root formation in many species, with the inhibition becom-ing greater with increasing concentrations of GA_3 from 1 μm and higher. There have also been reports, although a lesser number, which show that under specific conditions gibberellins enhance root formation (Batten and Goodwin 1978; Hartmann et al. 1990; Hansen 1987). It may be suggested that differ-ences in responsiveness could be due to the species used, however, both inhibition and stimulation have been shown in the same species. It appears that the responsiveness of a given plant to gibberellins is affected by a number factors mentioned earlier for all plant growth substances used in the promotion of rooting. Another complicating factor which is encountered when evaluating

Table 5.1.   The company, formulation, trade name and active ingredient for some commercial rooting formulations (Blazich 1989).

Company	Formulation	Trade name	Active ingredients
ACF Chemiefarma, The Netherlands	Powder	CHRYZOPON	0.1% IBA
		CHRYZOTEK	0.4% IBA
		CHRYZOSAN	0.6% IBA
		CHRYZOPLUS	0.8% IBA
	Powder	RHIZOPON A	0.5% IAA
		RHIZOPON A	0.7% IAA
		RHIZOPON A	1.0% IAA
		RHIZOPON B	0.1% NAA
		RHIZOPON B	0.2% NAA
		RHIZOPON AA	0.5% IBA
		RHIZOPON AA	1.0% IBA
		RHIZOPON AA	2.0% IBA
		RHIZOPON AA	4.0% IBA
		RHIZOPON AA	8.0% IBA
Brooker Chemical Corp., North Hollywood, CA	Powder	Hormex No. 1	0.1% IBA
		Hormex No. 3	0.3% IBA
		Hormex No. 8	0.8% IBA
		Hormex No. 16	1.6% IBA
		Hormex No. 30	3.0% IBA
		Hormex No. 45	4.5% IBA
MSD-AGVET, Rahway, NJ	Powder	Hormodin 1	0.1% IBA
		Hormodin 2	0.3% IBA
		Hormodin 3	0.8% IBA
Hortus Products, Newfoundland, NJ	Powder	Hormo-Root A	0.1% IBA
		Hormo-Root B	0.4% IBA
		Hormo-Root C	0.8% IBA
RHONE-POULENC, Research Triangle Park, NC	Powder	Rootone	0.1% IBA and 0.2% NAM
Coor Farm Supply, Smithfield, NC	Liquid	C-mone	1.0% IBA
		C-mone	2.0% IBA
ALPKEM Corp., Clackamas, OR	Liquid	DIP'N GROW	1.0% IBA and 0.5% NAA
Wilson Lab., Ontario, Canada	Liquid	Roots	0.4% IBA
Earth Science Products, Wilsonville, OR	Liquid	Wood's Rooting Compound	1.03% IBA and 0.51% NAA

the effects of gibberellins on rooting is that GA_3 is the only gibberellin that has been extensively studied with respect to rooting and only a limited number of studies have been conducted with other gibberellins (Hansen 1987). Considering that there are more than 90 gibberellins known today (Takahashi et al. 1991) it is very important to know if gibberellins have an effect in a given system and if they do which gibberellin has the largest promotive effect. In order to gain a better understanding of how gibberellins alone and in combination with other plant growth substances are involved in the rooting process additional research is necessary (Hansen 1987).

Abscisic Acid and Growth Retardants. Most of the research on abscisic acid (ABA) effects on adventitious rooting have utilized exogenous applications instead of measuring endogenous levels during root initiation and subsequent growth. In many cases ABA, which is a naturally occurring compound, has been shown to counteract the effects of gibberellins and inhibit shoot growth. There are conflicting reports on the effects of ABA on adventitious rooting (Hartmann et al. 1990; Davis 1989) depending upon a variety of factors mentioned earlier and the promotive effect on rooting has been shown to be to small and inconsistent to be of any commercial value at the present time (Davis 1989).

There are also a number of synthetic compounds which act as growth retardants including chlormequate chloride (CCC), paclobutrazol (PP333, Bonzi), XE-1019 (a triazole growth retardant related to PP333), morphactins, ancymidol (Arest), gonadotropins, daminozide (SADH, Alar), flurprimidol, and others (Hartmann et al. 1990; Davis 1989) which have been shown to influence adventitious rooting. Growth retardants are sometimes called antigibberellins because they inhibit gibberellin biosynthesis which reduces gibberellin levels and causes a decrease in shoot growth. Therefore, it has been suggested that their ability to promote rooting is due to their effect on the reduction of endogenous gibberellin levels and shoot growth (Hartmann et al. 1990). To date, growth retardants are not commercially used because of their small and typically inconsistent effects on rooting (Davis 1989).

Ethylene. In the 1930s scientists at the Boyce Thompson Institute for Plant Research discovered that ethylene and ethylene analogs stimulate adventitious root formation (Zimmerman et al. 1933; Zimmerman and Hitchcock 1933). This discovery was made prior to research by Thimann and Went (1934) which showed that auxin had a promotive effect on rooting. Auxins have also been shown to stimulate ethylene biosynthesis in many plant tissues (Arteca 1990), and it has been suggested that auxin-induced ethylene may induce adventitious root formation instead of auxin itself (Mudge 1989). There are a number of reports in the literature on the effects of ethylene on rooting with some showing a stimulation, inhibition, or no effect sometimes within the same genus and species (e.g., *Vigna radiata*) (Mudge 1989). Rooting promoted by ethylene has been reported more frequently in intact plants rather than cuttings, in herbaceous rather than woody plants, and in plants with preformed root initials rather than those without. At the present time it is not known if ethylene is directly involved in auxin-induced rooting of cuttings or flood-induced rooting in intact plants. This variability is reflected by the fact that ethylene is not widely used in commercial plant propagation. The reason for the large degree of variability of ethylene on root formation is probably for a wide range of reasons outlined in an earlier section of this chapter and is an inherent problem in studying the involvement of plant growth substances on adventitious rooting.

Brassinosteroids. Little is known about the effects of brassinosteroids (BR) on root initiation, growth, and development, while nothing is known about their occurrence in the roots. In general, BR and auxin have similar effects and when applied in combination can act synergistically in many test systems (Cutler et al. 1991). However, with respect to adventitious rooting in cuttings they both act very differently BR in most cases strongly inhibits while auxin promotes root formation (Roddick and Guan 1991). BR may directly inhibit root formation or indirectly affect this process by stimulating ethylene production (Arteca et al. 1988), which has been shown to suppress root formation in specific test systems (Mudge 1989). There are also a limited number of reports which show that BR can stimulate root growth (Sathiyamoorthy and Nakamura 1990; Cerana et al. 1983), although these results must be viewed with caution because of the very small increases observed. At the present time there is not a great deal of information on the effects of BR on root growth; however, the majority of the research thus far strongly supports that it has an inhibitory effect (Sakurai and Fujioka 1993).

Effects of Buds and Leaves

It was shown by Went (1934) that at least one bud on a pea cutting was required for root production. When all buds were removed from peas and treated with auxins no roots were formed, suggesting that specific factors other than auxins were involved in rooting. This work was later confirmed by others with pea cuttings (Mohammed 1975; Eriksen 1973) and in other test systems (Hartmann et al. 1990). Leaves have also been shown to have a strong stimulatory effect on root initiation (Reuveni and Raviv 1981).

Carbohydrates translocated from the leaves to the roots can strongly contribute to root formation; however, the promotive effects of leaves and buds are due to more specific factors (Breen 1974). Root-inducing factors contained in leaves and buds of easy-to-root cuttings have been extracted and applied to difficult-to-root plants, but there have been conflicting results (Hartmann et al. 1990); therefore, further studies are required to find the rooting plant growth substance (rhizocaline) proposed by Bouillene and Went (1933).

Co-factors and Endogenous Inhibitors

In order for rooting to occur, cuttings require the proper balance of plant-growth-promoting substances and a number of known and unknown cofactors. As mentioned in the previous section leaves are typically required for successful rooting to occur and are thought to contain the necessary components which promote or inhibit rooting. Sugars, nitrogenous compounds, phenolic compounds, and others have been shown to act as cofactors. Phenolic compounds are thought to stimulate the rooting process by protecting IAA from destruction by IAA oxidase. In addition to cofactors, it has also been demonstrated that the leaves also contain endogenous inhibitors other then ABA which inhibit the rooting process, however, work in this area and with cofactors is limited (Hess 1962, 1968).

Genetic Transformation to Enhance Rooting

Genetically manipulating plants to root by using *Agrobacterium rhizogenes* may play an important role in overcoming lack of rooting which occurs naturally in many plant species (Hartmann et al. 1990; Blakesley and Chaldecott 1993). Hairy root disease promoted by *A. rhizogenes* is characterized by prolific adventitious root development. With virulent strains of *A. rhizogenes*, a small portion of T-DNA, contained on a large extrachromosomal plasmid is transferred, integrated, and expressed into the plant genome. Virulent *A. rhizogenes* bacteria harbor an Ri (root-inducing) plasmid which contains genes involved in the biosynthesis of plant growth substances. In addition to promoting elevated levels of plant growth substances such as auxins and cytokinins, it has also been shown that specific regions of the Ri plasmid have the ability to confer increased sensitivity to auxin in plant tissues (Maurel et al. 1991; Shen et al. 1988; Blakesly and Chaldecott 1993). This technique in the future may make it possible to vegetatively propagate plant species which in the past could only be propagated by using seeds (Strobel and Nachmias 1989); however, before the use of *Agrobacterium* becomes a reality more research is necessary.

FACTORS AFFECTING ROOTING AND GROWTH OF PLANTS FROM CUTTINGS

Growth of Stock Plants for Cutting Production

The proper environmental conditions and physiological status of the stock plant is very important for successful rooting of cuttings (Figure 5.4). Stock plants should have the following optimized: water, temperature, light (intensity, photoperiod, quality), nutrition, and CO_2 enrichment.

Figure 5.4.   Effects of water stress on cuttings prior to planting, cuttings were water-stressed (bottom) prior to planting and unstressed (top) (three weeks after planting).

The water status of the plant should be maintained in order to avoid water stress by taking cuttings early in the day to maintain a turgid condition and also to avoid a buildup of ABA, ethylene, and other inhibitors which will adversely effect rooting (Moe and Anderson 1989). Stock plants should not be subjected to extremes in temperatures which will result in water stress because this will decrease the rooting potential of cuttings. Light duration, intensity, and wavelength have been shown to affect stock plant growth and subsequent rooting of cuttings. In fact, seasonal variation of rooting ability of cuttings has been linked to light levels which the stock plants are grown under. The effect of light levels under which stock plants are grown and the ability of cuttings to root are controversial, there are reports showing promotion, inhibition, or no effect of light on rooting (Hansen 1987; Moe and Anderson 1989), although it has generally been shown that when stock plants are grown under low light the new growth roots more easily (Hartmann et al. 1990). There are several reports in the literature which show that the photoperiod under which stock plants are grown influences the rooting ability of cuttings; however, it depended upon the species used (Moe and Anderson 1989; Hartmann et al. 1990). There are conflicting reports on the effects of light quality on stock plant growth and subsequent rooting of cuttings (Hartmann et al. 1990). The effects of carbon dioxide enrichment of stock plants has been shown to increase the number of cuttings which can be harvested from a single stock plant, but there is a lot of variation in the rooting response among species. The main reasons for increases in cutting yields are increased photosynthesis, higher relative growth rates, and greater lateral branching of the stock plants (Moe and Anderson 1989). However, without adequate light supplemental CO_2 enrichment is of minimal benefit (Molitor and von Hentig 1987).

The proper nutrition of stock plants has been shown to be very important. Very low nitrogen levels lead to reduced vigor, whereas high level of nitrogen leads to excessive vigor, either extreme is not good for rooting. At the present time it is difficult to interpret the effects of nutrition on rooting since there have only been a limited number of studies on the effects of various minerals during different stages of adventitious root formation (Blazich 1989)

Treatment of Cuttings

Handling of Cuttings. In order to maximize the rooting process it is very important to start with high-quality cuttings and this starts with high-quality stock plants. Cuttings are typically taken early in the day when they are still turgid. If they cannot be planted immediately, they are misted to reduce transpiration and held at 4°-8°C until they can be planted. There have been a number of studies on storage of unrooted cuttings (Hartmann et al. 1990); however, the key point is that once the cutting is taken from the parent plant it is very important to maintain its water status and minimize dry matter loss and pathogen attack.

Treatment of Cuttings with Auxins. The discovery that auxins stimulate adventitious rooting in cuttings was a major breakthrough for the propagation industry. Although IAA is the naturally occurring auxin found in plants, it is typically less effective for a number of reasons:

1. IAA oxidase found in plant tissues will break down IAA, but has little or no effect on synthetic auxins such as IBA or NAA.

2. Plants also possess the ability to conjugate IAA into inactive conjugates such as IAA aspartate and others, while synthetic auxins either cannot be conjugated or the conjugation occurs at a much slower rate.

3. When IAA solutions are kept under nonsterile conditions and/or sunlight they will rapidly be destroyed, this is not a problem with synthetic auxins. The acid form of auxin is only soluble in water at very low concentrations. In order to solubilize higher concentrations of auxin in its acid form it must first be dissolved in a few drops of alcohol or ammonium hydroxide prior to slowly adding it to water. In many instances auxins in their salt form are better because they are more soluble in water, thereby enhancing their effectiveness.

Unfortunately, auxins are not universal in the stimulation of rooting; in fact, there are a number of difficult-to-root species which do not respond to auxin. Both IBA and NAA are typically the principal auxins used for rooting cuttings because the majority of plant species are responsive to them, however, other auxins have also been shown to be effective (Figure 5.5). To determine the best auxin, the optimum concentration, etc. for rooting in a given species and set of conditions, simple trials are very important; whatever works best should be used.

Benefits of Auxin Treatments.   Although auxins are not effective in all plant species there are several direct benefits of using auxins. They are:

1. A higher percentage of cuttings produce roots.
2. Root initiation is typically quicker.
3. The number and quality of roots per cutting is increased.
4. Uniformity of rooting along the length of the cutting in increased (Blazich 1989).

Treatment Techniques.   The trade names of commercial rooting agents which are commonly used today are listed in Table 5.1. There are many methods for applying sufficient amounts of plant growth substances to a cutting, however, at the present time there are three methods which have come into widespread practical use. They are:

1. Application of auxin-talcum powder mixtures.
2. Dilute solution soaking.
3. Concentrated solution dip or quick dip (Hitchcock and Zimmerman 1939; Hartmann et al. 1990).

Other methods such as insertion of toothpicks soaked with auxin into severed roots, application of an auxin-lanolin mixture or auxin starch polyacrylate gel mixture to the roots and injection or vacuum infiltration into the roots, have been used but not on a commercial basis (Blazich 1989).

AUXIN TALCUM POWDER MIXTURES.   With this method the basal 1–2 cm of the cutting is either dipped or dusted with an auxin-talcum powder mixture.

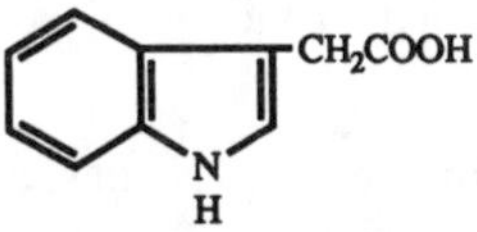

Indole-3-acetic acid
(IAA)

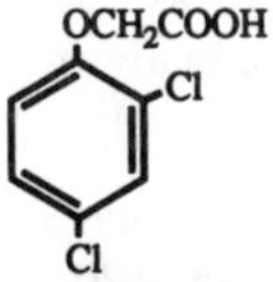

2,4-dichlorophenoxy
acetic acid
(2,4-D)

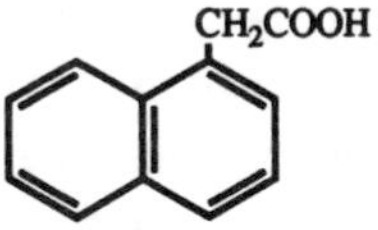

α-Naphthalene acetic acid
(α-NAA)

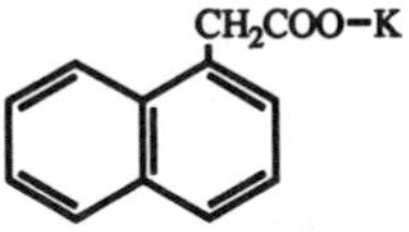

Potassium salt of α–NAA
(K-α-NAA)

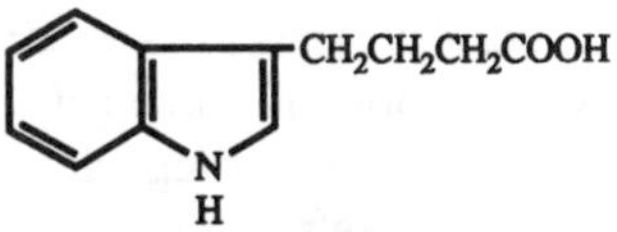

Indole-3-butyric acid
(IBA)

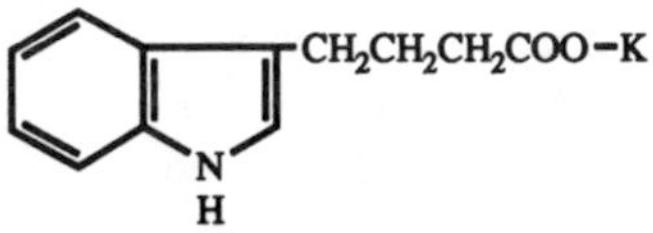

Potassium salt of IBA
(K-IBA)

Aryl ester of IBA
(P-IBA)

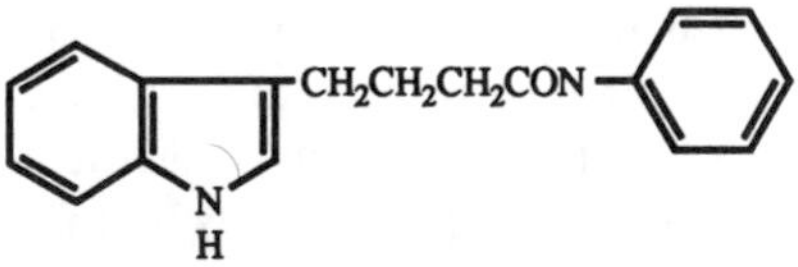

Aryl amide of IBA
(NP-IBA)

Figure 5.5. Structural formulas of auxins which are active in the promotion of adventitious root initiation in cuttings.

For the most effective application it is desirable to have the base of the cutting moist so the powder will adhere. The main advantages of this method are that commercial formulations are readily available and it is a quick and easy operation. However, the disadvantage is difficulty in obtaining uniform results due to potentially varying amounts of powder applied to individual cuttings (Blazich 1989).

DILUTE SOLUTION SOAKING.   The basal 1–2 cm of a cutting is soaked in a dilute solution (20–500 ppm) of one or more auxins for 2–24 hours and following soaking they are planted. This method is not very popular because it is slow, special equipment is required, cuttings must be kept under standardized environmental conditions, and there are potential problems with the spread of disease (Hartmann et al. 1990).

CONCENTRATED SOLUTION DIP OR QUICK DIP.   The basal 1–2 cm of the cutting is dipped for 1–5 seconds in a concentrated solution (500–30,000 ppm) of one or more auxins followed by planting. This method is very popular because it is economical, fast, and easy, and uniform treatment of bundled cuttings can be easily accomplished (Blazich 1989).

Nutrition and Use of Fungicides During the Rooting of Cuttings

Since rooting is a developmental process it has been difficult to determine the effect of nutrition on root primordia initiation vs. root primordia elongation. Although there have been a number of studies on nutrition of cuttings during rooting no general rule can be formulated and optimum levels of fertilization for rooting need to be determined for each specific use.

During the rooting process cuttings are subject to attack by microorganisms. Treatments with fungicides have been shown to give some protection resulting in both better survival and improved root quality. This enhancement of rooting is dependent on the method of application (drench, soak, incorporation into the propagation media), plant species, stage of growth, and environmental conditions. It should be noted that it is very important to conduct small-scale trials to assess which chemicals best fit the individual propagation system (Hartmann et al. 1990).

Environmental Conditions During Rooting

The presence of leaves on the cutting has been shown to be beneficial to adventitious root formation, however, if the proper water relations are not maintained the cutting will not survive. Therefore, it is important to maintain an atmosphere with a low evaporative demand on the plant to reduce transpirational loss, maintain temperatures which promote root initiation while avoiding heat stress in the leaves and maintain light levels, suitable for photosynthesis and carbohydrate production once root initiation has occurred (Hartmann et al. 1990).

REFERENCES

Arteca, R. (1990). "Hormonal stimulation of ethylene biosynthesis". In *Polyamines and Ethylene: Biochemistry, Physiology, and Interactions*, eds., H. E. Flores, R. N. Arteca, and J. C. Shannon. American Society of Plant Physiologists, Rockville, MD, pp. 216–223.

Arteca, R. N., Bachman, J. M., and Mandava, N. B. (1988). "Effects of indole-3-acetic acid and brassinosteroid on ethylene biosynthesis in etiolated mung bean hypocotyl segments". *J. Plant Physiol.* 133:430–435.

Batten, D. J. and Goodwin, P. B. (1978). "Phytohormones and the induction of adventitious roots". In *Phytohormones and Related Compounds—A Comprehensive Treatise. Vol. II*, eds., D. S. Letham, P. B. Goodwin, and T. J. V. Higgins. Elsevier/North-Holland Biomedical Press, Amsterdam, pp. 137–173.

Berry, J. B. (1984). "Rooting hormone formulations: A chance for advancement". *Proc. Int. Plant Prop. Soc.* 34:486–491.

Blakesley, D. and Chaldecott, M. A. (1993). "The role of endogenous auxin in root initiation. II. Sensitivity and evidence from studies on transgenic plant tissues". *Plant Growth Reg.* 13:77–84.

Blakesley, D., Weston, G. D., and Hall, J. F. (1991). "The role of endogenous auxin in root initiation. I. Evidence from studies on auxin application, and analysis of endogenous levels". *Plant Growth Reg.* 10:341–353.

Blazich, F. A. (1989). "Mineral nutrition and adventitious rooting". In *Adventitious Root Formation in Cuttings*, eds., T. D. Davis, B. E. Haissig, and N. Sankhla. Dioscorides Press, Portland, OR, pp. 61–69.

Bollmark, M. and Eliasson, L. (1986). "Effects of exogenous cytokinins on root formation in pea cuttings". *Physiol. Plant.* 68:662–666.

Bouillenne, R. and Went, F. W. (1933). "Recherches expérimentales sur la néo-formation des racines dans les plantules et les boutures des plantes supérieures". *Ann. Jard. Bot. Buitenzorg* 43:25–202.

Breen, P. J. (1974). "Effect of leaves and carbohydrate content and movement of ^{14}C-assimilate in plum cuttings". *J. Amer. Soc. Hort. Sci.* 99:326–332.

Cerana, R., Bonetti, A., Marre, M. T., Romani, G., Lado, P., and Marre, E. (1983). "Effects of a brassinosteroid on growth and electrogenic proton extrusion in Azuki bean epicotyls". *Physiol. Plant.* 59:23–27.

Cline, M. N. and Neely, D. (1983). "The histology and histochemistry of the wound healing process in geranium cuttings". *J. Amer. Soc. Hort. Sci.* 108:452–456.

Cutler, H. G., Yokota, T., and Adam, G. (1991). *Brassinosteroids: Chemistry, Bioactivity and Applications*, American Chemical Society, Washington, DC.

Davies, F. T. Jr, Lazarte, J. E., and Joiner, J. N., (1982). "Initiation and development of roots in juvenile and mature leaf bud cuttings of *Ficus pumila* L.". *Am. J. Botany* 69:804–811.

Davis, T. D. (1989). "Photosynthesis during adventitious rooting". In *Adventitious Root Formation in Cuttings*, eds., T. D. Davis, B. E. Haissig, and N. Sankhla. "Dioscorides Press, Portland, OR, pp. 79–87.

Davis, T.D., and Haissig, B.E. (1994). *Biology of Adventitious Root Formation*, Plenum Press, New York.

Dirr, M. A. (1981). "Rooting compounds and their use in plant propagation". *Proc. Intl. Plant Prop. Soc.* 31:472–479.

Eriksen, E. N. (1973). "Root formation in pea cuttings. I. Effects of decapitation and disbudding at different development stages". *Physiol. Plant.* 28:503–506.

Eriksen, E. N. (1974). "Root formation in pea cuttings. III. The influence of cytokinin at different developmental stages". *Physiol. Plant.* 30:163–167.

Flygh, G., Grönroos, R., Gulin, L., and von Arnold, S. (1993). "Early and late root formation in epicotyl cuttings of *Pinus sylvestris* after auxin treatment". *Tree Physiol.* 12:81–92.

Ginzburg, C. (1967). "Organization of the adventitious root apex in *Tamarix aphylla*". *Amer. J. Botany* 54:4–8.

Hansen, J. (1987). "Stock plant lighting and adventitious root formation". *HortScience* 22:746–749.

Hartmann, H. T., Kester, D. E., and Davies Jr., F. T. (1990). *Plant Propagation Principles and Practices. 5th Edition*, Prentice Hall, Englewood Cliffs, NJ.

Heide, O. M. (1965). "Interaction of temperature, auxin, and kinins in the regeneration ability of *Begonia* leaf cuttings". *Physiol. Plant.* 18:891–920.

Hess, C. E. (1962). "Characterization of the rooting co-factors extracted from *Hedera helix* L. and *Hibiscus rosa-sinensis* L.". *Proc. Intl. Hort. Cong.* 16:382–388.

Hess, C. E. (1968). "Internal and external factors regulating root initiation". In *Root Growth*, ed. W. J. Whittington. Butterworth, London, pp. 42–53.

Hitchcock, A. E. and Zimmerman, P. W. (1939). "Comparative activity of root inducing substances and methods for treating cuttings". *Contr. Boyce Thompson Inst.* 10:461–480.

Lovell, P. H. and White, J. (1986). "Anatomical changes during adventitious root formation". In *New root formation in plants and cuttings*, ed., M. B. Jackson, Martinus Nijhoff Publishers, Dordrecht, the Netherlands.

Maurel, C., Barbier-Brygoo, H., Brevet, J., Spena, A., Tempé, J., and Guern, J. (1991). "*Agrobacterium rhizogenes* T-DNA genes and sensitivity of plant protoplasts to auxins". In *Advances in Molecular Genetics of Plant-Microbe Interactions. Volume 1*, eds., H. Hennecke and D. P. S. Verma, Kluwer Academic Publishers, Dordrecht, The Netherlands pp. 343–351.

Meredith, W. C., Joiner, J. N., and Biggs, R. H. (1970). "Influences of indole-3-acetic acid and kinetin on rooting and indole metabolism of *Feijoa sellowiana*". *J. Am. Soc. Hort. Sci.* 95:49–52.

Moe, R. and Anderson, A. S. (1989) "Stock plant environment and subsequent adventitious rooting". In *Adventitious Root Formation in Cuttings*, eds., T. D. Davis, B. E. Haissig and N. Sankhla. Dioscorides Press, Portland, OR, pp. 214–234.

Mohammed, S. (1975). "Further investigations on the effects of decapitation and dibudding at different development stages of rooting in pea cuttings". *HortScience* 50:271–273.

Mok, D. W. S. and Mok, M. C. (1994). *Cytokinins. Chemistry, Activity, and Function,* CRC Press, Boca Raton, FL.

Molitor, H. D. and von Hentig, W. U. (1987). "Effect of carbon dioxide enrichment during stock plant cultivation". *HortScience* 22:741–746.

Mudge, K. W. (1989). "Effect of ethylene on rooting". In *Adventitious Root Formation in Cuttings*, eds., T. D. Davis, B. E. Haissig, and N. Sankhla. Dioscorides Press, Portland, OR, pp. 150–161.

Nemeth, G. (1979). "Benzyladenine-stimulated rooting in fruit-tree rootstocks cultured in vitro". *Z. Planzenphysiol.* 95:389–396.

Okoro, O. O. and Grace, J. (1978). "The physiology of rooting Populus cuttings. II. Cytokinins activity in leafless hardwood cuttings". *Physiol. Plant.* 44:167–170.

Priestley, J. H. and Swingle, C. F. (1929). "Vegetative propagation from the standpoint of plant anatomy". *USDA Tech. Bull.* 151.

Reuveni, O. and Raviv, M. (1981). "Importance of leaf retention to rooting avocado cuttings". *J. Amer. Soc. Hort. Sci.* 106:127–130.

Robinson, J. C. (1975). "The regeneration of plants from root cuttings with special reference to the apple". *Hort. Abst.* 45:305–15.

Roddick, J. G. and Guan, M. (1991). "Brassinosteroids and root development". In *Brassinosteroids. Chemistry, Bioactivity, and Applications*, eds., H. G. Cutler, T. Yokota, and G. Adam. American Chemical Society, Washington, DC, pp. 231–245.

Sakurai, A. and Fujioka, S. (1993). "The current status of physiology and biochemistry of brassinosteroids". *Plant Growth Reg.* 13:147–159.

Sathiyamoorthy, P. and Nakamura, S. (1990). "In vitro root induction by 24-epibrassinolide on hypocotyl segments of soybean (Glycine max (L.). "Merr.)". *Plant Growth Reg.* 9:73–76.

Shen, W. H., Petit, A., Guern, J., and Tempé, J. (1988). "Hairy roots are more sensitive to auxin than normal roots". *Proc. Natl. Acad. Sci. USA* 85:3417–3421.

Skoog, F. and Miller, C. O. (1957). "Chemical regulation of growth and organ formation in plant tissues culture in vitro". *Symp. Soc. Exp. Biol.* 11:118–131.

Strobel, G. A. and Nachmias, A. (1989). "*Agrobacterium rhizogenes*: A root inducing bacterium". In *Adventitious Root Formation in Cuttings*, eds., T. D. Davis, B. E. Haissig and Sankhla. Dioscorides Press, Portland, OR, pp. 284–288.

Takahashi, N., Phinney, B. O. and MacMillan, J. (1991). *Gibberellins*, Springer-Verlag, Berlin.

Thimann, K. V. (1935). "On an analysis of activity of two growth-promoting substances on plant tissues". *Proc. Kon. Ned. Akad. Wet.* 38:896–912.

Thimann, K. V. and Koepfli, J. B. (1935). "Identity of the growth-promoting and root-forming substances of plants". *Nature* 135:101–102.

Thimann, K. V. and Went, F. W. (1934). "On the chemical nature of the root forming hormone". *Proc. Kon. Ned. Akad. Wet.* 37:456–459.

Trewavas, A. J. (1981). "How do plant growth substances work? I". *Plant, Cell Env.* 4:203–228.

Trewavas, A. J. (1991). "How do plant growth substances work? II". *Plant, Cell Env.* 14:1–12.

Van Staden, J. and Harty, A. R. (1989). "Cytokinins and adventitious root formation". In *Adventitious Root Formation in Cuttings*, eds., T. D. Davis, B. E. Haissig, and N. Sankhla. Dioscorides Press, Portland, OR, pp. 185–201.

Von Sachs, J. (1882). "Stoff und Form der Pflanzenorgane. I". *Arb. Bot. Inst. Würzburg* 2:689–718.

Went, F. W. (1934). "On the pea test method for auxin, the plant growth hormone". *Proc. Kon. Ned. Akad. Wet.* 37:547–555.

Zimmerman, P. W., Crocker, W., and Hitchcock, A. E. (1933). "Initiation and stimulation of roots from exposure of plants to carbon monoxide gas". *Contr. Boyce Thompson Inst.* 5:1–17.

Zimmerman, P. W. and Hitchcock, A. E. (1933). "Initiation and stimulation of adventitious roots caused by unsaturated hydrocarbon gases". *Contr. Boyce Thompson Inst.* 5:351–369.

Zimmerman, P. W. and Wilcoxon, F. (1935). "Several chemical growth substances which cause initiation of roots and other responses in plants". *Contrib. Boyce Thomp Inst.* 7:209–229.

Dormancy

The definition of dormancy in a broad sense is, "a temporary suspension of visible growth of any plant structure containing a meristem," (Lang 1987). However, as mentioned earlier in this text there are three conditions which must be fulfilled in order to initiate germination:

1. The seed must be viable.
2. The seed must be subjected to appropriate environmental conditions.
3. Any primary dormancy must be removed.

In order to distinguish between dormancy imposed by internal and external conditions two terms have been used: quiescence, which is a condition where the seed or bud is under exogenous control (external conditions such as water supply, temperature, or other environmental conditions may be limiting), and rest which is the condition where the seed or bud is under *endogenous* control (internal factors prevent growth even though environmental conditions are favorable). Dormancy can be broken down to either primary or secondary dormancy and as the names imply primary dormancy typically occurs first followed by secondary dormancy. In certain instances even in the absence of primary dormancy seeds or buds exposed to adverse environmental conditions can develop secondary dormancy which further delays germination or bud break. Lang (1987) proposed a system of dormancy terminology which can be applied to any plant structure and if used has the potential to clarify some of the confusion surrounding dormancy. These terms are:

Ecodormancy—Dormancy due to one or more unsuitable factors of the environment which are nonspecific in their effect. In seeds this term is equivalent to quiescence.

Paradormancy—Dormancy due to the physical factors or biochemical signals originating external to the affected structure for the initial reaction, as in apical dominance or bud scale effects. In seeds, control would come from any of the enclosing structures surrounding the embryo, not restricted to biochemical signals. This category could be identified by more rapid germination and normal seedling growth following excision of the embryo.

Endodormancy—Dormancy regulated by physiological factors inside the affected structure such as the rest period in buds. In seeds, this type is present if embryo excision fails to produce either more rapid germination or normal seedling growth.

The development, maintenance, and release of dormancy is a very complicated process. One might ask if the function of a seed is to form a new plant and a bud to form a shoot then why does dormancy exist and why is it such a complicated process? The reason is simply in many cases it is not advantageous for a seed to germinate or a bud to break, therefore, dormancy acts as a survival mechanism, as will become evident in subsequent sections of this chapter. Dormancy is observable in many plant parts such as seeds, buds, tubers, and others. However, the bulk of the research in this area has utilized seeds as an experimental system probably because they are easier to work with then other plant parts. For this reason it may appear that this chapter is focusing on seed dormancy while in reality it is not, because seed and bud dormancy are equally important.

CATEGORIES OF SEED DORMANCY

Primary Seed Dormancy

Physical Dormancy (Seed-Coat Dormancy). This type of dormancy is caused by seed coverings which are impervious to water and falls under the category of paradormancy. It acts as a safety mechanism by preserving the seed in the dry state even under warm conditions. Germination can artificially be induced by disrupting the seed coat and allowing water to enter. In nature the seed coats are softened by the action of microorganisms, passing through the digestive tracts of birds or animals, or by mechanical abrasion, alternate freezing/thawing and in some species by fire. Examples of plant families which have the genetic characteristic of physical dormancy are: *Leguminoseae*, *Malvaceae*, *Cannaceae*, *Geraniaceae*, *Chenopodiaceae*, *Convolvulaceae*, and *Solanaceae* (Hartmann et al. 1990; Kelly et al. 1992; Bewley and Black 1984; Mayer and Poljakoff-Mayber 1989).

Mechanical Dormancy (Hard-Seed Dormancy). Mechanical dormancy is caused by the seed-enclosing structure being too strong to permit expansion of the embryo even though water can penetrate it and falls under the category of paradormancy. Germination can artificially be induced by cracking the structure covering the embryo or naturally by soil microorganisms. Examples of seeds experiencing this type of dormancy are walnut (*Juglans*), stone fruits (*Prunus*), and olive (*Olea*) (Hartmann et al. 1990; Bewley and Black 1984; Mayer and Poljakoff-Mayber 1989).

Chemical Dormancy (Inhibitor Dormancy). Chemical dormancy as the name implies is caused by germination inhibitors which accumulate in the fruit and seed coverings during development and falls under the category of *paradormancy*. This type of dormancy can sometimes be overcome by prolonged leaching of the seed, removal of the seed coat or both. Examples are *Polygonaceae*, *Chenopodiaceae* (*Atriplex*), *Portulaceae* (*Portulaca*), *Crucifereae* (Mustard), *Linaceae* (Flax), *Violaceae* (Violet), and *Labiteae* (*Lavendula*) (Hartmann et al. 1990; Bewley and Black 1984; Mayer and Poljakoff-Mayber 1989).

Morphological Dormancy (Rudimentary Embryo or Undeveloped Embryo). Morphological dormancy occurs when seeds are shed from the parent plant when their embryos are not fully developed and falls into two categories, paradormancy and endodormancy. The embryo begins to enlarge after the seed imbibes water and before germination begins. The cause of morphological dormancy can be due to two types of embryos: rudimentary embryos, which are proembryos embedded into a massive endosperm, such as *Ranunculaceae* (anemone, ranunculus), *Papaveraceae* (poppy), and *Araliaceae* (ginseng), or undeveloped embryos which are torpedo-shaped and fill up to one-half the size of the seed cavity, such as *Umbelliferaceae* (carrot), *Ericacea* (rhododendron), *Primulaceae* (cyclamen), and *Gentianaceae* (gentian) (Atwater 1980). Morphological dormancy can be overcome by subjecting seeds to temperatures which favor embryo enlargement or treatment with chemicals such as potassium nitrate or gibberellic acid. The temperature used and whether or not chemicals are effective depends upon the plant used (Hartmann et al. 1990; Bewley and Black 1984; Mayer and Poljakoff-Mayber 1989).

Physiological Dormancy.   Physiological dormancy refers to a general type of primary dormancy in freshly harvested seeds from herbaceous plants and falls into the paradormancy category. It is thought to be controlled by endogenous plant growth substances and environmental cues such as temperature and light. Freshly harvested seeds of lettuce and celery are inhibited by temperatures above 25°C; this type of dormancy is called thermodormancy. Seeds of many species which are temperature sensitive are also light sensitive. In fact, some plants including lettuce and many flower crops require either light or dark to germinate, these are said to have photodormancy. In order to break physiological dormancy seeds should be chilled, subjected to alternate temperatures or treated with potassium nitrate and/or gibberellic acid depending upon the plant used (Hartmann et al. 1990; Bewley and Black 1984; Mayer and Poljakoff-Mayber 1989).

Physiological Embryo Dormancy

Physiological dormancy is characterized by factors which directly affect the embryo as the name implies. Evidence for a dormant embryo is that the excised embryo will generally not germinate normally and when it does germinate it produces an abnormal seedling. Examples of plants with physiologically dormant embryos are in seeds from trees, shrubs, and some herbaceous plants from the temperate zone. Both the seed coat and endogenous conditions within the embryo affect physiological embryo dormancy, placing it into both paradormancy and endodormancy categories (Lang 1987). In order for afterripening to occur in seeds with physiological embryo dormancy, the proper moisture levels, aeration, chilling temperatures, and time are required; however, these conditions will vary with the plant used (Hartmann et al. 1990; Bewley and Black 1984; Mayer and Poljakoff-Mayber 1989).

Double Dormancy

Double dormancy combines two or more kinds of dormancy and falls under paradormancy and endodormancy. In order to promote germination all blocking conditions must be removed in the proper sequence. This type of dormancy is characteristic of species of trees and shrubs having seeds with hard seed coats (Hartmann et al. 1990; Bewley and Black 1984; Mayer and Poljakoff-Mayber 1989).

Secondary Seed Dormancy

Secondary seed dormancy acts as a safety mechanism for the seed by preventing germination of an imbibed seed if other environmental conditions are not favorable. Conditions which promote secondary seed dormancy include unfavorably high or low temperatures, prolonged darkness (skotodormancy), prolonged white light (photodormancy), prolonged far-red light, water stress and anoxia (Small and Gutterman 1992). Secondary seed dormancy can be in any of the three dormancy groups eco-, para-, or endodormancy. This type of dormancy can be overcome by chilling, light and sometimes with plant growth substances, particularly gibberellic acid; however, this depends on the plant and type of dormancy induced (Hartmann et al. 1990; Bewley and Black 1984; Mayer and Poljakoff-Mayber 1989).

BUD DORMANCY

Temperate zone perennial woody plants have evolved a dormancy mechanism allowing them to survive cold temperatures experienced during the winter months. For example, plants that would be killed by a frost can withstand temperatures far below freezing in their dormant state (Weiser 1970). The initiation, maintenance and release of dormancy in buds involves a complex interaction of factors that are genetic, chemical, and environmental. Bud dormancy imposed by external conditions such as the environment is called quiescence and dormancy controlled by internal factors such as genetics and/or chemicals is called rest, as is the case with seeds. Even though bud dormancy is extremely important in agriculture the amount of progress in this area has been slow as compared to the study of seed dormancy as mentioned earlier. One of the reasons for this slow progress is that perennial woody plants have a wide range of bud types that can become dormant including apical buds, adventitious buds and, in some cases axillary buds (Kozlowski 1971). Dormant buds can also be found in bulbs, tubers, rhizomes and turions (in aquatic vascular plants). The diversity of structures associated with bud dormancy is a contributing factor leading to its complexity and resulting in slow progress made in this area. In addition, the study of bud dormancy has more technical problems associated with it then seed dormancy. In order to study bud dormancy the tree or shrub used must be grown and maintained under conditions which mimic their annual growth cycle (Figure 6.1), which involves a considerable amount of time and effort as compared to the study of seed dormancy, which possess less technical problems.

Temperature and light are two of the most important environmental factors regulating bud dormancy, however, moisture, nutrients, and other factors have also been shown to control at least the initial steps in the development of dormant buds (Powell 1987). Environmental conditions promote changes in endogenous factors, which in turn serve as cues to the plant that seasons are changing. Plants generally undergo a cessation of plant growth due to quiescence prior to the beginning of physiological bud dormancy. Physiological bud dormancy is similar to seed dormancy in the following ways:

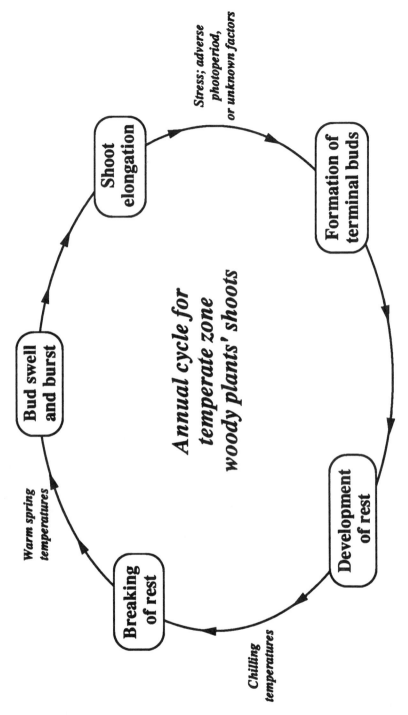

Figure 6.1. Series of events which take place over an annual growth cycle for shoots from temperate woody plants.

1. Buds are generally relatively dehydrated when dormant.
2. Reserve foods, particularly starch, are contained in dormant buds.
3. Inhibitors are high and promoters are typically low.

Once dormancy is broken there may be a brief period of quiescence followed by:

1. Rehydration of the bud leading to an increase in fresh weight.
2. Increased respiration.
3. Increased promoters.
4. Formation or activation of enzyme systems leading to a breakdown of storage materials.
5. Growth of the bud into a shoot.

As can be seen by the similarities between bud and seed dormancy many of the concepts discussed in the previous section on seed dormancy can also be applied to buds.

CONTROL OF SEED AND BUD DORMANCY

A considerable amount of evidence exists supporting the concept that endogenous plant growth substances are directly involved in the dormancy process, however, many of the details still remain a mystery (Powell 1987). In general there are four phases of dormancy:

1. Induction—This phase is characterized by a striking decrease in promoter and an increase in inhibitor activity. This phase may be triggered or affected by environmental factors such as temperature and light.
2. Maintenance—During this phase, general metabolism and endogenous promoters are very low while levels of inhibitors are very high. Fluctuations or changes in environmental conditions during this period will not have an immediate effect on ending this phase; however, environmental changes may shorten this period, causing seeds or buds to sprout prematurely, making them susceptible to injury due to environmental conditions such as a frost.
3. Trigger—This phase is especially sensitive to environmental cues or signals. During this time there is a dramatic decrease in inhibitors and a sharp rise in promoters.
4. Germination or bud break—This phase is characterized by high promoters and low inhibitors leading to increases in enzyme activity (Figure 6.2).

The length of each of these phases will depend on the genus and species plus many other factors.

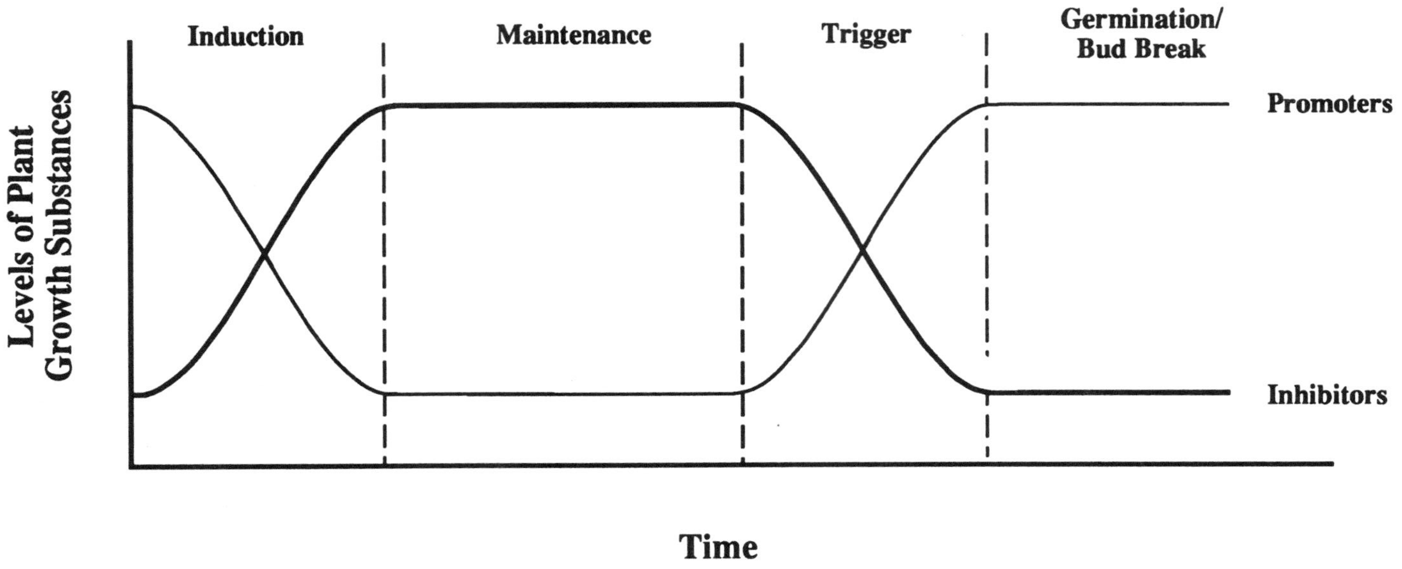

Figure 6.2. General changes in levels of plant growth substances which occur during the four stages of dormancy.

Effects of Plant Growth Substances

Experimental evidence supports the involvement of specific endogenous plant growth substances in the control of dormancy. Most of the information in the literature to date focuses on abscisic acid (ABA), gibberellins, and cytokinins with respect to their roles in dormancy; whereas ethylene, auxin, and other compounds have been shown to have little or no effect on the rest type of dormancy. Additional research on the identification and characterization of new inhibitors or promoters involved in dormancy has been slow. The focus of this section will be on the known plant growth substances and their involvement in seed and bud dormancy, but it must be kept in mind that others probably do exist and that an interaction between known and unknown compounds effects dormancy.

Abscisic Acid. Changes in endogenous levels of ABA in buds from the initiation through the breaking of rest have been reported in several species (Powell 1987; Rodriguez et al. 1991). However, based on this information the only conclusion which can be made is that ABA levels decline in buds during cold treatment, but it is still not clear if the cold treatment is responsible for this decline. Much of the evidence supporting the role of ABA in dormancy comes from seeds. As mentioned earlier, the mechanism of dormancy appears to be similar between seeds and buds; therefore, it seems safe to extrapolate many of the conclusions from one organ to another. At the present time there is a considerable amount of evidence in *Arabidopsis thaliana* indicating that the initiation of primary dormancy involves the action of ABA (Hilhorst and Karssen 1992), although the mechanism of dormancy in general remains unknown. Dormancy in wild type *A. thaliana* starts during maturation, but not in ABA-insensitive mutants (Karssen et al. 1983). In addition, there is a lesser degree of dormancy in ABA-insensitive mutants then the wild type (Koorneef et al. 1984). Interestingly, seeds from ABA-insensitive mutants contain higher levels of ABA than the wild type. It has been shown that dormancy is initiated only when the embryo itself can produce ABA (Karssen et al. 1983; Groot and Karssen 1992). This is further supported by studies using *Helianthus annus* which show that applications of fluridone (an ABA synthesis inhibitor) to developing seeds prevents ABA synthesis and development of embryo dormancy (Le Page-Degivry et al. 1990). However, fluridone does not prevent dormancy induction when it is applied after the rise in ABA levels. Evidence has also been presented that the induction of dormancy is not solely due to ABA levels but may also in part be due to the sensitivity of the seed to ABA (Walker-Simmons 1987; Trewavas 1991).

Even though applications of ABA have been shown to inhibit seed germination (Hilhorst and Karssen 1992; Ni and Bradford 1993) and delay bud break

(Davis and Jones 1991), they do not have the ability to induce dormancy (Karssen et al. 1983; Clutter 1978); although, at the present time strong arguments can be presented that ABA has a key role in the regulation of dormancy. The mechanism by which ABA regulates the induction of dormancy remains unknown. Studies at the molecular level have identified a number of ABA-induced genes (Bray 1991; Thomas et al. 1991; Kahn et al. 1993). In addition, there are a number of ABA-insensitive mutants which are now available (Ni and Bradford 1993). Future studies with these genes and mutants may shed some light as to whether the induction of dormancy is directly or indirectly controlled by ABA.

Gibberellins. There have been many studies evaluating changes in endogenous levels of gibberellins in seeds and buds (Takahashi et al. 1991; Clutter 1978; Wiebel et al. 1992). Studies evaluating exogenous applications of gibberellins have been shown to relieve certain types of dormancy including physiological dormancy, photodormancy, and thermodormancy (Hartmann et al. 1990). However, recent research with GA-deficient tomato and *Arabidopsis* mutants suggest that the synthesis of GA may not be involved in breaking dormancy and its action is only limited to the germination process (Hilhorst and Karssen 1992).

Cytokinins. The effectiveness of exogenous applications of cytokinins in the stimulation of growth in resting buds and seeds remains unclear. There have been reports showing that exogenous applications of cytokinins can overcome dormancy in buds and seeds, however, there are also reports showing that there is no effect. The relationship between chilling and the appearance of cytokinins has also been demonstrated; however, it has also been shown that little or no change occurs in response to chilling (Powell 1987; Thomas 1992; Thomas et al. 1992; Mok and Mok 1994). Additional work with mutant plants and molecular techniques is required to address the question are cytokinins involved in the dormancy process.

Ethylene. Ethylene will not promote germination in quiescent seeds nor will it alter the requirement for the rest period. In some seeds even though dormancy is broken no germination occurs because the embryo cannot overcome the physical restraint of the seed coat as mentioned in previous section on physical dormancy. Ethylene will overcome dormancy in many of the species where physical dormancy is imposed by stimulating the embryos to develop the force needed to penetrate the seed coat (Abeles et al. 1992). It has indirectly been shown that ethylene will promote lateral bud break; however, it is only effective in breaking bud dormancy after partial or full release from dormancy by chilling, suggesting that it does not have a direct regulatory role in bud dormancy (Powell 1987).

Others.   Analysis of IAA levels in *Pinus silvestris, Acer plantanoides,* and *Rosa rugosa* seeds have shown that there was no regulatory role of IAA in seed dormancy (Powell 1987). In *Corylus* buds during the winter months the IAA/ABA balance is in favor of ABA while the balance shifts toward IAA in the spring, coinciding with the breakage of bud dormancy and the initiation of shoot elongation (Rodriguez et al. 1991), indirectly suggesting the role of IAA in dormancy. However, at the present time there is little convincing evidence that auxins have any regulatory role in seed or bud rest. Thiourea has also been shown to overcome certain types of dormancy, such as deep embryo dormancy in *Prunus* seeds and high temperatures in lettuce seeds (Thomas 1992), but this effect has been suggested to be due to its cytokinin activity. Although there are many compounds which will inhibit germination or promote bud dormancy not all of these compounds can be regarded as dormancy inducing. Coumarin, a phenolic has been shown to induce light sensitivity in varieties of lettuce seeds not requiring light for germination (Nutile 1945). Due to its widespread distribution in plants and its strong inhibitory action it is considered to be one of the substances which may be a natural germination inhibitor (Mayer and Poljakoff-Mayber 1989).

EFFECTS OF
ENVIRONMENTAL FACTORS

Temperature and light are the two most important environmental factors in the control of dormancy. However, it should be noted that other environmental conditions should not be limiting to the extent that the plant is adversely affected.

Once shoot growth ceases a type of physiological dormancy or rest develops in the shoots which is similar to that in some seeds. Bud or seed rest varies in its intensity based on the season. Once primary dormancy is induced the chilling mechanism for buds and seeds are similar, both have optimum chilling temperatures from about 5° to 7°C, which are required for the termination of rest. Temperatures outside this range will not contribute to the chilling requirement, for example below 0°C is to cold and 12°C or above will be too warm. A description of how to calculate chilling units can be found in Richardson et al. (1974). The amount of chilling to break dormancy depends on the species and sometimes varieties within species (Powell 1987; Clutter 1978; Hartmann et al. 1990). The effects of temperature on dormancy in a variety of plant species is outlined in Clutter (1978). In addition to primary dormancy a secondary dormancy, can be induced if unfavorable temperatures which are too high or low are imposed (Hartmann et al. 1990).

Light

Garner and Allard (1923) showed that the length of the photoperiod was implicated in the control of bud dormancy. Since this time many investigators have found that short days promote the development of dormancy in trees and long days in some cases can release it (Clutter 1978; Nitsch 1957). However, in some woody plants shoot elongation stops and terminal buds are formed in the summer when day length is greatest (Powell 1987). A wide range of responses have been shown with respect to photoperiod, some plants respond to short days, long days, or dormancy develops irrespective of the photoperiod, these effects are outlined in a variety of species in Clutter (1978). Therefore, it appears that photoperiod alone is probably not the only factor involved in the initiation, maintenance, and breaking of bud dormancy. It has been shown that there is an interaction between temperature and photoperiod. In some cases long days can substitute for the chilling requirement and the effects of inductive photoperiods can be counteracted by low temperatures (Clutter 1978). In addition to primary dormancy, prolonged darkness (skotodormancy), white light (photodormancy), or far-red light (phytochrome response) can induce secondary dormancy in a number of plants (Thomas 1992; Hartmann et al. 1990; Small and Gutterman 1992) possibly acting as a safety mechanism.

REFERENCES

Abeles, F. B., Morgan, P. W., and Saltveit Jr., M. E. (1992). *Ethylene in Plant Biology. Second Edition*, Academic Press, San Diego, CA.

Atwater, B. R. (1980). "Germination, dormancy and morphology of the seeds of herbaceous ornamental plants". *Seed Sci. Tech.* 8:523–573.

Bewley, J. D. and Black, M. (1982). *Physiology and Biochemistry of Seeds in Relation to Germination. Vol. 2. Viability, Dormancy, and Environmental Control*, Springer-Verlag, Berlin.

Bray, E. A. (1991). "Regulation of gene expression by endogenous ABA during drought stress". In *Abscisic Acid, Physiology and Biochemistry*, eds., W. J. Davies and H. G. Jones. Bios Scientific Publishers, Oxford, U.K. pp. 81–97.

Clutter, M. E. (1978). *Dormancy and Developmental Arrest: Experimental Analysis in Plants and Animals*, Academic Press, New York.

Davis, W. J. and Jones, H. G. (1991). *Abscisic Acid: Physiology and Biochemistry*, Bios Scientific Publishers, Oxford, U.K.

Garner, W. W. and Allard, H. A. (1923). "Further studies in photoperoidism, the response of plants to relative length of day and night". *J. Agric. Res.* 23:871.

Groot, S. P. C. and Karssen, C. M. (1992). "Dormancy and germination of abscisic acid-deficient tomato seeds: studies with the *sitiens* mutant". *Plant Physiol.* 99:952–958.

Hartmann, H. T., Kester, D. E. and Davies Jr., F. T. (1990). *Plant Propagation: Principles and Practices. Fifth Edition,* Prentice Hall, Englewood Cliffs, NJ.

Hilhorst, H. W. M. and Karssen, C. M. (1992). "Seed dormancy and germination: The role of abscisic acid and gibberellins and the importance of hormone mutants". *Plant Growth Reg.* 11:225–238.

Kahn, T. L., Fender, S. E., Bray, E. A. and ÒConnell, M. A. (1993). "Characterization of expression of drought- and abscisic acid-regulated tomato genes in the drought-resistant species *Lycopersicon pennellii*". *Plant Physiol.* 103:597–605.

Karssen, C. M., Brinkhorst-van der Swan, D. L. C., Breekland, A. D., and Koornneff, M. (1983). "Induction of dormancy during seed development by endogenous abscisic acid: Studies on abscisic acid deficient genotypes of *Arabidopsis thaliana* L. Heynh". *Planta* 157:158–165.

Kelly, K. M., van Staden, J., and Bell, W. E. (1992). "Seed coat structure and dormancy". *Plant Growth Reg.* 11:201–209.

Koorneef, M., Reuling, G. and Karssen, C. M. (1984). "The isolation and characterization of abscisic acid-insensitive mutants of *Arabidopsis thaliana*". *Physiol. Plant.* 61:377–383.

Kozlowski, T. T. (1971). *Growth and Development of Trees. Vol. 1,* Academic Press, New York.

Lang, G. A. (1987). "Dormancy: A new universal terminology". *HortScience* 22:817–820.

Le Page-Degivry, M., Barthe, P., and Garello, G. (1990). "Involvement of endogenous abscisic acid in onset and release of *Helianthus annus* embryo dormancy". *Plant Physiol.* 92:1164–1168.

Mayer, A. M. and Poljakoff-Mayber, A. (1989). *The Germination of Seeds. Fourth Edition,* Pergamon Press, London.

Mok, D. W. S. and Mok, M. C. (1994). "*Cytokinins. Chemistry, Activity, and Function,* CRC Press, Boca Raton, FL.

Ni, B.R. and Bradford, K. J. (1993). "Germination and dormancy of abscisic acid- and gibberellin-deficient mutant tomato (*Lycopersicon esculentum*) seeds: Sensitivity of germination to abscisic acid, gibberellin, and water potential". *Plant Physiol.* 101:607–617.

Nitsch, J. P. (1957). "Photoperiodism in woody plants". *Proc. Amer. Soc. Hort. Sci.* 70:526–544.

Nutile, G. E. (1945). "Inducing dormancy in letuce seed with coumarin". *Plant Physiol.* 20:433.

Powell, L. E. (1987). "The hormonal control of bud and seed dormancy in woody plants". In *Plant Hormones and Their Role in Plant Growth and Development,* ed., P. J. Davies, Martinus Nijhoff Publishers, Dordrecht, The Netherlands, pp. 539–552.

Richardson, E. A., Seeley, S. D., and Walker, D. R. (1974). "A model for estimating the completion of rest for Redhaven and Elberta peachtrees". *HortScience* 9:331–332.

Rodríguez, A., Cañal, M. J., and Sánchez-Tamés, R. (1991). "Seasonal changes of plant growth regulators in *Corylus*". *J. Plant Physiol.* 138:29–32.

Small, J. G. C. and Gutterman, Y. (1992). "A comparison of thermo- and skotodormancy in seeds of *Lactuca serriola* in terms of induction, alleviation, respiration, ethylene, and protein synthesis". *Plant Growth Reg.* 11:301–310.

Takahashi, N., Phinney, B. O., and MacMillan, J. (1991). *Gibberellins*, Springer-Verlag, New York.

Thomas, T. H. (1992). "Some reflections on the relationship between endogenous hormones and light-mediated seed dormancy". *Plant Growth Reg.* 11:239–248.

Thomas, T. H., Kaminek, M., Mok, D. W. S. and Zazimalova, E. (1992). *Physiology and Biochemistry of Cytokinins in Plant*, SPB Academic Publishing, Hague, The Netherlands.

Thomas, T. L., Vivekananda, J., and Bogue, M. A. (1991). "ABA regulation of gene expression in embryos and mature plants". In *Abscisic Acid, Physiology and Biochemistry*, eds., W. J. Davies and H. G. Jones. Bios Scientific Publishers, Oxford, pp. 125–136.

Trewavas, A. (1991). "How do plant growth substances work? II." *Plant Cell Env.* 14:1–12.

Walker-Simmons, M. (1987). "ABA levels and sensitivity in developing wheat embryos of sprouting resistant and susceptible cultivars". *Plant Physiol.* 84:61–66.

Weiser, C. J. (1970). "Cold resistance and injury in woody plants". *Science* 169:1269.

Wiebel, J., Downton, W. J. S. and Chacko, E. K. (1992). "Influence of applied plant growth regulators on bud dormancy and growth of mangosteen (*Garcinia mangostana* L.)". *Scientia Hort.* 52:27–35.

Juvenility, Maturity and Senescence

During its life cycle the plant undergoes embryonic, juvenile, transitional (between juvenile and mature), and mature (adult) phases of growth and development followed by senescence and death. The juvenile phase in some species has a distinctive morphology of leaves, stems, and other structures which are no longer present when the plant becomes mature. Once the plant reaches maturity, flowering can be induced by appropriate external cues. The change from mature to senescent conditions typically involves the deterioration of many synthetic reactions leading to the death of the plant, thereby completing the cycle.

JUVENILITY

The juvenile phase of plant development can be defined as an initial period of growth when apical meristems will not typically respond to internal or external conditions to initiate flowers. This phase is characterized by exponential increases in size; absence of the ability to shift from vegetative growth to reproductive maturity leading to the formation of flowers; specific morphological and physiological traits, including leaf shape, thorniness, vigor, or disease resistance; and a greater ability to regenerate adventitious roots and shoots (Hartmann et al. 1990; Leopold and Kriedemann 1975). In general, juvenility is a period of complete inability to flower. An example of this is *Lunaria*, which cannot be induced to flower before seven weeks of age (Wellensiek 1958). However, it should be noted that when plants are in their juvenile phase they are not always devoid of flowering ability, for example, beets flower in response to cold at any age, but their responsiveness gradually increases as the plant grows older (Leopold and Kriedemann 1975).

Juvenile Morphology

Leaf form is one of the most readily observable expressions of juvenility. Differences in leaf form between juvenile and mature plants have been shown in many different plant species such as *Hedera helix, Sassafras albidum, Juniperus virginianum, Lycopersicon esculentum*, and others. *Hedra helix* and *Sassafras albidum* leaves in their juvenile phase are lobed, while in the mature phase they have a continuous leaf (Figure 7.1). In *Lycopersicon esculentum* the first non lobate leaves are followed by tri-, penta-, and heptalobate leaves, the latter accompanying flowering. *Juniperus virginium* have needlelike juvenile leaves, but when they become mature the leaves are scalelike (Figure 7.2).

There are many species where the stem and its growing point exhibit juvenile morphology. Woody plants are good examples of this, showing a juvenile branching pattern, with long whiplike branches and narrow branching angle (Blair et al. 1956). The presence of thorns on the stem is a characteristic of juvenile citrus and locust seedlings. As the tree matures, the new growth at the apex and the longest branch no longer develop thorns. It is thought that thorns on juvenile portions of the plant serve to protect the plant from browsing animals (Hartmann et al. 1990). The stem of *Hedera helix* changes from a creeping vine growth habit to an erect shrub when it passes out of the juvenile phase. Another change in stem morphology has been shown in brussels sprouts (Stokes and Verkerk 1951), where a juvenile seedling has a narrow pointed stem with a thin apex; however, at maturity the stem becomes wide and the apex broad and blunt. Enlargement of the apical meristem as the plant matures has been shown to be common in many herbaceous species (Millington and Fisk 1956).

Juvenile Physiology

Common physiological traits exhibited by plants in their juvenile stage of development are rapid growth rates, inability to flower, and high rooting

Figure 7.1. *Hedra helix* leaves are lobed in the juvenile phase (left) and in the mature phase they are continuous (right).

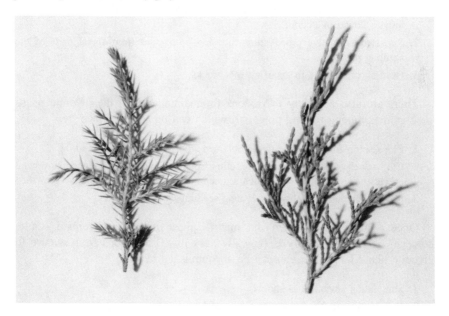

Figure 7.2. *Juniperus virginium* juvenile needlelike leaves (left) and mature scalelike leaves (right).

potential (Hartmann et al. 1990). A relationship between tree age and the rooting of cuttings was noted by Gardner (1929), before the concept of juvenility had become generally accepted. Since this time there have been reports showing that juvenile wood typically roots faster then mature wood (Trippi 1989). Loss or reduction of the ability to regenerate adventitious roots or stems is closely associated with maturity (Hartmann et al. 1990). As described in the previous chapter on rooting, auxins can generally stimulate root formation in cuttings, suggesting that the higher degree of rooting in juvenile versus mature cuttings is due to higher endogenous levels of auxin which the juvenile cuttings typically contain or due to a greater sensitivity in these tissues. Although this is an attractive theory it is not true in all cases since there are many difficult-to-root species which do not root in response to exogenous applications of auxin.

Modifying Juvenility

There are a variety of reasons why it may be necessary to induce early maturity. Practices to induce early maturation which promotes early flowering include:

1. Maximizing conditions that promote continuous vegetative growth allowing apical meristems to grow into the mature phase followed by appropriate treatments to initiate flower induction.
2. Repeated vegetative propagation choosing propagules from apical parts of the seedling plant.
3. Grafting or budding to dwarfing root stocks.

There are also a variety of reasons for maintenance of the juvenile phase such as maximizing rooting potential, which can be accomplished by:

1. Propagating from juvenile tissue obtained from the base of the seedling.
2. Cut back the plant to promote sprouting from the juvenile zone since many plant species have juvenile buds which are suppressed by normal vegetative growth.
3. In vitro cloning of embryos or young seedlings.

Once a plant has reached the mature phase it tends to remain stable; however, its rooting ability is typically very low. Reversions from mature to juvenile phase (rejuvenation) can be accomplished by:

1. Stimulating adventitious shoots.
2. Consecutive grafting to seedling rootstocks.
3. Consecutive subculturing of apical meristems in micropropagation.
4. Treating plants with gibberellic acid (Hartmann et al. 1990).

Nature of Juvenility

What is the biological basis for endogenous regulation of juvenile to mature phase changes? One of the first things which come to mind is that the duration of juvenility appears to be heritable. It has been suggested that individual cells in the apical meristem of juvenile and mature parts of the plant are epigenetically different and stable, but they change during cell division (Hackett 1985). As cells divide, one daughter cell may differentiate without further division, whereas the other continues to divide, causing gradients in juvenile:mature potential to develop with each subsequent cell division (Tilney-Bassett 1986). Another potential explanation for juvenile to mature phase changes is that the apical meristems are influenced by differentiated cells which surround them. Therefore, when cellular connections are severed between the two there is a reversion from mature to juvenile condition. Plant growth substances have been shown to be involved in the regulation of juvenile to mature growth phase changes. Evidence for this has come from experiments which show that when a mature shoot is grafted to a juvenile plant there is sometimes a change toward juvenile expression. This is thought to occur due to the transfer of plant growth substances from the roots and/or leaves of the juvenile plant to the mature shoot. Exogenous applications of gibberellins have been shown to cause a shift from mature to juvenile phenotype in *Hedera helix* and the opposite to occur in conifers while abscisic acid can reverse this change (Hackett 1985). When hard-to-root cherry clones are cocultivated with easy-to-root cultivars there is an increase in the ability of the hard-to-root clones to root. This effect can be duplicated by gibberellic acid treatments (Ranjit and Kester 1988).

MATURATION

Maturation refers to qualitative changes which allow the plant or organ to express its full reproductive potential. This is accomplished by a gradual transition of morphology, growth rate, and flowering capacity. The length of the maturation period can vary from days to years, depending on both genetics and the environment. One should be cautioned that maturation should not only be connected with flower formation but also with other reproductive structures such as tubers and bulbs. Although the period of maturation is genetically determined it can also be modified by plant growth substances and/or environmental conditions (Trippi 1989).

SENESCENCE

The process of senescence refers to endogenously controlled deteriorative changes, which are natural causes of death of cells, tissues, organs, or organisms (Leopold 1975; Noodén and Leopold 1978). Senescence may be thought of as a natural developmental process, namely, terminal differentiation. The concept that death may be actively induced by endogenous factors has been gaining acceptance. For example, it has been suggested that the difference between natural and accidental death is that natural death is due to internal factors which make the organism more vulnerable (Medawar 1957). In contrast to senescence, which is endogenously controlled, aging is caused by exogenous factors, wear and tear that accumulates over time (Leopold 1975; Noodén and Leopold 1978). Aging does not itself cause death but may cause decreased resistance to stress and disease, thereby increasing the probability of death. At the present time it is not easy to clearly distinguish between senescence and aging because the biochemical nature of the two is not well understood. Nonetheless, there are a number of clear examples of programmed senescence, including the rapid degeneration of petals following pollination or the coordinated senescence of leaves following fruit development in monocarpic plants; whereas, an example of aging is the loss of viability in stored seeds.

Why Study Senescence?

Senescence occurs at the whole plant, organ, tissue, and cell levels throughout the life cycle of an organism. It has an important function in xylem differentiation, development of leaf lobing patterns, and the breakdown of specialized cells in the embryo and female gametophyte. At the organ level it has a profound effect on leaves, flower parts, and fruits, while, at the whole plant level postreproductive senescence has a dramatic effect on the entire organism (Noodén and Leopold 1978). At the biochemical level much can be learned about many processes within the plant, for example, senescence involves the loss of self-maintenance; therefore, by studying senescence we can gain insight on how cells maintain themselves under nonsenescing conditions. At the whole plant level the study of senescence is helpful in understanding how the plant allocates resources from one site to another in order to cope with a changing environment or in order to increase chances of reproductive success.

Learning how to regulate senescence may have economic benefits. In monocarpic crop plants, there is a loss of assimilatory capacity as senescence progresses contributing to a limitation in yield (Grover 1993; Noodén and Leopold 1978). In fact, anything which speeds the senescence process such as development of metabolic sinks (e.g., fruits), unfavorable photoperiods, lengthy exposure to shade or excessive light, insufficient nutrients or water, extremes in temperature or other stresses such as ozone can dramatically reduce crop yields (Reddy et al. 1993). The keeping quality of flowers, vegetables, fruits, grains, and mushrooms is a function of senescence; therefore, the acceleration of this process contributes to postharvest losses which results in millions of dollars in loss due to waste annually (Burton 1982).

What Changes Occur During Senescence?

In order to better understand the senescence process it is necessary to monitor quantitative changes which occur during the event. The ability to quantify specific regulating components which are central to the senescence process would be ideal; however, at the present time this is not possible due to the lack of information in this area. For now it is necessary to use prominent components of the senescence process as measures. Components which have been used as measures of senescence include decreases in chlorophyll, total protein, nucleic acids, photosynthesis (reduction in photosynthetic enzymes such as ribulose bisphosphate carboxylase or phosphoenolpyruvate carboxylase), changes in plant growth substances, increased membrane permeability, and abscission. It must be noted that the study of senescence should be done with intact plants since there are many problems associated with the use of detached plant parts (Noodén and Leopold 1978; Grover 1993).

Effects of Plant Growth Substances on the Senescence Process

Senescence at the cell, tissue, organ, or whole plant level is kept under control by a number of internal as well as external factors such as environmental conditions. Plant growth substances appear to play a central role in regulating the senescence process. Since the regulation of senescence is of great economic importance, there have been many studies on the use of plant growth substance treatments on whole plants, attached organs, or detached plant parts. Despite these efforts it is difficult at the present time to form theories which meet the criteria of PESIGS rules: presence/parallel variation, excision, substitution, isolation, generality, and specificity (outlined in Chapter 1). The effects of plant growth substances on senescence have not been exhaustively studied in any one system; therefore, generality may not always be easily shown. Another complication is that senescence is probably controlled by an interaction of plant growth substances rather then by a single one. Many studies in this area involve exogenous plant growth substance applications, cynically termed spray, pray, then weigh. By comparison little work has been done on changes in endogenous levels of plant growth substances prior to and following the initiation of the senescence process, especially of natural senescence. In this section a summary on how some of the known plant growth substances are thought to be involved in the senescence process will be discussed.

Cytokinins. Cytokinins are involved in slowing many of the processes that contribute to plant senescence (Noodén and Leopold 1978; Kaminek et al. 1992; Mok and Mok 1994). This is supported mainly by two lines of research, first, the classical work of Richmond and Lang (1957), which showed that when kinetin was applied to detached leaves senescence was delayed. They showed that cytokinins are responsible for the maintenance of chlorophyll, protein, and RNA levels, all of which decline during senescence. Mothes and Engelbrecht (1961) showed that kinetin induced directed transport of substances in excised leaves in the dark. They showed that when a radioactive amino acid was applied to one leaf and kinetin to the adjacent leaf in the dark there was movement of the radioactive material toward the kinetin-treated leaf where it was shown to accumulate (Figure 7.3). When no kinetin was applied to the adjacent leaf there was no movement of radioactive material. Leopold and Kawase (1964) showed that when benzyladenine was applied to a single mature leaf it retarded senescence of that leaf while it inhibited growth and hastened senescence of the adjacent untreated leaf. Second, there have been a number of reports showing that rooting or rooting exudates can also delay senescence. Early investigators (Molisch 1938; Mothes 1960) showed that root formation delayed the senescence process in detached leaves and that kinetinlike substances produced by the roots also delayed this process (Kulaeva 1962).

In general, when cytokinins are applied at the right dose and the timing is correct they will delay senescence in most, but not all tissues. An example where cytokinin can promote senescence is in carnation (Woodson and Brandt 1991). In cases where cytokinins do not delay the senescence process other plant growth substances have been shown to be effective. Synthetic cytokinins are typically more effective in delaying senescence, probably due to their stability. It appears that exogenously applied cytokinins delay senescence by correcting an internal deficiency of cytokinins which occurs during the senescence process. There is a considerable amount of evidence showing that there is a decline in cytokinin levels prior to or during senescence (Noodén and Leopold 1978; Kaminek et al. 1992; Singh et al. 1992a, 1992b, 1992c; Mok and Mok 1994). However, most of these studies have been done with leaves and very few with flower parts or other organs.

Cytokinins have been shown to interact with other plant growth substances which promote or inhibit senescence. For example, cytokinins can counteract the promotive effects of abscisic acid on the senescence process. They may act in part by reducing the levels of endogenous ABA; however, ABA can also reduce cytokinin levels. Gibberellins have been shown to delay senescence. Kinetin has also been shown to cause an increase in gibberellinlike activity indicating that there may be a relationship between the two in delaying senescence (Noodén and Leopold 1978; Kaminek et al. 1992; Mok and Mok 1994).

In general cytokinins and ethylene have opposite effects on the senescence process; however, the relationship between the two is more complex than would appear. In most cases exogenous applications of cytokinins counteract the promotive effects of ethylene on the senescence process. However, in vegetative tissues cytokinins at low concentrations promote ethylene production

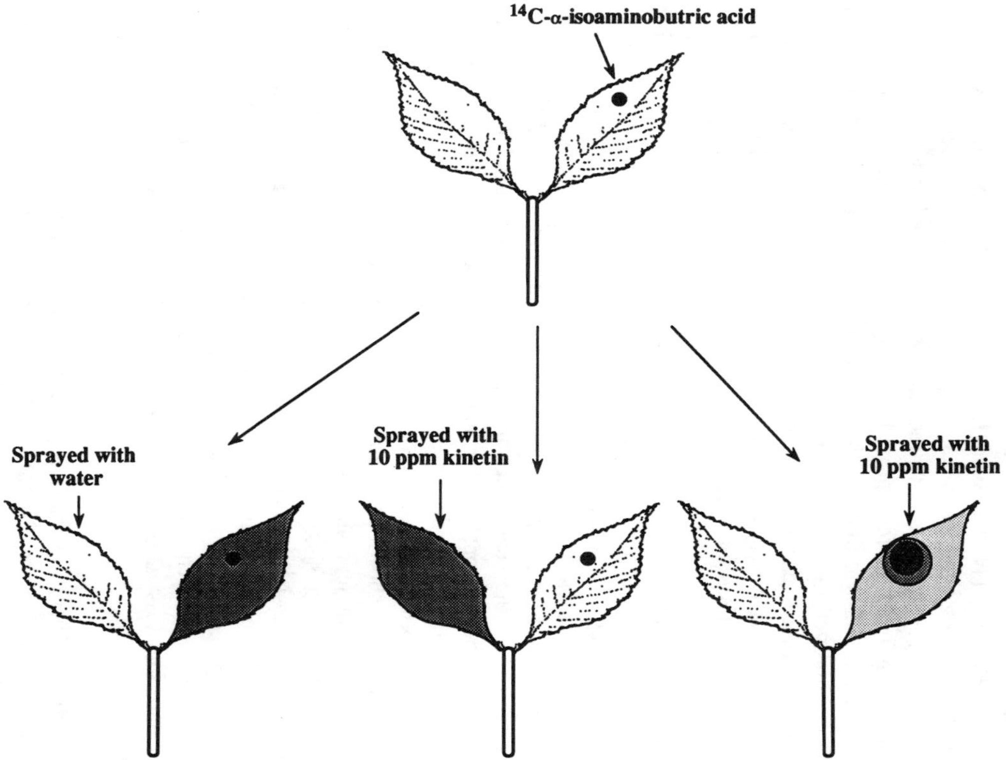

Figure 7.3. Movement of radioactive material toward the leaf sprayed with 10 ppm kinetin in the dark, showing that cytokinins create sinks (from Mothes and Engelbrecht (1961)).

Labels in figure:

¹⁴C-α-isoaminobutric acid

Sprayed with 10 ppm kinetin

Sprayed with 10 ppm kinetin

Sprayed with water

alone and in combination with other plant growth substances (Arteca 1990). This could account for the limited number of reports showing that cytokinins promote senescence. It has also been shown that cytokinins decrease tissue sensitivity to ethylene (Noodén and Leopold 1978). In summary, cytokinins generally delay senescence; however, they interact with other factors in doing so. Transgenic soybean plants have recently been produced in which cytokinin levels can be regulated, thereby providing a useful tool in better understanding the mode of action of cytokinins in the senescence process (Ainley et al. 1993). In addition, improved methodology for the detection of cytokinins will enable researchers to monitor changes in endogenous levels of cytokinins in organs other then leaves in a variety of test systems to better understand their involvement in senescence (Kaminek et al. 1992; Mok and Mok 1994).

Auxins.   Although there are considerably fewer reports on auxins than cytokinins, exogenous applications of both natural and synthetic auxins have been shown to delay senescence in a wide variety of tissues. There are also a number of reports in the literature where exogenous applications of auxin do not delay senescence and, in some cases, can actually promote it. When auxins are reported to promote senescence it is possible that auxin-induced ethylene production may be responsible for accelerated senescence. In general, the endogenous levels of auxins decrease before or during senescence, although in some cases they do not change (Noodén and Leopold 1978).

Auxins have been shown to interact with other plant growth substances in the promotion of ethylene in some tissues (Arteca 1990), whereas in fruit tissues auxin has been shown to inhibit ethylene production (Noodén and Leopold 1978). Auxin-induced ethylene production has been reported in a wide variety of experimental systems, however, the relationship between the two on the senescence process is still unclear (Abeles et al. 1992). In a limited number of cases auxins have been shown to counteract the promotive effect of ABA on senescence and to synergize with GA in delaying senescence (Noodén and Leopold 1978).

Gibberellins.   There are many references in the literature on the effects of gibberellins on senescence (Noodén and Leopold 1978; Saks and Vanstaden 1992; Takahashi et al. 1991), most of which describe the ability of GA to retard chlorophyll loss in leaves, fruit, pea shoot apices, cotyledons, and flower stalks. Gibberellins have also been shown to reduce RNA and protein degradation, delay senescence in petioles, and delay ripening (Noodén and Leopold 1978; Saks and van Staden 1992; Kaminek et al. 1992; Takahashi et al. 1991). Although GA typically delays the senescence process there are reports showing in some cases that it can speed it or have no effect depending on the age of the plant and a variety of other factors (Takahashi et al. 1991; Noodén and

Leopold 1978). Gibberellin levels have been found to decline prior to or during senescence in a wide variety of tissues. In fact senescing tissues appear to metabolize GA at a more rapid rate. In all tissues where GA levels decrease during senescence, exogenous applications of GA will delay the process. In many leaf tissues, both gibberellins and cytokinins will retard senescence; however, there are also reports where they act differently. Both ABA and auxin may interact with GA, with ABA typically having the opposite effect and auxin having the same effect (Noodén and Leopold 1978; Takahashi et al. 1991).

Ethylene, Abscisic Acid and Methyl Jasmonate. Although the response varies among plants, cytokinins, auxins, and gibberellins generally retard senescence. Ethylene and other compounds such as ABA and methyl jasmonate promote senescence (Abeles et al. 1992). Exogenous applications of ethylene have been shown to speed the rate of degradation parameters used to measure senescence such as chlorophyll, RNA, and protein in a variety of test systems. There are also a number of mRNAs which accumulate during leaf senescence in response to ethylene which are involved in the production of enzymes that promote degradative processes during senescence. It has been shown in a variety of test systems that ethylene production increases during senescence. Evidence based on chlorophyll degradation indicates that ethylene has a role in the senescence process in leaves. It has been shown that there is an inhibition of chlorophyll degradation when plants are treated with inhibitors which block ethylene biosynthesis or action. However, senescence can also occur without ethylene production. These observations indicate that senescence can be controlled by an increase in the rate of ethylene production or an increase in sensitivity to ethylene (Abeles et al. 1992).

Exogenous applications of ABA have been reported to promote a wide range of senescence-related processes in a variety of organs. ABA has been shown to decrease chlorophyll, protein, and nucleic acid synthesis and alter membrane structure. In general, endogenous levels of ABA or ABA-like substances have been shown to increase prior to or during senescence in a variety of tissues, both attached and detached. However, there are a number of reports which do not show a correlation between ABA level and senescence (Noodén and Leopold 1978; Davies and Jones 1991). Senescence, like many other processes, is probably regulated by an interaction of all known and unknown plant growth substances. There are reports in the literature which show that ABA can influence the endogenous levels of other plant growth substances and viceversa. ABA and ethylene act in a similar manner and both produce increases in the other. In gereral ABA leads to a reduction in levels of cytokinins, gibberellins, and auxins; however, as is always the case, there are exceptions to this rule (Noodén and Leopold 1978; Davies and Jones 1991).

Exogenous applications of methyl jasmonate have been shown to promote senescence in plants. Methyl jasmonate has the ability to stimulate ethylene biosynthesis at the step between ACC and ethylene by increasing ACC oxidase activity. Therefore it is possible that the promotive effect which methyl jasmonate has on senescence is due to its ability to stimulate the production of ethylene; however, more research is necessary in this area before definitive statements can be made (Noodén and Leopold 1978; Abeles et al. 1992; Sanz et al. 1993; Porat and Halevy 1993).

Use of Plant Growth Substances to Delay Senescence

Many vegetable, fruit, flower, and mushroom crops are now marketed fresh to increasingly distant markets. Therefore, the maintenance of quality during postharvest storage, shipping, and handling to reduce spoilage is of paramount importance. Cytokinins, plant growth retardants, auxins, gibberellins, and ethylene biosynthesis and action inhibitors have the ability to delay senescence. Some research has been directed toward the use of these compounds on vegetables, fruit, flowers, and mushrooms to delay deterioration after harvest. A common result of senescence is storage rot caused by the growth of bacteria and fungi on materials lost from senescing cells. This problem can be delayed or lessened by rapid, careful handling and proper storage conditions, but unfortunately this is not always possible.

Cytokinins, auxins, and plant growth retardants will generally delay senescence in vegetables, thereby reducing rot and potential losses. Cytokinins have been shown to be effective in delaying senescence in cabbage, lettuce, cauliflower, asparagus, broccoli, celery, Brussels sprouts and other vegetables such as endive, escarole, mustard greens, spinach, radish, carrot tops, parsley and green onion. Auxins have been reported to be effective in delaying senescence of cauliflower, broccoli, and Brussels sprouts, while plant growth retardants are effective in lettuce, asparagus, and broccoli (Weaver 1972).

Gibberellins have been shown to be effective in the retardation of senescence in fruits, an example of this is their use in navel oranges. When navel oranges approach maturity the rind changes in color from green to orange because there is a decrease in chlorophyll and an increase in carotenoid pigment. During this change the rind softens making the fruit susceptible to the following physiological disorders: rind staining (postharvest), decay (postharvest), sticky exudate (postharvest), water spotting (preharvest) and puffiness (pre- and postharvest). Applications with gibberellins have been shown to be very effective in overcoming problems with rind staining, water spotting, and decay, but only partially effective with puffiness and sticky exudate. Gibberellin applications have also been shown to be effective in delaying senescence in lemons, limes, grapefruit, and Valencia oranges. It should be noted that timing and concentration of GA applied are very important because if incorrectly done there may be problems with degreening (Weaver 1972).

Ethylene action (silver thiosulfate (STS)) or biosynthesis (aminoethoxyvinylglycine (AVG) or aminooxyacetic acid (AOA)) inhibitors, plant growth retardants, or cytokinins have been shown to be effective in delaying senescence in flowers. Ethylene biosynthesis or action inhibitors have been reported to be very effective in delaying senescence in many flower crops (Abeles et al. 1992). Plant growth retardants and cytokinins have been shown to be effective in delaying senescence in snapdragon or carnation and in carnation, daffodil, and stocks, respectively (Weaver 1972).

The plant growth retardant Alar has been reported to delay the senescence process in mushrooms, whereas cytokinin or cycocel were not effective (Weaver 1972).

Even though cytokinins, auxins, and plant growth retardants have been shown to delay senescence in a variety of systems none are commercially used for this purpose at the present time for a variety of reasons. Gibberellins are used in citrus to delay senescence commercially, however, this is presently its only use for this purpose. For many years it has been known that ethylene is involved in causing premature senescence in a variety of plant organs in a number of plant species. In fact, the ethylene biosynthesis inhibitors AVG and AOA were initially developed for commercial use, however, concerns of potential toxic effects stopped their further development and neither are registered for commercial use (Abeles et al. 1992). Silver thiosulfate is an ethylene action inhibitor which is commonly used to delay senescence in cut flower crops. However, neither AVG, AOA, or STS can be used to delay senescence in food crops. At the present time there is increasing concern about the use of heavy metals in greenhouses which may someday stop the use of STS on potted plants. Therefore genetically engineering plants to reduce ethylene production to delay senescence, an approach which has recently become available, has a great deal of potential. Prior to the biotechnology boom, folklore showed that sealing fruits or vegetables in a bag promoted ripening. Today we know the bag traps ethylene produced by the fruit, which speeds the ripening and aging process of fruits and vegetables. ACC synthase is one of the regulatory proteins involved in the production of ethylene, which has recently been exploited using molecular technology to regulate ripening in tomatoes. Researchers have expressed antisense ACC synthase (insertion of the gene for this enzyme in the backward orientation) in tomato plants and showed that fruit ripening was inhibited (Kende 1993; Theologis 1992). In the early 1990s three U.S. corporations (Calgene Inc., DNA Plant Technology, and Monsanto Co.) licensed this gene to reduce spoilage of fruit, vegetables, and flowers by preventing the production of ethylene. It is anticipated that genetically engineered tomatoes with the antisense ACC synthase gene will be on the market in the near future along with MacGregor tomatoes, which utilize softening genes to prevent spoilage. Recently, researchers have purified the plant protein ACC N-MTase (Guo et al. 1992 1993), which converts ACC (produced by ACC synthase) to an inactive end product and does not allow it to go to ethylene. This discovery will now allow this group to isolate the gene(s) for ACC N-MTase and to use this gene to genetically transform plants to reduce spoilage. In summary, there are now ethylene biosynthetic genes and genes which degrade the ethylene precursor ACC which can be used to reduce ethylene production, thereby delaying ripening and enabling growers to pick riper fruits without rot problems. In whole plants it is possible that the use of these genes can be used to block premature senescence which would enable plants to photosynthesize longer thereby, having the potential to increase crop yields.

REFERENCES

Abeles, F. B., Morgan, P. W., and Saltveit Jr., M. E. (1992). *Ethylene in Plant Biology. Second Edition*, Academic Press, San Diego, CA.

Ainley, W. M., McNeil, K. J., Hill, J. W., Lingle, W. L., Simpson, R. B., Brenner, M. L., Nagao, R. T., and Key, J. L. (1993). "Regulatable endogenous production of cytokinins up to toxic levels in transgenic plants and plant-tissues". *Plant Mol. Biol.* 22:13–23.

Arteca, R. (1990). "Hormonal stimulation of ethylene biosynthesis". In *Polyamines and Ethylene: Biochemistry, Physiology, and Interactions*, eds., H. E. Flores, R. N. Arteca and J. C. Shannon. American Society of Plant Physiologists, Rockville, MD, pp. 216–223.

Blair, D. S., MacArthur, M., and Nelson, S. H. (1956). "Observations in the growth phases of fruit trees". *Proc. Am. Soc. Hort. Sci.* 67:75–79.

Burton, W. G. (1982). *Postharvest Physiology of Food Crops*, Longman, London.

Davies, W. J. and Jones, H. G. (1991). *Abscisic Acid: Physiology and Biochemistry*, Bios Scientific Publishers, Oxford, U.K.

Gardner, F. E. (1929). "The relationship between tree age and the rooting of cuttings". *Proc. Am. Soc. Hort. Sci.* 26:101.

Grover, A. (1993). "How do senescing leaves lose photosynthetic activity". *Current Science* 64:226–234.

Guo, L., Arteca, R. N., Phillips, A. T., and Liu, Y. (1992). "Purification and characterization of ACC N-Malonyltransferase from etiolated mung bean hypocotyls". *Plant Physiol.* 100:2041–2045.

Guo, L. G., Phillips, A. T., and Arteca, R. N. (1993). "Amino acid N-malonyltransferases in mung beans: Action on 1-aminocyclopropane-1-carboxylic acid and D-phenylalanine". *J. Biol. Chem.* 268:25389–25894.

Hackett, W. (1985). "Juvenility, maturation and rejuvenation in woody plants". In *Horticultural Reviews*, ed.,J. Janick, AVI Publishing Co., Westport, CT, pp. 109–156.

Hartmann, H. T., Kester, D. E., and Davies Jr., F. T. (1990). *"Plant Propagation Principles and Practices. 5th Edition*, Prentice Hall, Englewood Cliffs, N. J.

Kamínek, M., Mok, D. S., and Zazimalova, F. (1992). *Physiology and Biochemistry of Cytokinins in Plants*, SPB Academic Publishing, Hague, The Netherlands.

Kende, H. (1993). "Ethylene biosynthesis". *Annu. Rev. Plant. Physiol. Plant Mol. Biol.* 44:283–307.

Kulaeva, O. N. (1962). "The effect of roots on leaf metabolism in relation to the action of kinetin on leaves". *Sov. Plant Physiol.* 9:182–189.

Leopold, A. C. (1975). "Aging, senescence and turnover in plants". *BioScience* 25:659–662.

Leopold, A. C. and Kawase, M. (1964). "Benzyladenine effects on bean leaf growth and senescence". *Am. J. Bot.* 51:294–298.

Leopold, A. C. and Kriedemann, P. E. (1975). *Plant Growth and Development. Second Edition*, McGraw-Hill Book Company, New York.

Medawar, P. B. (1957). *The Uniqueness of the Individual*, Basic Books, New York.

Millington, W. F. and Fisk, E. L. (1956). "Shoot development in *Xanthium pennsylvanicum*, L.". *Am. J. Bot.* 43:655–665.

Mok, D. W. S. and Mok, M. C. (1994). *Cytokinins: Chemistry, Activity, and Function*, CRC Press Inc., Boca Raton, FL.

Molisch, H. (1938). *The Longevity of Plants*, Science Press, Lancaster, PA.

Mothes, K. (1960). "Über das Altern der Blätter und die Moglichkeit ihrer Wiederverjüngung". *Naturwissenschaften* 47:337–351.

Mothes, K. and Engelbrecht, L. (1961). "Kinetin-induced directed transport of substances in excised leaves in the dark". *Phytochemistry* 1:58–61.

Noodén, L. D. and Leopold, A. C. (1978). "Photohormones and the endogenous regulation of senescence and abscission". In *Phytohormones and Related Compounds: A Comprehensive Treatise. Vol. 2: Phytohormones and the Development of Higher Plants*, eds., D. S. Letham, P. B. Goodwin and T. J. V. Higgins, Elsevier, Amsterdam, pp. 329–370.

Porat, R. and Halevy, A. H. (1993). "Enhancement of petunia and dendrobium flower senescence by jasmonic acid methyl ester is via the promotion of ethylene production". *Plant Growth Reg.* 13:297–301.

Ranjit, M. and Kester, D. E. (1988). "Micropropagation of cherry rootstocks: II. Invigoration and enhanced rooting of 46–1 Mazzard by co-culture with Colt". *J. Amer. Soc. Hort. Sci.* 113:150–154.

Reddy, G., Arteca, R. N., Dai, Y. –R. Flores, H. E., Negm, F. B., and Pell, E. J. (1993). "Changes in ethylene and polyamines in relation to mRNA levels of the large and small subunits of ribulose bisphosphate carboxylase/oxygenase in ozone-stressed potato foliage". *Plant Cell, Env.* 16:819–826.

Richmond, A. E. and Lang, A. (1957). "Effect of kinetin on protein content and survival of detached *Xanthium* leaves". *Science* 125:650–651.

Saks, Y. and Vanstaden, J. (1992). The role of gibberellic-acid in the senescence of carnation flowers". *J. Plant Physiol.* 139:484–488.

Sanz, L. C., Fernandezmaculet, J. C., Gomez, E., Vioque, B., and Olias, J. M. (1993). "Effect of methyl jasmonate on ethylene biosynthesis and stomatal closure in olive leaves". *Phytochemistry* 33:285–289.

Singh, S., Letham, D. S., and Palni, L. M. S. (1992a). "Cytokinin biochemistry in relation to leaf senescence. 7. Endogenous cytokinin levels and exogenous applications of cytokinins in relation to sequential leaf senescence of tobacco". *Physiol. Plant.* 86:388–397.

Singh, S., Letham, D. S., and Palni, L. M. S. (1992b). "Cytokinin biochemistry in relation to leave senescence. 8. Translocation, metabolism and biosynthesis of cytokinins in relation to sequential leaf senescence of tobacco". *Physiol. Plant.* 86:398–406.

Singh, S. T., Letham, D. S., Zhang, X. D., and Palni, L. M. S. (1992c). "Cytokinin biochemistry in relation to leaf senescence. 6. Effect of nitrogenous nutrients on cytokinin levels and senescence of tobacco leaves". *Physiol. Plant.* 84:262–268.

Stokes, P. and Verkerk, K. (1951). "Flower formation in brussels sprouts". *Meded. Landbouwhogesch. Wageningen* 50:141–160.

Takahashi, N., Phinney, B. O., and MacMillan, J. (1991). *Gibberellins*, Springer-Verlag, Berlin.

Theologis, A. (1992). "One rotten apple spoils the whole bushel: The role of ethylene in fruit ripening". *Cell* 70:181–184.

Tilney-Basset, R. A. E. (1986). *Plant Chimeras*, Edward Arnold, London.

Trippi, V. S. (1989). "Maturation and senescence: Types of aging". In *Plant Aging: Basic and Applied Approaches*, eds., R. Rodríguez, R. S. Tamés, and D. J. Durzan. Plenum Press, New York, pp. 11–18.

Weaver, R. J. (1972). *Plant Growth Substances in Agriculture*, W. H. Freeman and Company, San Francisco, CA.

Wellensiek, S. J. (1958). "Vernalization and age in *Lunaria*". *Proc. Kon. Ned. Akad. Wet.* C61:561–571.

Woodson, W. R. and Brandt, A. S. (1991). "Role of the gynoecium in cytokinin-induced carnation petal senescence". *J. Am. Soc. Hort. Sci.* 116:676–679.

Flowering

The transition from vegetative to reproductive development is clearly a very critical phase in the life cycle of higher plants. Although there has been a considerable amount of research on the physiological, biochemical, and molecular aspects of flowering, the actual mechanism by which the transition from vegetative to reproductive development occurs still remains unclear at the present time (Jordan 1993). Since there are many coordinated processes involved in reproductive development, background information for each of these stages will briefly be discussed prior to explaining the involvement of environmental factors and plant growth substances. The first stage is flower initiation which is an internal physiological change in the meristem which precedes any morphological change. The first noticeable morphological change indicating that a transition from vegetative to reproductive development is occurring is enhanced cell division in the central zone immediately below the apical part of the vegetative meristem. The divisions occurring here result in the differentiation of parenchyma cells which surround the meristem giving rise to flower primordia. The second stage is flower formation, the visible initiation of flower parts. The final stage is flower development which is the differentiation of the flower structure including events from flower formation to anthesis (flowering). Each of these stages like any other physiological process is determined by the genotype. However, in many species the start of reproductive development is regulated by environmental factors such as day length and temperature, which vary on a fairly regular basis throughout the year, and by a specific plant growth substance(s) or interactions between two or more. However, once again the response varies between species. In the following sections the involvement of environmental factors and plant growth substances on floral initiation will be discussed.

EFFECTS OF ENVIRONMENTAL FACTORS ON THE ONSET OF REPRODUCTIVE DEVELOPMENT

Photoperiodism

When flower initiation is determined solely by genotype and the plants have no specific light requirement they are called day-neutral plants, whereas, plants which flower in response to day length conditions are under the control of photoperiodism. Photoperiodically sensitive plants can be broken down into several categories:

1. Short-day plants flower only when the dark period is greater then a certain critical length, since plants measure the length of the dark period in order to flower.
2. Long-day plants flower only when the dark period is shorter then a certain critical length.
3. Although short-day and long-day plants make-up the majority of the photoperiodically sensitive plants there are also two other groups. One is short-long-day plants which flower only when subject to short days followed by long days, and the second group is long-short-day plants which require long days first followed by short days (Thomas 1993).

Day length is perceived by the plant leaves with a primary photochemical event which is the absorption of a photon by phytochrome. The mechanism of action of phytochrome has been discussed in detail elsewhere and will not be considered further in this text (Vince-Prue 1983). Although day length is perceived in the leaves the actual response occurs in the bud indicating that the signal is produced in the leaf and translocated to the site of action. This signaling substance has been called florigen, floral stimulus, or flower hormone. The existence of a signaling substance has been substantiated by grafting experiments conducted between induced and noninduced plants. When a leaf from a plant grown under conditions which promote flowering such as proper photoperiod or cold treatment is grafted on a plant under noninductive conditions flowering can be initiated. Based on results from successful grafting experiments it is very easy to speculate that the flowering stimulus is very similar in all plants. However, it would be risky to do so because there are also a number of grafting experiments which were unsuccessful in causing plants to flower (Zeevaart 1976). Although florigen may exist it has still not been isolated and characterized, which has led some to believe that flowering is initiated by an interaction between existing known and unknown plant growth substances which promote or inhibit flowering.

Vernalization

There are numerous plant species grown in temperate regions which will flower in response to low-temperature treatments. This phenomenon is called vernalization. In general, the effects of low-temperature treatments are observed when these plants are transferred to warmer growth-promoting temperatures. An exception to this is *Brassica oleracea* which will initiate flowers during exposure to low temperatures (Metzger 1987). Plants which will flower in response to low temperatures are called thermoinductive. Annual plants (which complete a life cycle in one growing season) can be broken down into two categories: one group called summer annuals which do not require low temperatures in order to flower; and the other group called winter annuals. In this group the cold requirement is facultative, meaning that flowering will eventually occur without low-temperature treatment; however, it takes longer. Biennial plants (which complete a life cycle in two years) require one full season of vegetative growth and must have a cold treatment in order to flower. There are also many perennial plants (which flowers and sets seed year after year without dying) which have a strict requirement for cold temperatures in order to flower. Differences among species with respect to their temperature requirements for flowering occur in annuals, biennials, and perennials (Thomas 1993).

The site of temperature perception occurs in the bud unlike the photoperiodic response, which is perceived in the leaves. There are also many plants which have dual requirements of low temperature and proper photoperiod in order to induce flowering. Once external signals are perceived by the plant they must be transmitted from one location to another resulting in molecular changes which promote flowering. The proposed mechanisms of signal transduction are summarized by Lumsden (1993); therefore, for more details on the signaling events involved in flowering, see this reference.

EFFECTS OF PLANT GROWTH SUBSTANCES ON FLOWER INITIATION, PROMOTION, AND INHIBITION

Flower Initiation

The role of plant growth substances in the juvenile to mature phase change is not well established at the present time and more research is necessary to better understand how this phase change occurs. It is known that plants will not typically flower during the juvenile phase as described in Chapter 7. However, once plants have completed their juvenile stage of development they have the ability to produce a flowering stimulus in response to the proper environmental cues. This stimulus initiates the transition of the apex from its vegetative stage to one which is committed to reproductive development. Based on grafting experiments it can be concluded that the floral stimulus is very similar or even identical in many cases. Since the existence of this flowering stimulus was first proposed by Chailakhyan (1936) there have been many attempts to identify and chemically characterize it. Unfortunately, after almost 60 years of research in this area the chemical nature of this flower stimulus remains unknown. To date research still suggests that the flower stimulus is a single substance or a few substances unique from the presently known plant growth substances and is very similar or even identical throughout the plant kingdom. However, it cannot be ruled out that there is an interaction between known plant growth substances which can directly trigger flowering or that they may be involved in triggering the production of a flower stimulus (Metzger 1987; Weaver 1972; Jordan 1993).

Flower Promotion and Inhibition

Early studies with plants in the *Bromeliaceae* family showed that auxins had the ability to promote flower formation. However, it is now known that auxin-induced ethylene production is responsible for the formation of flowers in plants from this family (Zeevaart 1978). In general exogenous applications of auxins such as IAA or NAA inhibit flower formation when applied under inductive conditions. The inhibitory effects of auxins are supported by studies with the auxin transport inhibitor 2,3,5-triiodobenzoic acid (TIBA), trade name Floraltone which has been shown to promote flower initiation in apples (*Plant Growth Regulator Handbook* 1981). It is possible that the inhibitory effect of auxin on flowering is due to auxin-induced ethylene production, however, at the present time there are only a limited number of conflicting reports evaluating changes in endogenous levels of IAA and ethylene during the flowering process. Therefore, until additional work is done in this area using modern quantitative techniques, the role(s) of auxin and its relationship with ethylene in flower formation will remain unclear.

Exogenous applications of ethylene either as a gas or with ethylene-releasing compounds such as ethrel inhibit or delay the promotion of flowering. It still remains unclear whether or not this inhibition is part of a natural regulatory mechanism (Abeles et al. 1992). Ethylene also has the ability to promote flowering in a limited number of plant species. The stimulatory effect of ethylene on flowering is best known for members of the pineapple family (De Greef et al. 1989) although ethylene effectively stimulates flowering in other plants as well. Today the promotion of flowering using either ethylene-releasing compounds or auxins which induce ethylene production have become an important horticultural practice (Abeles et al. 1992).

Cytokinins applied to the apex of plants under noninductive photoperiods have been shown to cause cell divisions which occur early in flower initiation (Bernier and Kinet 1985). However, in most of the cases where exogenous applications have been reported to promote flower formation, the plants had either been partially or fully induced. For example, when applied during floral initiation, BA increased the floret count and capitulum size in the short-day plant *Leucosperum cordifolium* (Napier et al. 1986) and flower count in the short-day plant *Schlumbergera truncata* (Ho et al. 1985; Runger and Poole 1985). Exogenous applications of cytokinins at times other than initiation stages either have no effect (Harkess and Lyons 1994), delay flowering, or increase branching (Runger and Poole 1985). The promotive effects of cytokinins on flowering may be indirect since it has been shown in *Pharbitis nil* that cytokinins can increase the translocation of a flower stimulus and assimilates from induced leaves (Ogawa and King 1979). There are a limited number of studies on the relationship between endogenous levels of cytokinins and flowering, therefore, at the present time the role of cytokinins in the promotion of flowering remains unclear (Kaminek et al. 1992; Mok and Mok 1994).

Exogenous applications of gibberellins under noninductive conditions have been shown to promote flowering in a variety of plant species (Zeevaart 1983; Harkess and Lyons 1994), including carrots as shown in Figure 8.1 (Lang 1957). Gibberellins are probably the only one of the eight classes of plant growth substances discussed in this text to have a significant effect on flowering (Metzger 1987; Takahashi et al. 1991). Although there have been reports showing that exogenous applications of gibberellins have an effect on flowering there are also conflicting reports showing that they have no effect. A review of the literature shows that attempting to correlate endogenous levels of GA with flower promotion is a complicated undertaking. Technical difficulties arise when trying to analyze the over 90 different gibberellins found in plants. In addition, it is difficult to link gibberellins with the flowering process, when a

Figure 8.1.   Effect of gibberellin on flower formation in carrot. Left, control received neither cold nor gibberellin treatment; center, received no cold but was treated with gibberellin; right, received cold treatment but no gibberellin (Lang 1957).

single GA may be mediating a physiological process in a given species, while others may be precursors or deactivation products (Takahashi et al. 1991). Although gibberellins have been shown to have a variety of roles in reproductive development, more research is necessary before a definite role for gibberellins can be assigned (Takahashi et al. 1991; Metzger 1987; Boyle et al. 1994).

Physiological evidence indicates that there is the existence of a graft-transmissible flower inhibitor. Following the discovery of ABA it was thought that this compound was responsible for causing inhibition of flowering. However, research has shown that ABA probably does not play a significant role in the regulation of flower formation as a graft-transmissible inhibitor (Metzger 1987; Davies and Jones 1991).

Salicylic acid has been shown to promote flowering in several plant species, however, its mechanism of action remains unknown (Raskin 1992). At the present time brassinosteroids and jasmonates have not been shown to be involved in the flowering process, however, more work is necessary before definitive statements can be made.

EFFECTS OF PLANT GROWTH SUBSTANCES ON STEM/INFLORESCENCE GROWTH IN ROSETTE PLANTS AND SEX EXPRESSION

Once flower primordia has been initiated there are many highly coordinated developmental steps which lead to flower opening and subsequent fertilization. In this section stem/inflorescence growth in rosette plants and sex expression will be discussed. The effects of plant growth substances in other aspects of reproductive development will be covered in Chapter 10.

Gibberellins and Stem/Inflorescence Growth

Both long-day and cold-requiring plants typically grow as rosettes until they are exposed to an inductive treatment. In these plants the formation of flowers is closely related to bolting (rapid stem elongation). Application of gibberellins to a wide variety of long-day and cold-requiring plants will induce flower formation under noninductive conditions, suggesting that gibberellins are limiting in noninduced plants. Gibberellins have also been shown to enhance flowering in short-day plants; however, they must be grown under inductive conditions otherwise treatments are ineffective (Weaver 1972; Takahashi et al. 1991). The formation of flowers in long-day and cold-requiring plants has been shown to be controlled by regulating endogenous levels of gibberellins using GA biosynthesis inhibitors (Jones and Zeevaart 1980 a, 1980 b; Metzger 1985). In addition, there have been reports showing that changes in endogenous levels of gibberellins occur during flowering, thereby supporting inhibitor studies. However, it is still difficult to make any general conclusions on the involvement of endogenous gibberellins on stem/inflorescence growth based on the limited number of studies with only a few of the more than 90 gibberellins known today (Metzger 1987; Takahashi et al. 1991).

Plant Growth Substances and Sex Expression

The study of sex in plants by man started many years ago with Empedocles (485–455 B.C.), Aristotle (384–322 B.C.), and Aristotle's student Theophrastus (370–322 B.C.). However, it was not until about 35 years ago that the effects of plant growth substances on the modification of sex expression in plants began to be studied (Chailakhyan and Khrianin 1987). Prior to discussing the effects of plant growth substances on sex expression a brief explanation of flowering terminology will be given in order to provide a better understanding of their involvement in this process. Flowers can be broken down into two different groups: those which are perfect, containing both pistils and stamens, or imperfect, containing either pistils or stamens. A monoecious plant such as maize has both male and female flowers on the same plant, while, dioecious plants such as spinach have male and female flowers on separate plants. The sex of imperfect flowers has a genetic basis, but environmental factors such as photoperiod, temperature, and nitrogen status influence this process (Metzger 1987). There have been numerous reports showing that exogenous applications of plant growth substances can modify the sex of flowers, suggesting that these substances mediate genetic and environmental control of sex expression (Chailakhyan and Khrianin 1987).

The effects of plant growth substances on sex expression have been extensively studied in *Cucumis*. Early studies showed that when auxin was applied to *Cucumis* there was an increase in the number of female flowers, while exogenous applications of gibberellins resulted in a stimulation of male flowers (Jones and Zeevaart 1980a, 1980b; Pharis and King 1985). It has since been shown that auxin-induced ethylene is responsible for the change in sex expression (Abeles et al. 1992). In many cases, changes in endogenous levels of either auxin or GA correlate with the expression of female or male flower formation. Overall, it is now generally accepted that sex expression in *Cucumis* is regulated by an internal balance of auxins acting through ethylene (Figure 8.2) and gibberellins. However, there have also been reports showing no correlation between endogenous levels of auxin or gibberellin with respect to sex expression, suggesting that plant growth substance balance may not be the only factor determining sex expression in *Cucumis*.

Exogenous applications of auxins, ethylene, or gibberellins have been shown to affect sex expression in *Cannabis sativa* in the same way as in *Cucumis*. In

Figure 8.2.   Female flower (left) from cucumber plant treated with ethephon and male flower (right) from untreated plant (Robinson et al. 1970).

Cannabis cytokinins have also been shown to promote femaleness. Studies evaluating endogenous levels of auxins, ethylene, gibberellins, or cytokinins in *Cannabis* have been shown to correlate with changes in sex expression induced by external treatments (Metzger 1987). Plant growth substances have also been reported to be involved in sex expression in begonia, hops, grape, muskmelon, squash, pumpkin, tomato, and cotton (Weaver 1972). The use of plant growth substances in the modification of sex expression in plants has a great deal of potential in seed production and in breeding programs (Chailakhyan and Khrianin 1987).

MOLECULAR BIOLOGY OF FLOWERING

As mentioned earlier there is a large body of descriptive information on the physiology and biochemistry of flowering; however, the molecular mechanisms which control flowering still remain unclear. There have been numerous recent advances in molecular biology of flowering which now provide an opportunity to study the molecular mechanisms involved in this process. The potential to genetically manipulate flower development will accelerate progress in research leading to agricultural applications. In a text by Jordan (1993), a group of scientists were assembled providing exciting information on recent advances on the molecular biology of flowering; therefore, for more information in this area see this text and references contained within.

REFERENCES

Abeles, F. B., Morgan, P. W. and Saltveit Jr., M. E. (1992). *Ethylene in Plant Biology. Second Edition*, Academic Press, San Diego, CA.

Bernier, G. and Kinet, J. M. (1985). "The control of flower initiation and development". In *Plant Growth Substances*, ed., M. Bopp, Springer-Verlag, Heidelberg, Germany, pp. 293–302.

Boyle, T. H., Marcotrigiano, M., and Hamlin, S. M. (1994). "Regulating vegetative growth and flowering with gibberellic acid in intact plants and cultured phylloclades of Crimson Giant easter cactus". *J. Amer. Soc. Hort. Sci.* 119:36–42.

Chailakhyan, M. K. (1936). "On the hormonal theory of plant development". *Dokl. Acad. Sci. USSR* 12:443–447.

Chailakhyan, M. K. and Khrianin, V. N. (1987). *Sexuality in Plants and its Hormonal Regulation*, Springer-Verlag, Heidelberg, Germany.

Davies, W. J. and Jones, H. G. (1991). *Abscisic Acid: Physiology and Biochemistry*, Bios Scientific Publishers, Oxford, U.K.

De Greef, J. A., De Proft, M. P., Mekers, O., Van Dijck, R., Jacobs, L., and Philippe, L. (1989). "Floral induction of bromeliads by ethylene". In *Biochemical and Physiological Aspects of Ethylene Production in Lower and Higher Plants*, eds., H. Clijster, M. De Proft, R. Marcelle, and M. Van Poucke, Kluwer Academic Publishers, Dordrecht, The Netherlands, pp. 313–322.

Harkess, R. L. and Lyons, R. E. (1994). "Gibberellin- and cytokinin-induced growth and flowering responses in *Rudbeckia hirta* L". *HortScience* 29:141–142.

Ho, Y., Sanderson, K. C., and Williams, J. C. (1985). "Effect of chemicals and photoperiod on the growth and flowering Thanksgiving cactus". *J. Am. Soc. Hort. Sci.* 110:658–662.

Jones, M. G. and Zeevaart, J. A. D. (1980a). "The effect of photoperiod on the levels of seven endogenous gibberellins in the long-day plant *Agrostemma githago* L.". *Planta* 149:274–279.

Jones, M. G. and Zeevaart, J. A. D. (1980b). "Gibberellins and the photoperiodic control of stem elongation in the long-day plant, *Agrostemma githago* L.". *Planta* 149:269–273.

Jordan, B. R. (1993). *Molecular Biology of Flowering*, CAB International, Sussex, England.

Kaminek, M., Mok, D. W. S., and Zazimalova, E. (1992). *Physiology and Biochemistry of Cytokinins in Plants*, SPB Academic Publishing, Hague, The Netherlands.

Lang, A. (1957). "The effects of gibberillin upon flower formation". *Proc. Natl. Acad. Sci. USA* 43:709–711.

Lumsden, P. J. (1993). "Mechanisms of signal transduction". In *Molecular Biology of Flowering*, ed., B. R. Jordan. C A B International, Sussex, England pp. 21–45.

Metzger, J. D. (1987). "Hormones and reproductive development". In *Plant Hormones and Their Role in Plant Growth and Development*, ed., P. J. Davies. Martinus Nijhoff Publishers, Boston.

Metzger, J. D. (1985). "Role of gibberellins in the environmental control of stem growth in *Thlaspi arvense* L.". *Plant Physiol.* 78:8–13.

Mok, D. W. S. and Mok, M. C. (1994). *Cytokinins. Chemistry, Activity, and Function*, CRC Press, Boca Raton, FL.

Napier, D. R., Jacobs, G., van Staden, J. and Forsyth, C. (1986). "Cytokinins and flower development in *Leucospermum*". *J. Am. Soc. Hort. Sci.* 111:776–780.

Ogawa, Y. and King, R. W. (1979). "Indirect action of benzyladenine and other chemicals on flowering of *Pharbitis nil* Chois". *Plant Physiol.* 63:643–649.

Pharis, R. P. and King, R. W. (1985). "Gibberellins and reproductive development in seed plants". *Annu. Rev. Plant Physiol* 36:517–568.

Plant Growth Regulator Handbook. Second Edition, (1981). Plant Growth Regulator Society of America, Lake Alfred, FL.

Raskin, I. (1992). "Role of salicylic acid in plants". *Annu. Rev. Plant Physiol. Plant Mol. Biol.* 43:439–463.

Robinson, R. W., Wilczynski, H., dela Guardia, M. D., and Shannon, S. (1990). "Chemical regulation of fruit ripening and sex expression", *New York Food and Life Science Quarterly* 3(1):10-11.

Runger, W. and Poole, R. T. (1985). "Schlumbergera". In *CRC handbook of Flowering*, eds., A. H. Halevy. CRC Press, Boca Raton, FL, pp. 277–282.

Takahashi, N., Phinney, B. O., and MacMillan, J. (1991). *Gibberellins*, Springer-Verlag, Berlin.

Thomas, B. (1993). "Internal and external controls on flowering". In *Molecular Biology of Flowering*, eds., B. R. Jordan. C A B International, Sussex, England pp. 1–19.

Vince-Prue, D. (1983). "Photomorphogenesis and flowering". In *Encyclopedia of Plant Physiology*, eds., W. Shropshire Jr. and H. Mohr. Springer-Verlag, Berlin, pp. 457–490.

Weaver, R. J. (1972). *Plant Growth Substances in Agriculture*, W. H. Freeman and Company, San Francisco.

Zeevaart, J. A. D. (1976). "Physiology of flower formation". *Annu. Rev. Plant. Physiol.* 27:321–348.

Zeevaart, J. A. D. (1978). "Phytohormones and flower formation". In *Plant hormones and Related Compounds*, eds., D. S. Letham, P. B. Goodwin and T. J. V. Higgins, Elsevier/North Holland, Amsterdam, pp. 291–327.

Zeevaart, J. A. D. (1983). "Gibberellins and flowering". In *The Biochemistry and Physiology of Gibberellins*, ed., A. Crozier. Praeger, New York, pp. 333–374.

Abscission

Many years ago the Greek philosopher Theophrastus, a student of Aristotle, described leaf abscission habits in higher plants. Since this time a great deal of research has been done in the area of abscission; in fact, manipulation of the abscission process is a common agricultural practice today. Abscission may be defined as the separation of a plant part, such as a leaf, flower, fruit, seed, stem, or others from the parent plant. Although we typically think of abscission occurring only in higher plants, it is also very common in lower plants. The occurrence of abscission is highly variable within species and cultivars of a species.

The most common example of abscission involves the separation of cells within specialized tissues in an abscission zone. In these cases the separation layer must be alive and have the ability to produce hydrolytic enzymes which promote abscission. This process can occur either very rapidly or slowly. In certain cases external mechanical forces such as wind are required to complete the abscission process. Although abscission occurs in a wide variety of plant parts both higher and lower, discussion will be limited to abscission habits in leaves, branches, flowers, fruits, and seeds in higher plants.

Leaves may live from weeks to years prior to senescence and abscission. Leaves can exhibit three patterns of abscission:

1. In many deciduous trees and shrubs grown in the temperate zone, seasonal abscission of leaves occurs within a very short period of time in response to environmental factors such as shortening photoperiod and cold.

2. Many deciduous trees and shrubs grown in tropical and subtropical regions shed their leaves in response to changes in growth and vigor of the plant. For example, stress due to drought conditions will promote abscission; however, once the plant receives water, leaf growth will be promoted. The cycle of abscission and formation of leaves can occur many times over the course of a year.

3. In evergreens grown in subtropical and temperate regions vernal leaf abscission of the previous season's growth occurs during the new flush of growth in the spring.

Branch abscission may also function in plants as a type of self-pruning habit. In some species branches abscise rapidly, suggesting that hydrolytic enzymes were produced and efficiently promoted the abscission process. However, there are also cases where only partial separation occurs, suggesting that the hydrolytic process was not complete. When this occurs factors such as wind or other mechanical forces assist in abscission.

Abscission of flowers, fruits, and seeds commonly occurs in higher plants, facilitating reproduction. In some species a large number of flowers are produced in order to assure that fruits are produced. In these species once some fruit are set and begin to develop, excess flowers are abscised in order to avoid the production of more fruit than can develop to maturity. In some species a large number of young fruits are produced, however, abscission of excess fruits occurs as a safety mechanism. Dispersal of seeds is extremely important for a plant species to survive. However, without abscission of seeds from the placenta and dehiscence, which is a form of abscission, seed dispersal would not occur. Today either the promotion or inhibition of abscission is very important in agriculture.

ANATOMY OF ABSCISSION

The abscission process has two very important parts which were first described by Von Mohl (1860a, 1860b, 1860c), separation and protection (wound healing). During leaf abscission both processes usually occur almost simultaneously, although there are cases where they occur separately. Abscission occurs in a variety of plant parts; however, only leaf abscission will be discussed since there is a large body of research in this area (for more information on the anatomy of abscission in leaves and other plant parts see Webster (1968); Addicott (1982); and Osborne (1989).

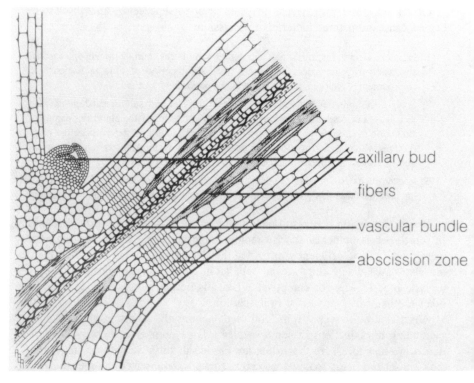

Figure 9.1.   The leaf abscission layer of cells (Addicott 1965).

Abscission of leaves, branches, flowers, fruits, and seeds in most plant species is preceded by the formation of an abscission zone or layer of specialized cells (Addicott 1982; Osborne 1989). The presence and even number of abscission zones differs among species, for example, coleus has one abscission zone at the base of the petiole, citrus and bean have two abscission zones, one at the base of the petiole and another at the base of the leaf, while tobacco has no abscission zones (Weaver 1972). In plant species with compound leaves, each leaflet has an abscission zone. An example of an abscission zone across the petiole near its junction with the stem is shown in Figure 9.1. The abscission zone is made up of one or more layers of thin-walled parenchyma cells extending across the petiole and excluding the vascular bundle (Figure 9.1). Prior to abscission, the middle lamella between cells located farthest from the stem (distal region) is digested. This results from de novo produced polysaccharide-hydrolyzing enzymes such as cellulases and pectinases, which are secreted from the cytoplasm into the cell wall. Along with the digestion of the cell wall, there is a burst of respiration and ethylene production in cells contained within the abscission zone closest to the stem (proximal region). The proximal cells of the abscission zone increase in length and diameter while cells located distally to this region do not. The combination of middle lamella degradation and mechanical forces generated by this differential growth results in separation.

PHYSIOLOGY OF ABSCISSION

Effects of Temperature, Oxygen, and Nutritional Factors

Temperature determines the rate at which abscission occurs. Fitting (1911) showed that with increasing temperatures there were accelerated rates of petal abscission. Since this time it has been shown that there is a temperature optimum for maximum abscission. In beans this optimum is 25°C, whereas in cotton it is 30°C. The Q_{10} (Figure 9.2) for this process is 2, which is characteristic for chemical reactions. Molisch (1886) showed that when branches were submerged in water the abscission process was retarded. From this experiment he concluded that oxygen is essential to the abscission process. It has since been shown that the rate of abscission is strongly affected by the percentage of oxygen in the atmosphere surrounding the explant. In cotton plants abscission follows a double sigmoid curve with increasing oxygen concentrations: up to

$$Q_{10} = \left(\frac{K_2}{K_1}\right)^{\frac{10}{T_2 - T_1}}$$

K_2 = rate at high temperature

K_1 = rate at lower temperature

T_2 = high temperature

T_1 = low temperature

Figure 9.2. Equation for calculating the Q_{10} for a given process.

10% O_2 there is a rapid increase in abscission, between 10% and 20% O_2 there is a plateau phase and a second sharp rise in abscission at about 25% O_2. At the present time there is no explanation for the acceleration of abscission by oxygen treatment above 25%. It has been suggested that the high levels of oxygen promote oxidative inactivation of IAA by enhancing IAA oxidase resulting in lower levels of IAA in the tissue. Another possibility is that oxygen stimulates the conversion of ACC to ethylene, which promotes abscission (Addicott 1982).

Nutritional factors such as carbohydrate, nitrogen, and mineral elements have been shown to have an effect on the abscission process. It has been shown that conditions which promote high carbohydrate such as high light which stimulates photosynthesis retard abscission, whereas conditions which promote low carbohydrates accelerate abscission. This is supported by experiments showing that the application of sucrose to explants retards abscission. It is thought that high carbohydrates in the plant contribute to the vigor of fruits and leaves. In general this increased vigor will enable these organs to more readily synthesize plant growth substances required for growth, development, and inhibition of abscission. It is generally accepted that plants given supplemental nitrogen retain their leaves longer and set more fruit than nitrogen-deficient plants. These plants typically have high levels of amino acids and other nitrogenous compounds which serve as building blocks for the synthesis of DNA, RNA, protein, and other factors which prevent abscission. It has also been shown that high-nitrogen plants also contain higher levels of auxins and cytokinins. Vigorously growing plants have the ability to delay abscission of their leaves and other plant organs by maintaining high levels of auxins and lower amounts of abscisic acid across the abscission zone. In addition, the higher levels of cytokinins may make leaves or other plant organs strong sinks promoting high rates of transport of carbohydrates and other materials which prevent abscission. Mineral elements are required for essential biochemical reactions in plants. If an essential element becomes limited abscission of leaves and/or other plant organs is likely to occur. Skoog (1940) showed that the proper levels of zinc nutrition are required in order to maintain normal levels of auxin within the plant, in fact, one of the first signs of zinc deficiency is reduced levels of auxin. It is now generally accepted that zinc deficiency is very effective in accelerating abscission. Calcium has also been shown to be involved in the abscission process. It is known that calcium pectate is the major constituent of the cell wall, particularly in the middle lamella. During abscission calcium is undetectable in the separation layer and adjacent cells. Plants which are deficient in calcium have been shown to readily abscise their leaves, whereas exogenous applications of calcium can retard the process possibly by making cell walls more resistant to hydrolases (Addicott 1982).

Effects of Plant Growth Substances

Ethylene and Auxins.   The ability of ethylene to promote abscission was shown many years ago (Wehmer 1917) and since this time there have been many reports on the ability of ethylene to stimulate abscission (Addicott 1982; Osborne 1989; Abeles et al. 1992). As mentioned earlier the physiology and biochemistry of abscission in leaves, fruits, flowers, and other organs is similar; however, not all plant organs have a preformed abscission zone nor are they all sensitive to ethylene (Abeles et al. 1992). In many dicotyledenous plants abscission can be induced by exogenous applications of ethylene (Figure 9.3). It is generally accepted that all plant parts produce ethylene and that ethylene generally increases in any ripening or senescing organ prior to and during abscission of that organ indicating a strong link between ethylene and abscission (Osborne 1989; Abeles et al. 1992; Addicott 1982). Exogenous applications of auxin made distally (located on the leaf side of the abscission zone) to the abscission zone have been shown to delay senescence, whereas proximal (located on the stem side of the abscission zone) applications accelerate the process. It has also been shown that when endogenous levels

Figure 9.3.   Petal abscission in regal *Pelargonium* plants following treatment with ethylene (courtesy of C. F. Deneke).

of auxin in the leaves or other plant organs are reduced, abscission typically occurs (Addicott 1982; Guinn and Brummett 1988; Osborne 1989; Abeles et al. 1992). Therefore, it is generally accepted that abscission is enhanced when proximal quantities of auxin are equal to or larger than distal quantities. High distal concentrations of auxin are effective in inhibiting abscission only during a critical period of time. When abscission zones are in stage 1, abscission is inhibited by auxin, while in stage 2 auxins stimulate abscission. It has been shown that exogenous ethylene applications made during stage 1 are ineffective in stimulating abscission probably due to the sensitivity of the tissue, whereas, during stage 2 ethylene strongly promotes abscission. The length of time for an explant to complete stage 1 depends on the species and environment. Young leaves are rich in auxin but as the leaves age and become less productive as a result of stress or shading, auxin levels decline. Eventually the abscission zones lose sensitivity to auxin and become sensitive to ethylene. Hall (1952) was the first to propose that auxin was involved in controlling the aging process and that aging regulated the sensitivity of the tissue to ethylene. This theory has since been supported by others who have shown that auxins and cytokinins, which slow the aging process, also retard leaf (Abeles et al. 1967), fruit (Griggs et al. 1970), and flower abscission (Roberts et al. 1984; Tanaka et al. 1985). It was suggested that auxin blocked ethylene action by delaying the aging process. After the aging process had been initiated auxins no longer delayed abscission but instead stimulated abscission by enhancing ethylene production (Rubinstein and Abeles 1965).

Abscisic Acid.   Pioneering studies leading to the discovery of ABA suggested that it was the plant growth substance directly responsible for abscission. In fact, ABA was first isolated and characterized using a cotton explant abscission-accelerating bioassay to evaluate activity (Davies and Jones 1991). Since this time there have been a number of studies attempting to correlate endogenous levels of ABA with abscission. However, it has been difficult to show that ABA is directly involved in the abscission process since in many cases endogenous levels of ABA were shown to increase prior to the initiation of cell separation (Osborne 1989; Davies and Jones 1991; Addicott 1982), suggesting that ABA was indirectly involved in the promotion of abscission. It is generally accepted that under stress conditions ABA levels are increased, thereby accelerating the senescence process which eventually leads to abscission. Since ABA accelerates the senescence process it becomes difficult to separate enhanced ethylene production which normally occurs in senescing tissues and ABA effects.

ABA has been shown to speed the rate of abscission in most explant systems which have been tested. Initial studies suggested that ABA directly enhanced ethylene production by plant tissue. Upon closer evaluation it was shown that ABA promoted premature senescence, resulting in the production of ethylene, which in turn promoted abscission. It has also been shown that aminoethoxyvinylglycine (AVG) (an inhibitor of ethylene biosynthesis) had the ability to block the accelerated abscission promoted by ABA and that exogenous application of ethylene promoted abscission even in the presence of AVG, indicating that ABA had no effect independent of its promotion of ethylene production. There have been numerous attempts to remove ethylene from ABA-treated tissues, thereby providing proof that ABA is directly involved in the promotion of abscission. However, in all of these reports the air surrounding the tissue had lower levels of ethylene, but endogenous levels could not be depleted sufficiently to provide unequivocal doubt that ethylene was not causing an effect. One must remember that ethylene is a very active plant growth substance required in only trace amounts to be effective. Further studies with ABA and/or ethylene mutants may clarify this matter in the future (Osborne 1989).

Other Plant Growth Substances.   Gibberellins have been shown to promote abscission; however, in these cases it was reported that the response was due to a stimulation of ethylene production rather than due to GA itself (Morgan and Durhan 1975; Wittenbach and Bukovac 1973). There have also been reports showing that GA can delay abscission by increasing the ability of that organ to act as a sink (Addicott 1982). Cytokinins have also been reported to have a potential role in abscission although information in this area is very limited. It has been shown that when cytokinins are applied to young fruits there is a stimulation in growth and a retardation of abscission. As mentioned earlier in the chapter on senescence, cytokinins delay leaf senescence in a variety of plant species; this effect indirectly results in delaying the abscission process (Addicott 1982). Therefore, it has been suggested that when cytokinins are applied to different plant organs it makes them a strong sink diverting nutrients and other materials to this area, thereby delaying the abscission process. Although, cytokinins can retard abscission, the site of application will determine whether it retards or promotes it. When cytokinins are applied at sites distant from the abscission zone there tends to be an acceleration of the abscission process, whereas applications directly to or above the abscission zone inhibits it. Exogenous applications of jasmonates have been shown to promote chlorophyll degradation and many other effects associated with the senescence process. It is thought that jasmonates promote leaf abscission by accelerating senescence. It is tempting to suggest that jasmonates may be directly involved in abscission, however, their effect on senescence and subsequent leaf abscission may be indirect since it has been shown that methyl jasmonate stimulates ethylene biosynthesis, a plant growth substance known to be involved in abscission. The effects of jasmonates on the abscission process may also be due to their structural similarity to ABA (Sembdner and Parthier 1993).

In a review by Osborne (1989) ethylene is identified as the signal which initiates the abscission process in dicotyledonous plants. Evidence supporting this theory includes the ability of the following treatments to delay abscission:

1. Hypobaric conditions around the abscission zone (removes ethylene).
2. Suppression of ethylene production with inhibitors of ethylene biosynthesis such as AOA, AVG, Co^{2+} or anoxia.
3. Blocking ethylene action with silver thiosulfate or norbornadiene.

Although there is a considerable amount of evidence implicating ethylene in the abscission process it is still possible that an uncharacterized material termed senescence factor may be the direct cause of abscission (Addicott 1982) or that it may be due to an interaction between known and unknown plant growth substances.

AGRICULTURAL ABSCISSION

Most if not all cultivated plants do not always cooperate with respect to their abscission habits, thereby presenting a problem in agriculture. Until fairly recently, methods for manipulating the abscission process in agricultural crops were limited. Prior to this time growers had to thin fruits, flowers, leaves, and other organs by hand, which is a very tedious and cost-ineffective process. Cultural conditions such as optimal mineral nutrition, water supply, and elimination of abiotic and biotic stresses may be used to retard abscission. In this section agricultural practices designed for promoting or delaying the abscission process in leaves will be discussed. In Chapter 10 agricultural practices to delay or promote abscission in flowers and fruits will be discussed.

Leaf Abscission

Promotion. The use of chemicals to defoliate plants for agricultural purposes began when it was found that foliar applications of calcium cyanamid could cause leaves to abscise from cotton plants, thereby facilitating mechanical harvesting of the bolls. Other chemicals such as thidiazuron (Droop), ammonium nitrate, endothall, paraquat, sodium cacodylate, sodium chlorate, tributyl phosphorotrithioate, and tributyl phosphorotrithioite have been found to be effective (Addicott 1982; Plant 1981). Ethephon (an ethylene-releasing compound) has been evaluated as a cotton defoliant; however, others are cheaper and have fewer side effects. Thidiazuron has been shown to be effective as a cotton defoliant when applied as a preharvest spray. Chemicals such as paraquat can also be used either as defoliants or desiccants depending on the timing and concentration used. When using chemicals as defoliants they must be applied seven to 14 days prior to harvest so abscission can be induced. When used as desiccants they are applied at higher concentrations one or two days prior to harvest and cause the foliage to rapidly lose water and abscise. If the water loss occurs too rapidly and the abscission layer does not have adequate time to form, drying leaves will remain attached to the plant, causing problems with harvest. The advantage of desiccants over defoliants is that they can be used later than defoliants, thereby allowing the leaves to photosynthesize for longer periods which maximizes yields.

Defoliation of nursery plants is done in some areas of the United States prior to digging and shipment. Ethephon has been shown to be effective in defoliation of deciduous nursery stock (Abeles et al. 1992). An inexpensive way to defoliate plants on a small scale is to place buckets of ripening apples in a closed room with plants to be defoliated. The ethylene evolved from these apples will be enough to promote abscission. This practice is commonly done on small farms where expense is a critical factor.

Delaying.   The ability to retard or prevent leaf abscission is important in floriculture and nursery crops especially during shipping. NAA applications have been shown to prevent abscission of leaves and berries of holly during shipment (Roberts and Ticknor 1970). Another example where NAA is effective in delaying abscission is its ability to delay needle abscission during shipment of conifers used as Christmas trees (Worley and Grogan 1941). Although there are other reports showing that plant growth regulators delay leaf abscission, the best way to prevent it is through good cultural practices such as:

1. Providing optimal mineral nutrition, especially nitrogen.
2. Avoiding water stress and other types of stress which lead to the production of ethylene or ABA.
3. Avoiding the accumulation of ethylene and keeping respiration rates low during shipment or storage.
4. When the production of ethylene cannot be avoided ethylene scrubbers, biosynthesis inhibitors or action inhibitors have been shown to be useful (Addicott 1982; Abeles et al. 1992).

REFERENCES

Abeles, F. B., Holm, R. E., and Gahagan, H. E. (1967). "Abscission: The role of aging". *Plant Physiol.* 42:1351–1356.

Abeles, F. B., Morgan, P. W., and Saltveit Jr., M. E. (1992). *Ethylene in Plant Biology. Second Edition*, Academic Press, San Diego, CA.

Addicott, F.T. (1965). "Physiology of abscission". In *Encyclopedia of Plant Physiology*, Springer-Verlag, Berlin.

Addicott, F. T. (1982). *Abscission*, University of California Press, Berkeley, CA.

Davies, W. J. and Jones, H. G. (1991). *Abscisic Acid: Physiology and Biochemistry*, Bios Scientific Publishers, Oxford U.K.

Fitting, H. (1911). "Untersuchungen über die vorzeitige Entblätterung von Blüten". *Jahrb. Wiss. Bot.* 49:187–263.

Griggs, W. H., Iwakiri, B. T., Fridley, R. B. and Mehlschau, J. (1970). "Effect of 2-chloroethylphosphonic acid and cycloheximide on abscission and ripening of Bartlett pears". *HortScience* 5:264–266.

Guinn, G. and Brummett, D. L. (1988). "Changes in abscisic acid and indoleacetic acid before and after anthesis relative to changes in abscission rates of cotton fruiting forms". *Plant Physiol.* 87:629–631.

Hall, W. C. (1952). "Evidence on the auxin-ethylene balance hypothesis of foliar abscission". *Bot. Gaz.* 113:310–322.

Molisch, H. (1886). "Untersuchungen über Laubfall". *Sitzungsber. Akad. Wiss. Wien, Math.-Naturw. Kl., Abt. I.* 93:148–184.

Morgan, P. W. and Durhan, J. I. (1975). "Ethylene-induced leaf abscission is promoted by gibberellic acid". *Plant Physiol.* 55:308–311.

Osborne, D. J. (1989). "Abscission". *Crit. Rev. Plant Sci.* 8:103–129.

Plant Growth Regulator Handbook, (1981). Plant Growth Regulators Society of America, Lake Alfred, FL.

Roberts, A. N. and Ticknor, R. L. (1970). "Commercial production of English holly in the Pacific Northwest". *Am. Hortic. Mag.* 49:301–314.

Roberts, J. A., Schindler, C. B., and Tucker, G. A. (1984). "Ethylene-promoted tomato flower abscission and the possible involvement of an inhibitor". *Planta* 160:159–163.

Rubinstein, B. and Abeles, F. B. (1965). "Relationship between ethylene evolution and leaf abscission". *Bot. Gaz.* 126:255–259.

Sembdner, G. and Parthier, B. (1993). "The biochemistry and the physiology and molecular actions of jasmonates". *Annu. Rev. Plant Physiol. Plant Mol. Biol.* 44:569–589.

Skoog, F. (1940). "Relationships between zinc and auxin in the growth of higher plants". *Am. J. Bot.* 27:939–951.

Tanaka, H., Denpoya, K., and Hashimoto, T. (1985). "Relationship between flower abscission and ethylene production in some flowering ornamentals treated with plant growth regulators and STS". *Bull. Fac. Agric. Tamagawa Univ.* 25:72–78.

Von Mohl, H. (1860a). "Einige nachträgliche Bemerkungen zu meinem Aufsatz über den Blattfall". *Bot. Zeitg.* 18:132–133.

Von Mohl, H. (1860b). "Ueber die anatomischen Veränderungen des Blattgelenkes, welche das Abfallen der Blätter herbeiführen". *Bot. Zeitg.* 18:1–7, 10–17.

Von Mohl, H. (1860c). "Ueber den Ablösungsprocess saftiger Pflanzenorgane". *Bot. Zeitg.* 18:273–277.

Weaver, R. J. (1972). *Plant Growth Substances in Agriculture*, W. H. Freeman and Company, San Francisco.

Webster, B. D. (1968). "Anatomical aspects of abscission". *Plant Physiol.* 43:1512–1544.

Wehmer, C. (1917). "Leuchtgaswirkung auf Pflanzen. 2. Wirkung des gases auf gruene Pflanzen". *Ber. Deut. Bot. Ges.* 35:318–322.

Wittenbach, V. A. and Bukovac, M. J. (1973). "Cherry fruit abscission: Effect of growth substances, metabolic inhibitors and environmental factors". *J. Amer. Soc. Hort. Sci.* 98:348–351.

Worley, C. L. and Grogan, R. G. (1941). "Defoliation of certain species as affected by α-naphthaleneacetic acid treatment". *J. Tenn. Acad. Sci.* 16:326–328.

Physiology of Fruit Set, Growth, Development, Ripening, Premature Drop, and Abscission

Regulation of fruit set, growth, development, ripening, premature fruit drop, and subsequent abscission is very important in agriculture. Prior to discussing the regulation of these processes background information starting with pollination, which is the transfer of pollen from the anther to the stigma, will be provided. Once pollination has occurred the pollen tube grows down the style into the ovary until it reaches the embryo sac within the ovule. Two male gametes from the pollen tube are inserted into the embryo sac, one of which unites with the female gamete, a process known as fertilization, to produce a zygote which divides to become the embryo. The other unites with two polar nuclei to produce the endosperm. The ovary gives rise to the fruit, which may be defined as the structure which results from the development of tissues which support the ovules of the plant (Nitsch 1965) and the ovule leads to the seed (Figure 10.1).

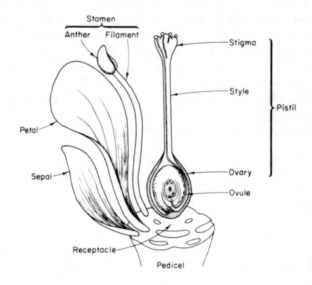

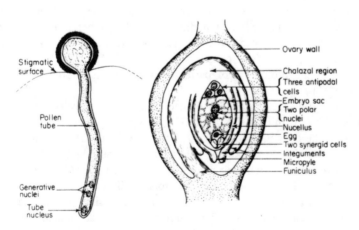

Figure 10.1.   Angiosperm plant flower structure (from Hartmann et al. (1990)).

PHYSIOLOGY OF FRUIT SET

Fruit set may be defined as the rapid growth of the ovary which usually follows pollination and fertilization. Both pollination and fertilization must normally take place to produce viable seed. However, there are cases where development of fruits occur without pollination or fertilization resulting in seedlessness; this process is called parthenocarpy. There are two types of parthenocarpy which have been found, one is vegetative parthenocarpy, which is characterized by the development of fruit without pollination (e.g., pineapple and Washington Navel orange). The other is stimulative parthenocarpy which requires the stimulus of pollen without subsequent fertilization in order to set fruit (e.g., Black Corinth grape) (Weaver 1972).

Effects of Auxins and Gibberellins on Fruit Set

Synthetic auxins such as NAA have been shown to be most effective in promoting set in fruits having multiple ovules such as strawberry, squash, fig, tomato, rose, tobacco, eggplant and others. IAA is much less effective in promoting fruit set than the synthetic auxins, probably because it is unstable in the light and can be broken down by IAA oxidase within the plant or converted to inactive conjugates (Weaver 1972). Gibberellins also have the ability to promote fruit set in all plants which are responsive to auxins plus a number of others such as blueberries, citrus, grapes, and stone fruits (Coggins et al. 1966; Crane et al. 1960, 1961) where auxins were shown to be ineffective (Weaver 1972). Gibberellic acid (GA_3) has been shown to be the effective gibberellin in many cases. The induction of parthenocarpy by gibberellins has been reported in a number of plants including grape, peach, apricot, almond, and tomato; however, there is a considerable amount of specificity among the gibberellins in the promotion of parthenocarpy and also varietal differences (Weaver 1972).

Bukovac and Nakagawa (1967) treated emasculated flowers from Wealthy apple trees with GA_1 through GA_{10} and GA_{14} by applying lanolin paste containing 5 mM GA to the cut style and adjacent receptacle tissue. Unpollinated controls treated with lanolin alone did not enlarge and abscised within two or three weeks after treatment. After four weeks, styles treated with GA_4 and GA_7 had growth rates equal to the pollinated controls, while other gibberellins tested exhibited low growth rates at this time. It is not surprising that GA_4, and GA_7 are very active since they have been found to be the predominate gibberellins in apple seeds (Dennis and Nitsch 1966).

Effects of Cytokinins, Abscisic Acid, and Ethylene on Fruit Set

Synthetic cytokinins have been shown to be effective in increasing fruit set in grapes (Weaver 1972), figs (Crane 1965), and muskmelon (Jones 1965). It has been suggested that the ability of cytokinins to mobilize assimilates to the area of application is responsible for increased fruit set. However, if this were true cytokinins should be effective in increasing fruit set in a wide range of plants and this has not been found. Abscisic acid and ethylene cause abscission of flowers and young fruits so their effects on fruit set are negative.

PHYSIOLOGY OF FRUIT GROWTH AND DEVELOPMENT

Fruits undergo dramatic increases in size following anthesis. An example of this is the apple which may increase in volume up to 6,000 times during a 20-week growth period (Luckwill 1957). Fruit growth typically follows two distinct growth curves. One type is a smooth sigmoid curve which is exhibited by many plants including apples, pears, tomatoes, cucumbers and strawberries (Figure 10.2). The second type of growth curve is represented by a double sigmoid which can be viewed as two successive sigmoid curves, examples of this type of fruit growth are blueberries, grapes, figs and many stone fruits including cherries, olives and peaches. This type of growth pattern has two periods of rapid growth separated by an intermediate period where little or no growth occurs. During the first stage of rapid growth the ovary and its contents grow rapidly except for the embryo and endosperm. Stage 2, which appears to be a plateau, is characterized by growth of the embryo and endosperm, lignification of the endocarp, and a small amount of growth of the ovary wall. In stage 3, rapid growth of the mesocarp occurs promoting a rapid increase in fruit size followed by maturation (Figure 10.2).

Effects of Plant Growth Substances on Fruit Growth and Development

The use of plant growth substances to control fruit set, size, shape, and maturation has become important in agriculture today because they have the ability to increase fruit size, color, and shape, thereby increasing marketability. In addition, by hastening or delaying maturation the grower can utilize peak demands, avoid unfavorable environmental conditions, and extend the market period.

Auxins and Fruit Growth. Two lines of evidence have been presented showing that auxins are involved in fruit growth. First, there is a correlation

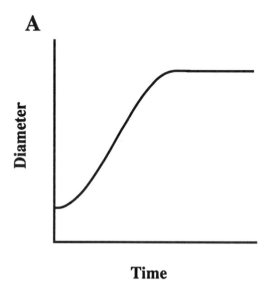

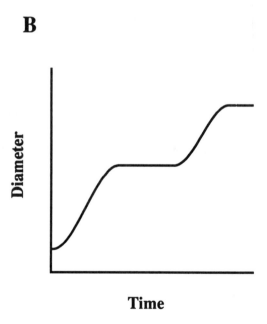

Figure 10.2. Typical fruit growth curves: (a) Smooth sigmoid curve; and (b) double sigmoid curve.

between seed development and final size and shape of the fruit. Second, exogenous applications of auxins to certain fruits at particular stages of their development induces a growth response.

Fruit size and shape is closely correlated with seed number and seed distribution in many fruits. Muller-Thurgau (1898) was the first to report in grapes that there was a direct correlation between size of the berry and number of seeds. Since this time it has been shown that the endosperm and embryo in the seed produces auxin, which moves outward and stimulates growth of the endosperm (Nitsch 1950; Dreher and Poovaiah 1982; Mudge et al. 1981; Southwick and Poovaiah 1987).

The location of the seeds within the fruit has a profound effect on its shape. This phenomenon has been extensively studied in strawberries because the achenes surround the fleshy receptacle on the outside and are easy to remove. Each achene induces growth of the receptacle tissue around it, therefore, when varying numbers of achenes are removed from young fruits different shapes can be produced (Nitsch 1950; Mudge et al. 1981). When insufficient pollination of strawberry flowers occurs a small number of achenes develop distorted fruits. Nitsch (1950), working with strawberries, was the first to convincingly show that developing achenes have a dramatic effect on fruit growth. He showed that when achenes were completely removed from the receptacle, growth ceased. When several achenes were left attached to the receptacle, growth occurred directly below the achene. These experiments suggested the possibility that auxins which are contained in the achenes were affecting receptacle growth. Evidence that auxins were involved in regulating fruit growth was presented by Nitsch (1950). He showed that when all achenes were removed and replaced with lanolin paste alone there was no growth. However, when all achenes were removed and lanolin paste containing 100 ppm β-NOA was added, the receptacle grew the same as the strawberry fruit, where no achenes had been removed (Figure 10.3). Since this time other researchers have obtained similar results with B-NOA and other auxins (Mudge et al. 1981).

Although there have been a number of studies showing a close correlation between the number of seeds and final fruit size and between seed distribution and shape of the fruit there is generally not a close correlation between the total amount of auxin produced in the seeds and fruit growth (Nitsch 1952 1955; Dreher and Poovaiah 1982) (Figure 10.4).

Most of the molecular studies on the mechanisms of auxin action have focused on auxin-induced cell elongation, while little is known about how auxins regulate physiological processes such as fruit growth and development which involves cell division, cell elongation, and differentiation. Recently, researchers using strawberry as a model system have begun work on the elucidation of auxin involvement in fruit growth and development at the molecular level (Reddy and Poovaiah 1990). They have isolated and characterized

Figure 10.3. Effect of the removal of fertilized achenes and auxin application on growth of the strawberry receptacle: (a) Control; (b) all achenes removed and lanolin paste applied to the receptacle; (c) all achenes removed, but lanolin paste containing 100 ppm β-napthoxyacetic acid was applied to the receptacle; and (d) growth occurring under three fertilized achenes (magnified three times) (Nitsch 1950).

a cDNA (λSAR5) for an auxin-repressed mRNA and have shown a positive correlation between repression of mRNA corresponding to λSAR5 and fruit growth. During the development of pollinated fruits λSAR5 was dramatically reduced. They also showed that when pollinated fruits were deachened, the fruits did not grow and the auxin-repressed mRNA level increased. However, when exogenous applications of auxins were made to deachened fruits, growth occurred and a reduction in λSAR5 was observed. Variant genotypes of strawberries which will not grow in the absence of exogenous auxin showed very high levels of expression of λSAR5. However, when exogenous auxin applications were made λSAR5 was reduced dramatically (Reddy and Poovaiah 1990). These studies provide a foundation for research in this area which may lead to a better understanding of how auxins regulate fruit growth and development at the molecular level.

Gibberellins and Fruit Growth. Seeds are rich sources of gibberellins; therefore, as with auxin it was suspected that they were somehow involved in fruit growth. The ability of exogenous applications of gibberellins to induce parthenocarpic fruits suggests that they were involved in fruit growth. It has been shown that growth patterns of parthenocarpic fruits produced by exogenous gibberellin application closely paralleled patterns induced under normal conditions in peaches (Crane et al. 1960), grapes (Weaver and McCune 1960), and other crops (Weaver 1972). The time of GA applications has a

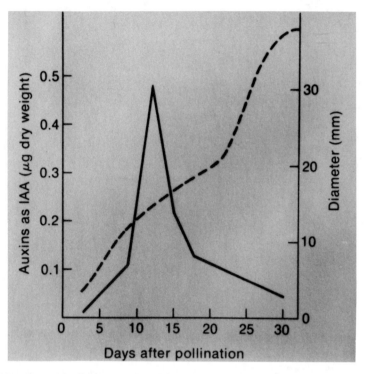

Figure 10.4.   Changes in auxin levels (solid line) and fruit diameter (dashed line) in Marshall strawberry (Nitsch 1952, 1955).

profound effect on the ability to promote enlargement of fruits and berry shape (Christodoulou et al. 1968; Zuluaga 1968). As shown in Figure 10.5, gibberellin treatments made at the time of full bloom elongate the berries and later applications produce larger berries (Christodoulou et al. 1968). The enhancement of fruit growth promoted by gibberellins in seeded varieties of grapes is generally minimal. When there is an increase in the size of seeded varieties of grape it is generally associated with a lower number of seeds found within the fruit, suggesting that endogenous levels of GA are affecting fruit size and when they are high enough exogenous applications have no effect. Gibberellins and auxins have been shown to have a synergistic effect on the growth of tomato fruits (Luckwill 1959). There is also evidence in the literature that gibberellins will promote a stimulation of endogenous auxin levels (Sastry and Muir 1963), further suggesting an interaction between the two.

   Gibberellins have been shown to affect fruit shape. Asymmetric fruit growth can be induced in seeded and parthenocarpic apple and Japanese pear by localized GA_4 and GA_7 application to one side of fruits two weeks after bloom

Figure 10.5.   Effects of gibberellin treatments on grape berry size and shape.

(Nakagawa et al. 1968; Bukovac and Nakagawa 1968) (Figure 10.6). The number and size of cells on the treated side increases while the untreated side remains unchanged. Both seeded and seedless fruit respond to applications of GA$_3$ by producing longer fruit (Westwood and Bjornstad 1968).

There is little correlation between endogenous levels of gibberellins within the fruit and fruit growth as is the case with auxins. The concentrations of gibberellinlike substances in the seed, endocarp, and mesocarp of apricot fruits have been shown to correlate with the growth rate of these specific tissues between anthesis and maturity but not with total fruit growth (Jackson and Coombe 1966). This lack of correlation between endogenous levels of gibberellins and fruit growth has also been shown in grapes (Iwahori et al. 1968), apples (Luckwill et al. 1969), peaches (Jackson 1968), and others (Weaver 1972).

Cytokinins and Fruit Growth.   Young developing fruits have rapidly dividing cells, and the seeds contained within them have been shown to be rich sources of cytokinins (Letham 1967; Kaminek et al. 1992), suggesting that they play an important role in fruit growth. Exogenous applications of cytokinins on grapes have shown mixed results. In seedless Black Corinth grapes berry size was increased (Weaver and van Overbeek 1963), but berry size in Thompson

Figure 10.6. Effects of localized GA_4 treatments as indicated by arrows two weeks after full bloom in Wealthy apple: seeded nontreated (left) seeded treated with GA_4 (center) and parthenocarpic treated with GA_4 (Bukovac and Nakagawa 1968).

Seedless grapes was only slightly increased (Weaver et al. 1966), and in seeded varieties cytokinins actually reduced berry size. Available results indicate that cytokinins exert most of their effect on fruit set, while gibberellins affect mostly fruit growth and auxins affect fruit set and growth.

Exogenous applications of N-(purine-6-yl)-α-phenylglycine, BPA, BA, and zeatin have been shown to promote apple fruit enlargement and development of calyx lobes in Delicious apples (Williams and Stahly 1969). Working with Cox Orange Pippin apples it was shown that exogenous applications of zeatin suppressed fruit elongation and weight (Letham 1968). However, it is possible that factors such as concentration, type of cytokinin used, and timing may have been responsible for the negative effect. Gibberellins have been reported to induce asymmetric growth of apple fruits (Figure 10.6) (Bukovac and Nakagawa 1968). Both cytokinins and gibberellins together have been shown to stimulate the development of elongated fruits with well-developed calyx lobes (Figure 10.7) (Williams and Stahly 1969). Unrath (1974) suggested the commercial implications of a cytokinin and GA_4 and GA_7 combination for the control of apple shape. Today, Promalin, a product from Abbott Laboratories containing a mixture of BA + GA_4 and GA_7 is commercially used in 'Delicious' apples to regulate shape, size, and weight and to increase per acre yield of fruits (*Plant Growth Regulator Handbook* 1981).

Chemical Thinning of Flowers and Fruits

Thinning of certain species and varieties of fruit trees is a necessary commercial practice. In fruit trees especially, most varieties of apple thinning eliminates problems with biennial bearing and physical damage to the tree and also enhances fruit size, shape, color, and overall quality.

Figure 10.7.   Effects of cytokinin plus gibberellin treatments on the stimulation of elongated fruits with developed calyx lobes.

Auchter and Roberts (1934) tested several chemicals in an attempt to find one which would prevent fruit set in some varieties of apples. They found that tar oil distillates would kill flower buds when applied at the cluster bud stage. A summary of thinning sprays from the early 1930s through 1964 is outlined in reviews by Batjer (1964, 1965). Since this time, various chemicals given in Table 10.1 have been shown to be effective in fruit thinning.

However, the timing and concentration used is critical. NAA, NAAm, Sevin, Accel, and ethephon (plus other ethylene releasing compounds) are effective postbloom thinners (Southwick et al. 1964; McKee and Forshey 1966). In order to be effective in thinning apples, NAA or NAAm should be applied after full bloom and in pears five to seven days after petal fall (*Plant Growth Regulator Handbook* 1981). One or two applications of Accel should be made when king

Table 10.1.   Chemicals used for thinning apples.

Common name	Chemical name	Trade name
Ethephon	(2-chloroethyl)phosphonic acid CEPA, Amchem 66-329	Ethephon, Ethrel,
Silaid	(2-chloroethyl)methylbis (phenylmethoxy)silane	Silaid
Alsol	(2-chloroethyl)tris (2-methoxyethoxy)silane	Alsol, Etacelasil
DNOC	Sodium 4,6-dinitro-o-cresylate	Elgetol
DNOC	4,6-dinitro-o-cresol	Dinitro-dry
NAA	Naphthaleneacetic acid	Fruitone-N, Fruit Fix-800, Fruit Fix-200, Fruit Set, Stafast, Kling-Tite
NAAm	Naphthaleneacetamide	Amide-Thin W, Anna-Amide
Carbaryl	1-naphthyl N-methyl carbamate	Sevin
Oxamyl	Methyl N'N'-dimethyl-N-[(methyl carbamoyl)oxy]-1-thiooxamimidate	Vydate
Cytokinin + Giberellin	N-(phyenylmethyl-H-purine 6-amine and GA$_4$ and GA$_4$	Accel
Silvex/ Fenoprop	2-(2,4,5-trichlorophenoxy) propanoic acid	Fruitone T

fruitlets are approximately 10 mm in diameter, this period is between seven and 21 days after full bloom. Ethrel should be applied 10–20 days after full bloom and if used for thinning may not be used for fruit loosening in the fall. Sevin has been shown to be effective in thinning apples when applied 15–27 days after full bloom (Batjer and Thompson 1961). Other carbamates used, such as Vydate, have also been shown to be effective in postbloom thinning. Postbloom thinning provides the grower with an opportunity to evaluate the degree of fruit set before applying sprays and by delaying application, the danger of frost is reduced prior to spraying. DNOC is also effective in thinning apples, it must be applied at full bloom or petal fall because it is a caustic agent that burns stigmatic surfaces preventing pollination and therefore is not effective as a postbloom thinning agent (Weaver 1972).

The proper concentration of thinning agent to use varies with weather conditions, tree vigor, variety, and other factors. Therefore, it is very important to always get specific recommendations from local authorities prior to spraying.

The incidence of rot in grape species or varieties that produce tight clusters is very high because after rain, tight clusters dry very slowly. In addition, tight clusters also cause berries within the cluster to be crushed, thereby providing an excellent media for decay-causing organisms. Thinning of grapes with sprays presents certain difficulties not encountered in thinning fruit trees. The flower clusters of grapes are very small until the shoots are 3–4 inches long. Hand-thinning is not practical and thinning with chemical sprays has met with limited success. As an alternative to thinning, gibberellins have been used to elongate the bunch of a number of varieties, thereby removing the need for thinning. However, once again the timing and concentration is critical because when the recommended dose of gibberellin is exceeded, the vine may be injured.

Fruit Ripening

The manipulation of fruit ripening is of major economic importance. Ethylene has been shown to be involved in the ripening of many fruits. In fact, the term climacteric refers to fruits which will ripen in response to ethylene (e.g., tomato and banana) and non-climacteric refers to fruits which will not ripen in response to ethylene (e.g., grape and strawberry) (Abeles et al. 1992). There are several lines of evidence showing that ethylene is involved in the ripening, of climacteric fruits. First, it has been shown that exogenous applications of ethylene will promote ripening. Second, prior to fruit ripening endogenous levels of ethylene are very low, however, once the ripening process is initiated there is a dramatic increase in ethylene production and subsequent ripening. Third, ripening can be delayed by hypobaric storage, treatment with ethylene biosynthesis inhibitors such AVG or AOA, or treatment with ethylene action inhibitors such as silver or carbon dioxide. Present knowledge on the involvement of ethylene in the fruit ripening process can be used in two ways. The first, way is to promote faster, more uniform ripening to meet market demands and to facilitate mechanical harvesting. At the present time ethephon can be used to speed and also promote uniform ripening in cherries, apples, boysenberries, pineapples, blueberries, coffee, cranberries, and figs (Abeles et al. 1992; *Plant Growth Regulator Handbook* 1981). The second way is to delay ripening in order enhance the fruits shelf life. Increased shelf life can be accomplished by blocking ethylene biosynthesis or action or to use conditions which remove ethylene from the area surrounding the fruit. However, these approaches are quite expensive and do not prevent ripening from occurring. Ethylene is thought to regulate the ripening process by facilitating the expression of genes which are responsible for enhancing respiration rates, autocatalytic ethylene production, chlorophyll production, carotenoid synthesis, conversion of starch to sugars, and increased cell-wall degrading enzymes (Gray et al.

1992). Recently, genes induced during fruit ripening and genes involved in ethylene biosynthesis have allowed scientists to construct ripening mutants in tomatoes using reverse genetics. Since gene replacement technology in plants was not available, researchers used antisense RNA technology as a tool to modify fruit ripening. Pioneering studies attempting to inhibit softening in tomatoes utilized antisense polygalacturonase (PG) RNA. PG was thought to be the enzyme responsible for cell-wall hydrolysis during ripening (Smith et al. 1988; Sheehy et al. 1988). The use of PG antisense RNA dramatically inhibited PG mRNA accumulation and enzyme activity, but softening still occurred in the transgenic fruits suggesting that PG was not the sole determinant of cell-wall hydrolysis. Two approaches to prevent fruit ripening by inhibiting ethylene production were to metabolize ACC by overproducing the *Pseudomonas* ACC deaminase (Klee et al. 1991), or to inhibit ACC oxidase activity with antisense RNA (Hamilton et al. 1990). Both were only partially effective in delaying fruit ripening because they did not sufficiently decrease ethylene production in order to effectively inhibit the ripening process. The most successful approach to date has been to inhibit ACC synthase with antisense RNA (Oeller et al. 1991). By using this approach researchers were able to inhibit ethylene production to less then 0.1 nl/g per hour which is as low as levels found in naturally occurring nonripening mutants (Theologis 1992).

During the tomato fruit ripening process, two ACC synthase genes LE-ACC2 and LE-ACC4 are expressed (Olson et al. 1991; Rottmann et al. 1991). By using antisense constructs derived from LE-ACC2 mRNA, both LE-ACC2 and LE-ACC4 were almost completely inhibited (Oeller et al. 1991) and fruits from these plants never ripened, whereas, control fruits kept in air began to produce ethylene 50 days after pollination and were fully ripe 10 days later. Red color resulting from chlorophyll degradation and lycopene biosynthesis is inhibited in tomato fruits containing antisense RNA. Ethylene or propylene treatments for a period of six days reversed the antisense phenotype, while shorter exposures did not produce the fully ripe phenotype. Fruits treated with ethylene for six days were indistinguishable from naturally ripened fruits with respect to texture, color, aroma, and compressibility. Interestingly, overripening could be prevented when antisense fruits were removed from an ethylene atmosphere after they were fully ripe. From these findings it was concluded that:

1. Ethylene-mediated ripening requires continuous transcription of the necessary genes, which may reflect a short half-life of induced gene products.
2. Ethylene is autocatalytically regulated.
3. Ethylene acts as a rheostat rather than a switch for controlling the ripening process.
4. Ethylene is the key regulatory molecule for fruit ripening and senescence, not the by-product of ripening (Theologis 1992).

Recently, ACC N-MTase, which is the enzyme responsible for the conversion of ACC to MACC, which is an inactive end-product, was purified and characterized by Guo et al. (1992, 1993). This recent development will enable these researchers to isolate the gene(s) for ACC N-MTase and to use this plant-derived gene in the same manner as with the deaminase gene (Klee et al. 1991); however, this time a plant gene will be used instead of a bacterial gene which should be more effective. In summary, the use of antisense technology is only the first step in controlling fruit ripening. The development of gene transplacement by homologous recombination should enable researchers to produce nonleaky ripening mutants (Theologis 1992). For details on ripening in fruits of commercial interest see a summary in Abeles et al. (1992).

Plant growth substances other than ethylene such as auxins, cytokinins, and gibberellins have also been shown to reduce or delay various aspects of ripening. While these plant growth substances play other roles, they each share the ability to delay senescence, thereby reducing the sensitivity of the fruit to ethylene (Abeles et al. 1992).

Prevention of Fruit Drop

Problems with mature apples, pears, citrus, and other types of fruit falling off the tree prior to harvest occurs frequently. This is a problem in agriculture because when fruit falls from the tree they are often injured and much less valuable. In the past, growers used to spread straw under the trees of certain cultivars or fruit types to reduce injury, a practice not used today because of the many problems associated with it. Preharvest fruit drop also means that harvesting may need to be started earlier than desired to assure optimum fruit quality.

Gardner et al. (1939) found that certain growth regulators applied to fruit trees delayed drop. Shortly after this discovery the use of these materials was put into commercial practice. Today, NAA trade names Fruitone-N, Fruit Fix 860, Fruit Fix 200, Fruit Set, Stafast, Kling-Tite, and others are commercially used for the control of apple and pear preharvest drop. NAA is applied as a foliar spray at the first sign of sound mature fruit drop (*Plant Growth Regulator Handbook* 1981). 2-(2,4,5-trichlorophenoxy) propanoic acid, common name silvex or fenoprop and trade names Fruitone T and Fruitone T Double strength, is applied as a foliar spray two weeks prior to harvest for most varieties of apples except McIntosh. For McIntosh apples to be marketed immediately after harvest they should be sprayed seven to eight days before harvest drop normally begins and for apples to be stored applications should be made four to five days before harvest drop normally begins.

Daminozide trade name Alar has been shown to be effective in reducing harvest drop in apples when applied as a preharvest spray. It has also been reported to have a number of other beneficial effects such as reducing water core, storage scald, maintaining fruit firmness, and increasing fruit color. Timing and concentration have been shown to be very important in order to achieve the beneficial effects. Adverse effects caused by improper timing and concentration are reduced fruit size, flattening of fruits by decreasing the size of the calyx lobes, and delayed harvest by two to four weeks (Weaver 1972). At the present time Alar has been removed from the market due to potential health risks; therefore, its future use remains in question.

Induction of Fruit Abscission

As shown in the previous section there are established procedures for the prevention of fruit drop. However, in some cases these treatments cause fruits to be retained by the plant so tightly that their removal becomes very difficult, thereby making harvesting extremely difficult. Shaking trees, vines, and other plants where fruit is held too tightly can cause a reduction in yield because more fruit is left on the plant. In addition, bruising, crushing of fruit, excessive breakage of limbs, and other plant parts is a problem.

Ethylene-releasing compounds (Figure 10.8) have been shown to be very effective in the promotion of abscission in cherries, apples, blackberries, cantaloupes, walnuts, macadamia nuts, and tangerines, and in cotton speeds fruit dehiscence, which also allows for an earlier more concentrated harvest. Undesirable side effects from ethylene-releasing compounds such as Ethrel include leaf abscission, gummosis, reduced yields in successive years, and declining tree vigor. In order to avoid these problems timing and concentration is extremely important. A study by Bukovac et al. (1969) with tart cherries as an example illustrates the benefits and also potential problems with ethylene-releasing compounds to promote fruit abscission. In this study ethephon was applied at four concentrations and a control receiving no treatment was used, fruit removal force and other factors were then evaluated. As shown in Figure 10.9, 500 and 1,000 ppm ethephon promoted a dramatic reduction in the force required to remove tart cherry fruit with minimal phytotoxic effects, thereby making this method commercially acceptable. Higher concentrations of ethephon, such as 2,000 and 4,000 ppm were shown to cause defoliation and excessive gummosis, promotion of a gummy exudate from the stem Figure 10.10. Therefore, even though higher concentrations of ethephon reduced the fruit removal force they were not commercially acceptable because of damage to the tree. In all plants where ethylene-releasing compounds are effective, the timing and concentration is extremely important in avoiding potential problems.

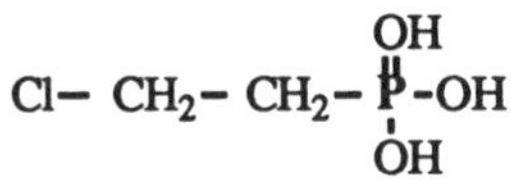

$$Cl - CH_2 - CH_2 - \overset{\overset{\displaystyle OH}{\|}}{\underset{\underset{\displaystyle OH}{|}}{P}} - OH$$

Ethephon
(2-chloroethyl)phophoric acid

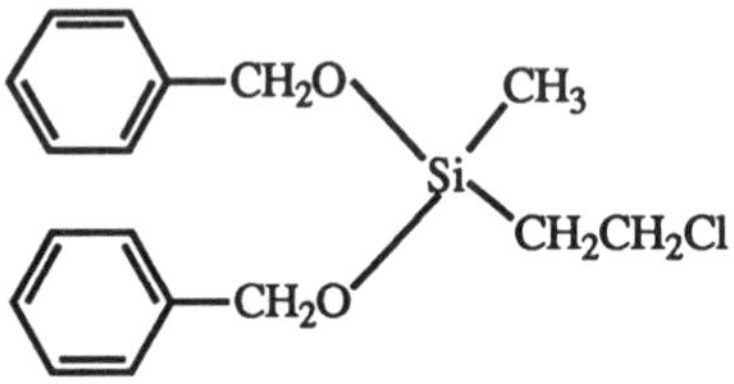

Silaid
(2-chloroethyl)methylbis(phenylmethoxy)silane

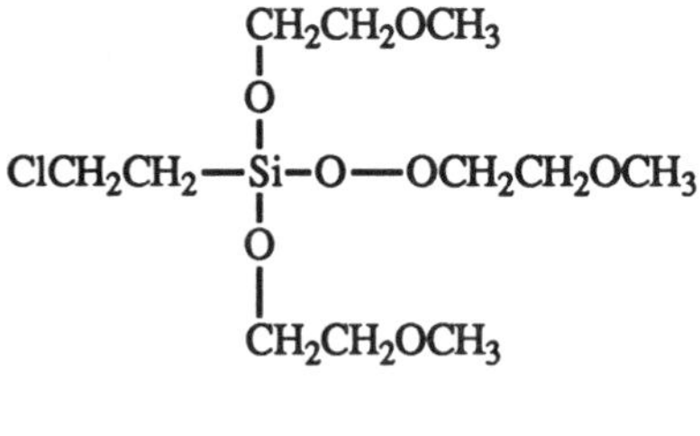

$$ClCH_2CH_2 - \overset{\overset{\displaystyle CH_2CH_2OCH_3}{\overset{|}{O}{|}}}{\underset{\underset{\displaystyle CH_2CH_2OCH_3}{\underset{|}{O}{|}}}{Si}} - O - OCH_2CH_2OCH_3$$

Alsol
(2-chloroethyl)tris(2-methoxyethoxy)silane

Figure 10.8. Structural formulas for the commercially used ethylene-releasing compounds: ethephon (2-chloroethyl)phosphonic acid), silaid (2-chloroethyl)methylbis-(phenylmethoxy)silane), and alsol (2-chloroethyl)tris(2-methoxyethoxy)silane).

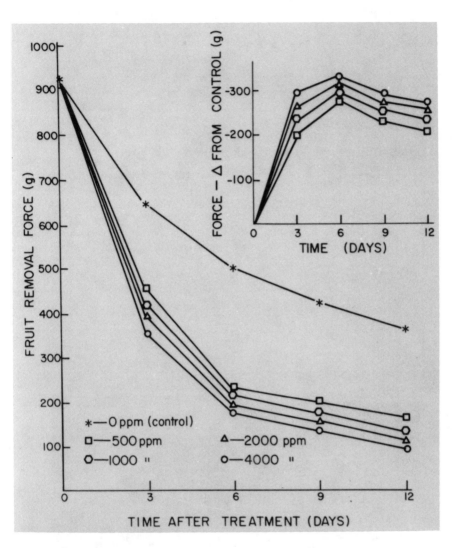

Figure 10.9.   Effects of various concentrations of ethephon on fruit removal force in tart cherries (Bukovac et al. 1969).

Figure 10.10.   Effect of ethephon on tart cherries top, dieback which occurs center, gummosis on the current season's wood and bottom, gummosis on old wood (Bukovac et al. 1969).

REFERENCES

Abeles, F. B., Morgan, P. W., and Saltveit Jr., M. E. (1992). *Ethylene in Plant Biology. Second Edition*, Academic Press, Inc., San Diego, C.A.

Auchter, E. C. and Roberts, J. W. (1934). "Experiments in spraying apples for the prevention of fruit set". *Proc. Amer. Soc. Hort. Sci.* 30:22–25.

Batjer, L. P. (1964). "Apple thinning with chemical sprays". *WA Agr. Exptl. Sta. Bull.* 1:651.

Batjer, L. P. (1965). "Fruit thinning with chemicals". *USDA Res. Ser. Ag. Information Bull.* 289.

Batjer, L. P. and Thompson, B. J. (1961). "Effects of 1-naphthyl *N*-methylcarbamate (Sevin) on thinning apples". *Proc. Amer. Soc. Hort. Sci.* 77:1–8.

Bukovac, M. J. and Nakagawa, S. (1967). "Comparative potency of gibberellins in inducing parthenocarpic fruit growth in *Malus sylvestris* Mill". *Experientia* 23:865.

Bukovac, M. J. and Nakagawa, S. (1968). "Gibberellin-induced asymmetric growth of apple fruits". *HortScience* 3:172–173.

Bukovac, M. J., Zucconi, F., Larsen, R. P. and Kesner, C. D. (1969). "Chemical promotion of fruit abscission in cherries and plums with special reference to 2-chloroethylphosphonic acid". *J. Amer. Soc. Hort. Sci.* 94:226–230.

Christodoulou, A., Weaver, R. J. and Pool, R. M. (1968). "Relation of gibberellin treatment to fruit-set, berry development, and cluster compactness in *Vitis vinifera* grapes". *Proc. Amer. Soc. Hort. Sci.* 92:301–310.

Coggins, Jr., C. W. , Hield, H. Z., Burns, R. M., Eaks, I. L. and Lewis, L. N. (1966). "Gibberellin research with citrus". *Calif. Agr.* 20:12–13.

Crane, J. C. (1965). "The chemical induction of parthenocarpy in the *Calimyrna* fig and its physiological significance". *Plant Physiol.* 40:606–610.

Crane, J. C., Primer, P. E. and Campbell, R. C. (1960). "Gibberellin-induced parthenocarpy in *Prunus*". *Proc. Amer. Soc. Hort. Sci.* 75:129–137.

Crane, J. C., Rebeiz, C. A. and Campbell, R. C. (1961). "Gibberellin-induced parthenocarpy in the J. H. Hale peach and the probably cause of Button production". *Proc. Amer. Soc. Hort. Sci.* 78:111–118.

Dennis, Jr. F. G. , and Nitsch, J. P. (1966). "Identification of gibberellin A_4 and A_7 in immature apple seeds". *Nature* 211:781–782.

Dreher, T. W. and Poovaiah, B. W. (1982). "Changes in auxin content during development in strawberry fruits". *J. Plant Growth Reg.* 1:267–276.

Gardner, F. E., Marth, P. C., and Batjer, L. P. (1939). "Spraying with plant-growth substances for control of the pre-harvest drop of apples". *Proc. Amer. Soc. Hort. Sci.* 37:415–428.

Gray, J., Pictor, S., Shabbeer, J., Schuch, W. and Grierson, D. (1992). "Molecular biology of fruit ripening and its manipulation with antisense genes". *Plant Mol. Biol.* 19:69–87.

Guo, L., Arteca, R. N., Phillips, A. T., and Liu, Y. (1992). "Purification and characterization of 1-aminocyclopropane-1-carboxylate N-malonyltransferase from etiolated mung bean hypocotyls". *Plant Physiol.* 100:2041–2045.

Guo, L. G., Phillips, A. T., and Arteca, R. N. (1993). "Amino acid N-malonyltransferases in mung beans: Action on 1-aminocyclopropane-1-carboxylic acid and D-phenylalanine". *J. Biol. Chem.* 268:25,389–25,394.

Hamilton, A. J., Lycett, G. W., and Grierson, D. (1990). "Antisense gene that inhibits synthesis of the hormone ethylene in transgenic plants". *Nature* 346:284–287.

Iwahori, S., Weaver, R. J. and Pool, R. M. (1968). "Gibberellin-like activity in berries of seeded and seedless Tokay grapes". *Plant Physiol.* 43:333–337.

Jackson, D. I. (1968). "Gibberellin and the growth of peach and apricot fruits". *Australian J. Biol. Sci.* 21:209–215.

Jackson, D. I. and Coombe, B. G. (1966). "Gibberellin-like substances in the developing apricot fruit". *Science* 154:277–278.

Jones, C. M. (1965). "Effects of benzyladenine on fruit set in muskmelon". *Proc. Amer. Soc. Hort. Sci.* 87:335–340.

Kaminek, M., Mok, D. W. S., and Zazimalova, E. (1992). *Physiology and Biochemistry of Cytokinins in Plants*, SPB Academic Publishing, Hague, The Netherlands.

Klee, H. J., Hayford, M. B., Kretzmer, K. A., Barry, G. F. and Kishmore, G. M. (1991). "Control of ethylene synthesis by expression of a bacterial enzyme in transgenic tomato plants". *Plant Cell* 3:1187–1193.

Letham, D. S. (1967). "Chemistry and physiology of kinetin-like compounds". *Ann. Rev. Plant Physiol.* 18:349–364.

Letham, D. S. (1968). "A new cytokinin bioassay and the naturally occurring cytokinin complex. In *Biochemistry and Physiology of Plant Growth Substances*, eds., F. Wightman and G. Setterfield. Runge Press, Ottawa, pp. 19–31.

Luckwill, L. C. (1959). "Fruit growth in relation to internal and external chemical stimuli. In *Cell, Organism, and Milieu*, ed., D. Rudnick. Ronald Press, New York, pp. 223–251.

Luckwill, L. C. (1957). "Hormonal aspects of fruit development in higher plants." In *The Biological Action of Growth Substances*, ed., H. K. Porter. Cambridge University Press, Cambridge, pp. 63–85.

Luckwill, L. C., Weaver, P. and MacMillan, J. (1969). "Gibberellins and other growth hormones in apple seeds". *J. Hort. Sci.* 44:413–424.

McKee, M. W. and Forshey, C. G. (1966). "Effects of chemical thinning on repeat bloom of McIntosh apple trees". *Proc. Amer. Soc. Hort. Sci.* 88:25–32.

Mudge, K. W., Narayanan, K. R. and Poovaiah, B. W. (1981). "Control of strawberry fruit set and development with auxins". *J. Amer. Soc. Hort. Sci.* 106:80–84.

Müller-Thurgau, H. (1898). "Abhängigkeit der ausbildung der traubenbeeren und einiger anderer Früchte von der entwicklung der samen." *Landw. Jahrb. Schweiz* 12:135–205.

Nakagawa, S., Bukovac, M. J., Hirata, N. and Kurooka, H. (1968). "Morphological studies of gibberellin-induced parthenocarpic and asymetric growth in apple and Japanese pear fruits". *J. Jap. Soc. Hort. Sci.* 37:9–19.

Nitsch, J. P. (1965). Physiology of flower and fruit development. In *Encyclopedia of plant physiology*, ed., W. Ruhland. Springer-Verlag, Berlin, pp. 1537–1647.

Nitsch, J. P. (1952). "Plant hormones in the development of fruits". *Quarterly Rev. Biol.* 27:33–57.

Nitsch, J. P. (1950). "Growth and morphogenesis of the strawberry as related to auxin". *Amer. J. Botany* 37:211–215.

Oeller, P. W., Min-Wong, L., Taylor, L. P., Pike, D. A. and Theologis, A. (1991). "Reversible inhibition of tomato fruit senescence by antisense RNA". *Science* 254:437–439.

Olson, D. C., White, J. A., Edelman, L., Harkins, R. N. and Kende, H. (1991). "Differential expression of two genes for 1-aminocyclopropane-1-carboxylate synthase in tomato fruits". *Proc. Natl. Acad. Sci. USA* 88:5340–5344.

Plant growth regulator handbook, (1981). Plant Growth Regulators Society of America, Lake Alfred, FL.

Reddy, A. S. N. and Poovaiah, B. W. (1990). "Molecular cloning and sequencing of a cDNA for an auxin-repressed mRNA: correlation between fruit growth and repression of the auxin-regulated gene". *Plant Mol. Biol.* 14:127–136.

Rottmann, W. E., Peter, G. F., Oeller, P. W., Keller, J. A., Shen, N. F., Nagy, B. P., Taylor, L. P., Campbell, A. D. and Theologis, A. (1991). "1-Aminocyclopropane-1-carboxylate synthase in tomato is encoded by a multigene family whose transcription is induced during fruit and floral senescence". *J. Mol. Biol.* 222:937–961.

Sastry, K. K. S. and Muir, R. M. (1963). "Gibberellin: effect of diffusible auxin in fruit development". *Science* 140:494–495.

Sheehy, R. E., Kramer, M., and Hiatt, W. R. (1988). "Reduction of polygalacturonase activity in tomato fruit by antisense RNA". *Proc. Natl. Acad. Sci. USA* 85:8805–8809.

Smith, C. J. S., Watson, C. F., Ray, J., Bird, C. R., Morris, P. C., Schuch, W., and Grierson, D. (1988). "Antisense RNA inhibition of polygalacturonase gene expression in transgenic tomatoes". *Nature* 334:724–726.

Southwick, S. M. and Poovaiah, B. W. (1987). "Auxin movement in strawberry fruit corresponds to its growth-promoting activity". *J. Am. Soc. Hort. Sci.* 112:139–142.

Southwick, F. W., Weeks, W. D., and Olanyk, G. W. (1964). "The effect of naphthaleneacetic acid type materials and 1-naphthyl *N*-methylcarbamate (Sevin) on the fruiting, flowering, and keeping quality of apples". *Proc. Amer. Soc. Hort. Sci.* 84:14–24.

Theologis, A. (1992). "One rotten apple spoils the whole bushel: The role of ethylene in fruit ripening". *Cell* 70:181–184.

Unrath, C. R. (1974). "The commercial implication of gibberellin A_4A_7 plus benzyladenine for improving shape and yield of Delicious apples". *J. Amer. Soc. Hort. Sci.* 99:381–384.

Weaver, R. J. (1972). *Plant Growth Substances in Agriculture*, W. H. Freeman and Company, San Francisco.

Weaver, R. J. and McCune, S. B. (1960). "Further studies with gibberellin on *Vitis vinifera* grapes". *Bot. Gaz.* 121:151–162.

Weaver, R. J. and van Overbeek, J. (1963). "Kinins stimulate grape growth". *Calif. Agr.* 17:12.

Weaver, R. J., van Overbeek, J., and Pool, R. M. (1966). "Effect of kinins on fruit set and development in *Vitis vinifera*". *Hilgardia* 37:181–201.

Westwood, M. N. and Bjornstad, H. O. (1968). "Effects of gibberellin A_3 on fruit shape and subsequent seed dormancy". *HortScience* 3:19–20.

Williams, W. M. and Stahly, E. A. (1969). "Effect of cytokinins and gibberellins on shape of Delicious apple fruits". *J. Amer. Soc. Hort. Sci.* 94:17–19.

Zuluaga, E. M., Lumelli, J., and Christensen, J. H. (1968). "Influence of growth regulators on the characteristics of berries of *Vitis vinifera* L.". *Phyton* 25:35–48.

CHAPTER 11

Tuberization

The initiation and subsequent growth of tubers in potato plants results from a series of biochemical and morphological changes which occur above and below the ground. Early investigators attributed the process of tuber formation to the existence of surplus carbohydrates. Kraus and Kraybill (1918) established the nutritional theory. They studied changes in the C/N ratio in relationship to the growth of tomato plants. The first person to introduce a form of the nutritional theory as it applies to tuberization was Wellinsiek (1929). He suggested that the causal factor of tuberization was the concentration of metabolites from photosynthesis, especially the carbohydrate-to-nitrogen ratio. Others have supported this theory (Werner 1934; Milthorpe 1963). Under unfavorable conditions such as high temperature and low light intensity large amounts of assimilates are used for shoot and root growth, thereby inhibiting tuberization. If tubers are already present during unfavorable conditions they can be totally reabsorbed if these conditions persist. More recently, a considerable amount of attention has been directed toward the effects of plant growth substances on tuberization, now called the hormonal theory. The first report suggesting the existence of a specific factor produced under favorable photoperiods was by Driver and Hawkes (1943). This factor was proposed to be transmitted to the stolons where it induced tuberization. It was not until Gregory's classical grafting experiments (Gregory 1954, 1956) that experimental evidence was presented supporting the hormonal theory of tuberization. Based on grafting experiments he postulated the existence of a specific tuber-forming substance which was produced under certain conditions of photoperiod and temperature and transmissible through graft unions and was not a major metabolite, such as a carbohydrate. Since this time there have been a number of other studies supporting this theory (Chapman 1958; Kumar and Wareing 1973, 1974; Ewing 1987).

Prior to discussing the biochemical and physiological changes which occur during the tuberization process, some background information will be given. The potato tuber is morphologically a modified stem with nodes and internodes. Potato tubers are initiated on underground rhizomes commonly called stolons which may be defined as lateral shoots. The tuber originates just below the stolon tip by the process of tuber initiation and swelling occurs acropetally in internodes which were preexisting at the time of tuber initiation (Cutter 1978). The series of events which take place during tuber initiation and subsequent growth of the tuber are:

1. Elongation of the stolon stops and there is a change in polarity of growth, during the early stages of tuber initiation radial cell elongation occurs followed by cell division (Koda and Okazawa 1983a, 1983b).

2. The first biochemical sign of tuberization is starch deposition which is accompanied by the development of high levels of patatin, a glycoprotein found in tubers (Paiva et al. 1983).

3. During growth and development of the tuber both cell division and enlargement occur until maturity, once the tuber is mature there is a dramatic decrease in cell division, while cell enlargement continues.

In whole plants, tubers form underground, however, when leaf/bud cuttings are taken tubers have been shown to originate from axillary buds on aboveground portions of the plant (Figure 11.1). These tubers originate from axillary buds rather than stolons and are typically smaller and contain chlorophyll. Leaf/bud cuttings taken from tuberizing plants usually develop tubers if the buds are buried in the soil. This method has been a useful tool to study the effects of nutrition, environment, and plant growth substances on the tuberization process (Ewing 1985). Tuberization is a complex phenomenon and is influenced by a myriad of variables such as nutrition, environment, mother tuber, genetics, leaf area, partitioning of assimilates, and others. In order to reduce these variables, scientists have developed in vitro systems allowing for

Figure 11.1.   Tuberization of axillary buds of single-node leaf cuttings. (a) Leaf cutting treated with water for seven days; (b) effects of pretreating leaf cuttings for 15 hours with water (1, 2) or 5 mM EGTA (3, 4) and grown for seven days in water (1, 3) or in 1 mM $CaCl_2$ + 1 mM $MgCl_2$ (2, 4); (c) effect of pretreating with 5 mM EGTA + 50 mM Ca ionophore A23187 for 15 hours and grown for seven days in $CaCl_2$ + 1 mM $MgCl_2$ (1) or in water (2) (from Balamani et al. (1986)).

Figure 11.2.   Tuberization in nodal sections taken from etiolated sprouts (*Solanum tuberosum L.* var. Russett Burbank) and cultured in vitro.

the greatest control of growing conditions (Figure 11.2). The effects of plant growth substances on tuberization in whole plants, leaf/bud cuttings, and in vitro will be discussed in subsequent sections.

CONTROL OF TUBERIZATION BY PLANT GROWTH SUBSTANCES

According to the hormonal theory of tuberization there are two factors which affect tuberization in a sequential fashion: environmental factors and plant growth substances. Environmental factors, especially photoperiod and temperature, play a major role in the initiation of tubers (Jolivet 1969; Seabrook et al. 1993). Although there are a number of factors which influence tuberization, all of them suggest that the process is under the control of plant growth substances. Since the pioneering work of Gregory (1954, 1956) and others (Chapman 1958; Kumar and Wareing 1973, 1974) there has been a considerable amount of work supporting the hormonal theory using both naturally occurring and synthetic plant growth regulators. The experimental evidence can be broken down into two categories. First, studies on the correlation between endogenous levels of plant growth substances and tuberization, and second, the effects of exogenous applications of plant growth substances on tuberization in whole plant, cuttings and in vitro.

ENDOGENOUS LEVELS OF PLANT GROWTH SUBSTANCES

Gibberellins

Potato tubers have been shown to contain gibberellinlike substances (Okazawa 1959). Gibberellin levels vary in different organs and change during plant growth and development. In general gibberellin activity is highest during the phase of rapid growth and lowest in mature dormant tubers (Bialek 1974). High nitrogen levels which promote rapid vegetative growth and high levels of gibberellins in the shoots inhibit tuberization, whereas a reduction in nitrogen levels favors tuberization (Krauss and Marschner 1982; McGrady and Ewing 1990; Zrust and Mica 1992). Most studies reported thus far indicate that gibberellins have a promotive effect on stolon elongation, however, they are inhibitory to the tuberization process (Ewing 1987).

Cytokinins

There have been several reports showing that there are elevated levels of cytokinins in response to conditions which promote tuberization (Forsline and

Langille 1975; Okazawa 1970; Langille and Forsline 1974; Mauk and Langille 1978; Arteca et al. 1980). The major cytokinin found in potato leaves is zeatin riboside (Mauk and Langille 1978; Arteca et al. 1980), although, zeatin has also been found in potato tubers (Okazawa 1970), leaves and roots (Arteca et al. 1980). It still remains unclear as to the direct involvement of cytokinins in the tuberization process since at the present time there are only a limited number of reports on changes in their endogenous levels during tuberization.

Inhibitors

There have been several studies on the role of endogenous inhibitors on tuberization, one of which was identified as abscisic acid (ABA), while the others remain unidentified (Kumar and Wareing 1974; Smith and Rappaport 1969). The involvement of ABA in the induction of tuberization remains unclear since its levels were shown to increase under inductive conditions (Krauss and Marschner 1982; Arteca et al. 1980); however, there are also reports where there is not a good correlation between endogenous levels of ABA and tuberization (Kumar and Wareing 1974; Wareing and Jennings 1980).

Other Plant Growth Substances

In dahlia it has been shown that there are changes in endogenous levels of ethylene which correlate with tuberization (Biran et al. 1972); however, this has not been reported in potato. Indole-3-acetic acid (IAA) has been shown to increase in the leaves of potato plants when the roots were exposed to high levels of CO_2 in order to induce the tuberization process (Arteca et al. 1980), and it has since been suggested that IAA may play an important role in the tuberization process (Melis and van Staden 1984). Changes in polyamine biosynthesis during the initial stages of tuberization have been shown to occur. The polyamines spermidine and spermine along with their biosynthetic enzymes were all shown to increase during the initial stages of potato tuberization and then decreased as the tuber increased in size. In addition, the levels of putrescine were shown to fall continuously in the stolon tip and in tubers as tuberization proceeded. In other organs such as root, leaf, and stems the levels of polyamines and their related enzymes were much lower (Taylor et al. 1993). At the present time there are a limited number of reports on changes in plant growth substances during the tuberization process. As mentioned in earlier chapters of this book, analytical techniques for plant growth substances have greatly improved in recent years. Therefore, future research on changes in plant growth substances during tuberization using this new methodology may shed some light on their involvement in this important process. In addition to known plant growth substances the search for a specific tuberizing substance continues in the same way as florigen.

EXOGENOUS APPLICATIONS OF PLANT GROWTH SUBSTANCES

At the present time studies on changes in endogenous levels of plant growth substances have not led to a clear understanding of their involvement in tuberization. There have been numerous studies evaluating the effects of exogenous applications of plant growth substances in whole plants, cuttings, and in vitro which has provided us with much of our knowledge on their involvement in the tuberization process.

Gibberellins

There are many studies on the effects of exogenous applications of gibberellins on tuberization and in all cases they have an inhibitory effect. Foliar applications of GA_3 have been shown to have a strong inhibitory effect on tuberization in whole plants (Okazawa 1960; Lovell and Booth 1967). In addition, GA_1, GA_3, GA_4, GA_7, or GA_9 applied as a foliar spray to cuttings also inhibits tuberization (Tizio 1971; Kumar and Wareing 1974). When GA_3 was added to the culture medium of stem sections cultured in vitro, it was shown to have either an inhibitory effect (Harmey et al. 1966; Koda and Okazawa 1983b) or a delay in tuberization by promoting the elongation of stolons (Figure 11.3) (Tizio 1964). More recently it has been shown that there is an interaction between glucose and gibberellins on tuberization of potato plantlets cultured in vitro (Martinez and Tizio 1991); however, further studies are required to better elucidate their relationship.

Early studies showed that the growth-retardant 2-chloroethyl-trimethylammonium chloride (CCC) has the ability to either accelerate (Dutta and Kaley 1968; Dyson 1965; Dyson and Humphries 1966) or promote (Gifford and Moorby 1967) tuberization. Since this time a number of other growth retardants have also been shown to be effective in the stimulation of tuberization both in vivo and in vitro (Levy et al. 1993; Simko 1993; Simko 1991; Harvey et al. 1991). The beneficial effect of plant growth retardants on tuberization may be due to their general depression of vegetative growth or due to a reduction in gibberellin levels; however, more research is necessary in order to better understand the mechanism by which plant growth retardants effect tuberization.

Cytokinins

In whole plant and cutting studies the effects of exogenous applications of cytokinins on tuber initiation has been negative (Ewing 1987). In vitro studies with cytokinins have shown that they induce tuberization (Figure 11.4) (Palmer and Smith 1969,1970; Smith and Palmer 1970; Pelacho and Mingo-Castel

Figure 11.3.   Effects of gibberellic acid on elongation of stolons from nodal sections taken from etiolated sprouts (*Solanum tuberosum L.* var. Russett Burbank) and cultured in vitro.

Figure 11.4.   Kinetin-induced tuberization in nodal sections taken from etiolated sprouts (*Solanum tuberosum* var. Norland) and cultured in vitro.

1991). However, as is always the case when using plant growth substances there have been reports showing that kinetin has no effect on tuberization in vitro (Simko 1993).

Salicylates and Jasmonates

Exogenous applications of salicylic acid have been shown to stimulate tuberization of potato plants both in vivo and in vitro. A comparison of compounds related to salicylic acid were evaluated for their ability to induce tuberization. It was found that benzoic acids with a free carboxyl group and a substituent at the C-2 position of the benzene ring have the ability to induce tuberization, with salicylic acid having the highest activity of all compounds tested. The ability of salicylic acid to stimulate tuberization in potato is much less then jasmonic acid (1R,2S-jasmonic acid) (Koda et al. 1992a). In a subsequent study salicylic acid was not found in the leaves of potato plants grown under tuber-inducing conditions. This suggests that salicylic acid is not responsible for the induction of tuberization under natural conditions (Koda et al. 1992a); whereas, jasmonic acid has been identified in Jerusalem artichoke (Matsuura et al. 1993) and yam plants (Koda and Kikuta 1991) and exogenous applications of it have also been shown to stimulate tuberization in these plants (Matsuura et al. 1993; Koda and Kikuta 1991; Koda et al. 1992a, 1992b; Pelacho and Mingo-Castel 1991).

Other Plant Growth Regulating Substances

El-Antably et al. (1967) reported an increase in tuberization by foliar application of abscisic acid on potato plants grown under long days. This stimulation was probably due to a general depression of overall growth rather than a direct effect on tuberization. Abscisic acid has since been shown to be ineffective in the promotion of tuberization in cuttings, shoot sections, and sprouts (Ewing 1987; Claver 1970; Palmer and Smith 1969; Smith and Rappaport 1969; Tizio and Maneschi 1973). Catchpole and Hillman (1969) claimed that ethylene promoted an increase in tuberization of stolons on potato sprouts. Subsequent studies have shown that ethylene either had no effect on tuberization (Singh 1970; Palmer and Barker 1973) or decreased it (Langille 1972). There have been a limited number of studies on the effects of auxins in the initiation of tubers. These studies show that auxin may accelerate tuber formation (Harmey et al. 1966; Lawrence and Barker 1963); however, conflicting reports by Booth (1963) and Okazawa (1967) indicate that auxin is probably involved in stolon growth rather than tuber initiation.

It has been shown that when the root systems of potato plants are treated with high levels of CO_2 for a 12-hour period there is an increase in dry matter content as early as two days after treatment. When the treated plants were

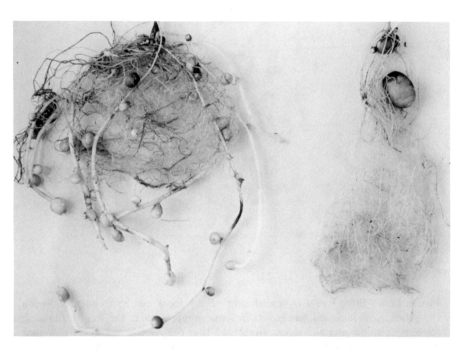

Figure 11.5.   Effects of short-term CO_2 enrichment of the root zone on stolon elongation, tuber initiation, and growth in *Solanum tuberosum L.* var. Russett Burbank. Photograph was taken three weeks following CO_2 enrichment. Control is on the right and CO_2-treated sample is on the left (from Arteca et al. (1979)).

allowed to grow for a period of three to six weeks there was a large increase in tuberization. CO_2 treated plants exhibited an increase in stolon length, number of tubers per stolon and overall dry weight (Figures 11.5 and 11.6). This work is in agreement with the findings of Patterson (1975) who showed that a single 12-hour treatment with CO_2 to the roots of potato plants grown in sand culture resulted in significantly more tubers than controls. In a subsequent study it was shown that CO_2 applied to the root zone of potato plants modifies endogenous plant growth substances (trans-zeatin, zeatin riboside, IAA and ABA) (Arteca et al. 1980), which could have possibly acted as a trigger resulting in increases in dry matter content and tuberization (Arteca et al. 1979). Mingo-Castel et al. (1974 1976) have reported that CO_2 stimulated tuberization of potato stolons cultured in vitro and that the stimulatory effect was inhibited by ethylene which was by itself also inhibitory. In vitro studies have indicated that polyamines have a role in the tuber formation process. In a study where high levels of sucrose together with kinetin and CCC were used to induce tuberization, specific inhibitors of polyamine biosynthesis were evalu-

Figure 11.6. Effects of CO_2 enrichment of the shoot and root zones on stolon elongation, tuber initiation, and growth in *Solanum tuberosum L.* var. Russett Burbank; ambient air to the root and shoots (right), 12 hour CO_2 enrichment to the root zone (middle) and continuous CO_2 enrichment of the shoots during the light period for six weeks (left). Photograph was taken six weeks following CO_2 enrichment (from Arteca et al. (1979)).

ated for their ability to reduce tuberization (Protacio and Flores 1992). It was found that polyamines derived via the ornithine decarboxylase-mediated pathway are necessary for tuber formation in vitro, probably at an early phase of morphogenesis involving cell division. Coumarins have been shown to induce tuberization in potato stolons cultured in vitro (Stallknecht 1972a, 1972b 1973). The mode of action of coumarin-induced tuberization appears to be different than kinetin-induced tuberization because high levels of nitrogen are inhibitory to coumarin-induced tuberization but not to kinetin-induced tuberization. In addition, Alar, CCC, ethrel, and IAA stimulate coumarin-induced tuberization (Stallknecht 1972b), while ABA and GA inhibit. Other factors which have been shown to stimulate tuberization are high magnesium in the nutrient solution (Coutrez-Geernick 1969), vitamin C (van Schreven 1956), magnetism (Pittman 1972), methyl-b-D-glucopuranosyl tubernate, methyl-b-D-glucopuranosyl helianthenate A and B (Matsuura et al. 1993), and calcium (Balamani et al. 1986).

MOLECULAR STUDIES
ON THE TUBERIZATION PROCESS

In recent years, of molecular techniques have been used to better understand the tuberization process at the molecular level. Physiological studies have shown that cytokinins are involved in the tuberization process; therefore, early molecular studies utilized cytokinin genes to evaluate their effects on the tuberization process. These studies showed that there was precocious tuberization and enhanced cytokinin levels in *Agrobacterium* transformed potatoes (Ooms and Lenton 1985). Since this time the involvement of plant growth substances, with the exception of gibberellins, has remained unclear. A molecular approach using differential screening has been used to identify several tuberization-related genes (Jackson et al. 1993; Taylor et al. 1992a, 1992b, 1992c; Bevan et al. 1988). A partial cDNA clone has been isolated which represents a gene that is induced in potato (*Solanum tuberosum L.*) leaves during tuber initiation. This gene is also present in tobacco (*Nicotiana tobacum L.*), where it is induced upon flowering and in tomato (*Lycopersicon esculentum L.*), where it is induced during fruit formation. These studies provide further evidence that there is a similarity between the flower- and tuber-inducing pathways. However, it is also possible that it could be involved in a secondary process induced during tuberization, flowering, and fruit formation (Jackson et al. 1993). Other tuberization-related genes have also been identified in potato by differential screening (Taylor et al. 1992b; Bevan et al. 1988). Two of these cDNA clones were shown to be increased 25–30- fold during the early stages of tuberization. These genes were also shown to be expressed in leaves, stems, and roots with their expression pattern changing in the same way as tuberization (Taylor et al. 1992a, 1992c). The previously mentioned work provides the foundation for further study in the area of tuberization which may lead to breakthroughs in better understanding this process.

PLANT GROWTH SUBSTANCE INVOLVEMENT
IN THE TUBERIZATION PROCESS

When summarizing the literature on any given subject we typically try to find a general explanation of the mechanism of action for a specific physiological process. In spite of many years of research in the area of tuberization, we are still in a serious state of ignorance. Although we have learned about many factors influencing tuberization, they are just parts in a huge puzzle which have to be fit together. Environmental factors such as temperature and photoperiod which trigger changes in endogenous levels in plant growth substances are undoubtably linked in the tuberization process. Among the eight classes of plant growth substances discussed in this book, gibberellins clearly have been shown to be involved in the inhibition of the tuberization process. Evidence supporting the inhibitory effect of gibberellins is:

1. Environmental conditions which promote tuberization cause a reduction in endogenous gibberellin levels.
2. Exogenous applications of gibberellins inhibit or delay tuberization.
3. Growth retardants have been shown to promote tuberization, probably by decreasing endogenous levels of gibberellins.

Other plant growth substances have been shown to promote, inhibit, or have no effect on tuberization, and at the present time a clear role on individual plant growth substances remains unclear. The effects of plant growth substances on tuberization are probably due to an interaction between those which are known and unknown. It is still possible that a single tuberizing plant growth substance exists; however, at the present time there is no strong support for its existence. Future studies using improved analytical techniques for evaluating changes in plant growth substances both known and unknown may shed some light on this process. In addition, the use of molecular technology to identify specific tuberization genes may provide researchers with the key to the mystery of tuberization.

REFERENCES

Arteca, R. N., Poovaiah, B. W., and Smith, O. E. (1979). "Changes in carbon fixation, tuberization and growth induced by CO_2 application to the root zone of potato plants". *Science* 205:1279–1280.

Arteca, R. N., Poovaiah, B. W., and Smith, O. E. (1980). "Use of high performance liquid chromatography for the determination of endogenous hormone levels in *Solanum tuberosum* L. subjected to CO_2 enrichment of the root zone". *Plant Physiol.* 65:1216–1219.

Balamani, V., Veluthambi, K., and Poovaiah, B. W. (1986). "Effect of calcium on tuberization in potato (*Solanum tuberosum* L.)". *Plant Physiol.* 80:856–858.

Bevan, M., Goldsborough, A., Jefferson, R., Sheerman, S., Iturriaga, G., and Atkinson, E. (1988). "Transcriptional regulation of genes during potato tuberization". *Heredity* 61:280.

Bialek, K. (1974). "Preliminary study of the activity of gibberellin-like substances in potato tubers". *Z. Pflanzenphysiol.* 71:370–372.

Biran, I., Gur, I., and Halevy, A. H. (1972). "The relationship between exogenous growth inhibitors and endogenous levels of ethylene and tuberization of dahlias". *Physiol. Plant.* 27:226–230.

Booth, A. (1963). "The role of growth substances in the development of stolons". In *The Growth of the Potato*, eds., J. D. Ivins and F. L. Milthorpe. Butterworths, London, pp. 99–113.

Catchpole, A. H. and Hillman, J. (1969). "Effect of ethylene on tuber initiation in *Solanum tuberosum* L". *Nature* 223:1387.

Chapman, H. W. (1958). "Tuberization in the potato plant". *Physiol. Plant.* 11:215–224.

Claver, F. K. (1970). "The effects of abscisic acid on tuberization of potato sprouts in vitro". *Phyton Rev. Int. Bot. Exp.* 27:25–29.

Coutrez-Geerinck, D. (1969). "Tuberization of potatoes under different culture conditions". *Ann. Physiol. Veg. Univ. Bruxelles* 3:63–84.

Cutter, E. G. (1978). "Structure and development of the potato plant". In *The Potato Crop*, ed., P. M. Harris. Halsted Press, John Wiley & Sons, New York, pp. 70–152.

Driver, C. M. and Hawkes, J. G. (1943). "Photoperiodism in the potato. Part I. General. Part II. The photoperiodic reactions of some South American potatoes". In *Tech. Comm. Imp. Bur. Plant Breeding and Genetics*, London School Agric., Cambridge, U.K.

Dutta, T. R. and Kaley, D. M. (1968). "Effects of 2-chloroethyltrimethyl ammonium chloride and gibberellic acid on potato growing during winter at Simla". *Ind. J. Agric. Sci.* 38:140–148.

Dyson, P. W. (1965). "Effects of gibberellic acid and (2-chloroethyl)-trimethylammonium chloride on potato growth and development". *J Sci. Food and Agric.* 16:542–549.

Dyson, P. W. and Humphries, E. C. (1966). "Modification of growth habit of Majestic potato by growth regulators applied at different times". *Applied Biol.* 58:171–182.

El-Antably, M. M. M., Wareing, P. F., and Hillman, J. (1967). "Some physiological responses to D,L abscisin (dormin)". *Planta* 73:74–90.

Ewing, E. E. (1985). "Cuttings as simplified models of the potato plant". In *Potato Physiology*, ed., P. H. Li. Academic Press, New York, pp. 153–207.

Ewing, E. E. (1987). "The role of hormones in potato (*Solanum tuberosum* L.) tuberization". In *Plant Hormones and Their Role in Plant Growth and Development*, ed., P. J. Davies. Martinus Nijhoff Publishers, Boston, pp. 515–538.

Forsline, P. L. and Langille, A. R. (1975). "Endogenous cytokinins in *Solanum tuberosum* as influenced by photoperiod and temperature". *Physiol. Plant.* 34:75–77.

Gifford, R. M. and Moorby, J. (1967). "The effect of CCC on the initiation of potato tubers". *Eur. Potato J.* 10:235–238.

Gregory, L. E. (1954). "Some factors controlling tuber formation in the potato plant". *Ph.D. Dissertation,* Univ. of Calif., Los Angeles.

Gregory, L. E. (1956). "Some factors for tuberization in the potato". *Ann. Bot.* 41:281–288.

Harmey, M. A., Crowley, M. P., and Clinch, P. E. M. (1966). "The effect of growth regulators on tuberization of cultured stem pieces of *Solanum tuberosum* L.". *Eur. Potato J.* 9:146–151.

Harvey, B. M. R., Crothers, S. H., Evans, N. E., and Selby, C. (1991). "The use of growth-retardants to improve microtuber formation by potato (*Solanum tuberosum*)". *Plant Cell, Tissue Organ Culture* 27:59–64.

Jackson, S. D., Sonnewald, U., and Willmitzer, L. (1993). "Characterization of a gene that is expressed in leaves at higher levels upon tuberization in potato and upon flowering in tobacco". *Planta* 189:593–596.

Jolivet, E. (1969). "Physiologie de la tuberisation". *Ann. Physiol. Vég.* 11:265–301.

Koda, Y. and Kikuta, Y. (1991). "Possible involvement of jasmonic acid in tuberization of yam plants". *Plant Cell Physiol.* 32:629–633.

Koda, Y., Kikuta, Y., Kitahara, T., Nishi, T. and Mori, K. (1992b). "Comparisons of various biological activities of stereoisomers of methyl jasmonate". *Phytochemistry* 31:1111–1114.

Koda, Y. and Okazawa, Y. (1983a). "Characteristic changes in the levels of endogenous plant hormones in relation to the onset of potato tuberization". *Japan. J. Crop Sci.* 52:592–597.

Koda, Y. and Okazawa, Y. (1983b). "Influences of environmental, hormonal, and nutritional factors on potato tuberization in vitro". *Japan. J. Crop Sci.* 52:582–591.

Koda, Y., Takahashi, K., and Kikuta, Y. (1992a). "Potato tuber-inducing activities of salicyclic acid and related compounds". *J. Plant Growth Regulation* 11:215–219.

Kraus, E. J. and Kraybill, H. R. (1918). "Vegetation and reproduction with special reference to the tomato". *Ore. Agr. Exptl. Sta. Bull.* 149.

Krauss, A. and Marschner, H. (1982). "Influence of nitrogen nutrition, daylength, and temperature on contents of gibberellic and abscisic acid and on tuberization in potato plants". *Potato Res.* 25:13–21.

Kumar, D. and Wareing, P. F. (1973). "Studies on tuberization in *Solanum andigene*. I. Evidence for the existence and movement of a specific tuberization stimulus". *New Phytol.* 72:283–287.

Kumar, D. and Wareing, P. F. (1974). "Studies on tuberization of *Solanum andigena*. II. Growth hormones and tuberization". *New Phytol.* 73:833–840.

Langille, A. R. (1972). "Effects of (2-chloroethyl) phosphonic acid on rhizome and tuber formation in the potato, *Solanum tuberosum*". *J. Amer. Soc. Hort. Sci.* 97:305–308.

Langille, A. R. and Forsline, P. L. (1974). "Influence of temperature and photoperiod on cytokinin pools in the potato *Solanum tuberosum* L.". *Plant Sci. Lett.* 2:189–191.

Lawrence, C. H. and Barker, W. G. (1963). "A study of tuberization in the potato *Solanum tuberosum*". *Am. Potato J.* 40:349–356.

Levy, D., Seabrook, J. E. A., and Coleman, S. (1993). "Enhancement of tuberization of axillary shoot buds of potato (*Solanum tuberosum* L.) cultivars cultured in vitro". *J. Exp. Botany* 44:381–386.

Lovell, P. and Booth, A. (1967). "Effects of gibberellic acid on growth, tuber formation and carbohydrate distribution in *Solanum tuberosum* L.". *New Phytol.* 66:525–537.

Martinez, L. and Tizio, R. (1991). "Interaction of glucose and gibberellins on tuberization of potato plantlets (*Solanum tuberosum* L.) cultivated in vitro". *Phyton-International J. Exp. Botany* 52:83–88.

Matsuura, H., Yoshihara, T., Ichihara, A., Kikuta, Y., and Koda, Y. (1993). "Tuber-forming substances in Jerusalem artichoke (*Helianthus tuberosus* L.)". *Bioscience Biotech. Biochem.* 57:1253–1256.

Mauk, C. S. and Langille, A. R. (1978). "Physiology of tuberization in *Solanum tuberosum* L.: *cis*-zeatin riboside in the potato plant — Its identification and changes in endogenous levels as influenced by temperature and photoperiod". *Plant Physiol.* 62:438–442.

McGrady, J. J. and Ewing, E. E. (1990). "Potato cuttings as models to study maturation and senescence". *Potato Research* 33:97–108.

Melis, R. J. M. and van Staden, J. (1984). "Tuberization and hormones". *Z. Pflanzenphysiol.* 133:271–283.

Milthorpe, F. L. (1963). "Some aspects of plant growth. An introductory survey". In *The Growth of the Potato*, eds., J. D. Ivins and F. L. Milthorpe. Butterworths, London, pp. 3.

Mingo-Castel, A. M., Negm, F. B., and Smith, O. E. (1974). "Effect of carbon dioxide and ethylene on tuberization of isolated potato stolons cultured in vitro". *Plant Physiol.* 53:789–801.

Mingo-Castel, A. M., Smith, O. E. and Kumamoto, J. (1976). "Studies on the carbon dioxide promotion and ethylene inhibition of tuberization in potato explants cultured in vitro". *Plant Physiol.* 57:480–485.

Okazawa, Y. (1959). "Studies on the occurrence of natural gibberellin and its effect on the tuber formation of potato plants". *Proc. Crop Sci. Soc. Japan* 28:129–133.

Okazawa, Y. (1960). "Studies on the relation between the tuber formation of potato and its natural gebberellin content". *Proc. Crop Sci. Soc. Japan* 29:121–124.

Okazawa, Y. (1967). "Physiological studies on the tuberization of potato plants". *J. Fac. Agr. Hokkaido Univ. Sapporo* 55:267–336.

Okazawa, Y. (1970). "Physiological significance of enodgenous cytokinin occurring in potato tubers during their developmental period". *Nippon Sakumotsu Gakkai Kiji* 39:171–176.

Ooms, G. and Lenton, J. R. (1985). "T-DNA Genes to study plant development. Precocious tuberization and enhanced cytokinins in *A. tumefaciens* transformed potato". *Plant Mol. Biol.* 5:205–212.

Paiva, E., Lister, R. M., and Park, W. D. (1983). "Induction and accumulation of major tuber proteins of potato in stems and petioles". *Plant Physiol.* 71:161–168.

Palmer, C. E. and Barker, W. G. (1973). "Influence of ethylene and kinetin on tuberization and enzyme activity in *Solanum tuberosum* L. stolons cultured in vitro". *Ann. Bot.* 37:85–93.

Palmer, C. E. and Smith, O. E. (1969). "Cytokinins and tuber initiation in the potato *Solanum tuberosum* L.". *Nature* 221:279–280.

Palmer, C. E. and Smith, O. E. (1970). "Effect of kinitin on tuber formation on isolated stolons of *Solanum tuberosum* L. cultured in vitro". *Plant Cell Physiol.* 11:303–314.

Patterson, D. R. (1975). "Effect of CO_2 enriched atmosphere on tuberization and growth of the potato". *J. Amer. Soc. Hort. Sci.* 100:431–434.

Pelacho, A. M. and Mingo-Castel, A. M. (1991). "Jasmonic acid induces tuberization of potato stolons cultured in vitro". *Plant Physiol.* 97:1253–1255.

Pittman, U. J. (1972). "Biomagnetic responses in potatoes". *Can. J. Plant Sci.* 52:727–733.

Protacio, C. M. and Flores, H. E. (1992). "The role of polyamines in potato tuber formation". *In Vitro Cellular and Dev. Biol.* 28:81–86.

Seabrook, J. E. A., Coleman, S., and Levy, D. (1993). "Effect of photoperiod on in vitro tuberization of potato (*Solanum tuberosum* L.)". *Plant Cell Tissue Organ Culture* 34:43–51.

Simko, I. (1991). "In vitro potato tuberization after the treatment with paclobutrazol". *Biologia* 46:251–256.

Simko, I. (1993). "Effects of kinetin, paclobutrazol, and their interactions on the microtuberization of potato stem segments cultured in vitro in the light". *Plant Growth Regulation* 12:23–27.

Singh, G. (1970). "Influence of ethrel on growth and yield of potatoes". *Res. Life Sci.* 18:38–43.

Smith, O. E. and Palmer, C. E. (1970). "Cytokinin-induced tuber formation on stolons of *Solanum tuberosum* L.". *Physiol. Plant.* 23:599–606.

Smith, O. E. and Rappaport, L. (1969). "Gibberellins, inhibitors, and tuber formation in the potato *Solanum tuberosum*". *Am. Potato. J.* 46:185–191.

Stallknecht, G. F. (1972a). "Coumarin-induced tuber formation on excised shoots of *Solanum tuberosum* L. cultivated in vitro". *Plant Physiol.* 50:412–413.

Stallknecht, G. F. (1972b). "The effect of growth regulating chemicals applied to Russet Burbank potatoes grown in the greenhouse". *Amer. Potato J.* 49:356.

Stallknecht, G. F. (1973). "Effect of inhibitors on coumarin-induced tuberization of axillary shoots of *Solanum tuberosum* L. grown in vitro". *Amer. Potato J.* 50:381.

Taylor, M. A., Arif, S. A. M., Kumar, A., Davies, H. V., Scobie, L. A., Pearce, S. R., and Flavell, A. J. (1992a). "Expression and sequence analysis of cDNAs induced during the early stages of tuberization in different organs of the potato plant (*Solanum tuberosum* L.)". *Plant Mol. Biol.* 20:641–651.

Taylor, M. A., Arif, S. A. M., Pearce, S. R., Davies, H. V., Kumar, A. and George, L. A. (1992b). "Differential expression and sequence analysis of ribosomal protein genes induced in stolon tips of potato (*Solanum tuberosum* L.) during the early stages of tuberization". *Plant Physiol.* 100:1171–1176.

Taylor, M. A., Burch, L. R., and Davies, H. V. (1993). "Changes in polyamine biosynthesis during the initial stages of tuberization in potato (*Solanum tuberosum* L.)". *J. Plant Physiol.* 141:370–372.

Taylor, M. A., Kumar, A., George, L. A., and Davies, H. V. (1992c). "Isolation and molecular characterization of a tuberization-related cDNA clone from potato (*Solanum tuberosum* L.)". *Plant Cell Reports* 11:623–626.

Tizio, R. (1964). "Action de l'acide gibbèrellique sur la tubèrisation de la Pomme de terre." C. R. Acad. Sci. Paris 259:1187–1190.

Tizio, R. (1971). "Action et rôle probable de certaines gibbérellines (A_1, A_3, A_4, A_5, A_7, et A_{13}) sur la croissance des stolons et la tubérisation de la Pomme de terre (*Solanum tuberosum* L.)". *Potato Res.* 14:132–204.

Tizio, R. and Maneschi, E. (1973). "Different mechanisms for tuber initiation and dormancy in the potato (*Solanum tuberosum* L.)". *Phyton Rev. Int. Bot. Exp.* 31:51–62.

van Schreven, D. A. (1956). "On the physiology of tuber formation in potatoes. IV. Influence of vitamin C on tuber and sprout formation of potatoes asceptically grown in the dark". *Plant Soil* 8:87–94.

Wareing, P. F. and Jennings, A. M. V. (1980). "The hormonal control of tuberization in potato". In *Plant Growth Substances*, ed., F. Skoog. Springer-Verlag, New York, pp. 293–300.

Wellensiek, S. J. (1929). "The physiology of tuber formation in *Solanum tuberosum* L.". *Meded. Landb. Hoogesch. Wagen.* 33:6.

Werner, H. O. (1934). "The effect of a controlled nitrogen supply with different temperatures and photoperiods upon the development of the potato plant". *Res. Bull. Neb. Agr. Exp. Stn.* No. 75.

Zrust, J. and Mica, B. (1992). "Stolon and tuber initiation and development in potatoes at various levels of nitrogen nutrition". *Rostlinna Vyroba* 38:1045–1052.

CHAPTER

12

Manipulation of Growth and Photosynthetic Processes by Plant Growth Regulators

The purpose of most plant-related research is to explore possible ways to manipulate growth and increase productivity of plants. Throughout this text it has been shown that plant growth regulators are involved in many aspects of plant growth and development such as flowering, rooting, and other processes. This chapter will focus on restricting plant size and the manipulation of photosynthetic processes. The chemical control of plant growth to reduce size through the use of plant growth retardants is a common practice to make a plant more compact, which in many cases is more commercially acceptable. Plant growth regulators have also been shown to be involved in the regulation of photosynthesis and the movement of photosynthetic products from their site of synthesis in the leaf (source) to their sites of accumulation (sink). Regulation of photosynthesis and movement of photosynthetic products can occur at numerous points, thereby increasing or decreasing plant size.

PLANT GROWTH RETARDANTS

There are a number of plant growth retardants which are currently available for restricting growth. The most commonly used and best understood group of plant growth retardants consists of those which inhibit gibberellin biosynthesis (Figure 12.1). However, there are also a number of miscellaneous compounds which also have the ability to retard plant growth by means other than blocking gibberellin biosynthesis. In this section, plant growth retardants which inhibit gibberellin biosynthesis will be discussed first, followed by other miscellaneous growth-retarding compounds.

Gibberellin Biosynthesis Inhibitors

Onium Compounds. There are a number of onium compounds including chlormequate chloride (2-chloroethyltrimethylammonium chloride) (Cycocel, CCC), mepiquate chloride (1,1-dimethyl-piperidinum chloride), AMO-1618 [2′-isopropyl-4′-(trimethylammonium chloride)-5′-methylphenyl piperidine-1-carboxylate], phosphon D [tributyl(2,4-dichlorobenzyl) phosphonium chloride]

Figure 12.1. Effects of flurprimidol on the height of chrysanthymum plants (courtesy of E. J. Holcomb).

Chlormequate chloride   Mepiquat chloride

Figure 12.2. Structural formulas for two onium growth retardants.

and piperidium bromide [1-allyl-1,3,7-dimethyloctyl)-piperidium bromide]. The most commonly used onium compounds are Cycocel and mepiquate chloride (Figure 12.2). For more details on structural requirements for biological activity on this class of compounds see Gausman (1986).

The inhibition of the cyclization of geranylgeranyl pyrophosphate to copallyl pyrophosphate is the primary mode of action of onium growth retardants leading to the inhibition of gibberellin formation. Plants treated with onium compounds have shortened internodes and thicker greener leaves than untreated controls. When plant growth is restricted by onium compounds a number of other benefits have been obtained. There have been reports showing that onium compounds can enhance net photosynthesis; however, the basis for this enhancement remains unclear. In addition, plants treated with onium compounds have been shown to be able to better tolerate drought conditions than untreated plants. Although it is still not clear how onium compounds promote drought tolerance, several possibilities have been suggested. One is that a reduction in the leaf area caused by onium compounds reduces the transpirational surface, which in turn reduces water loss (Davis and Curry 1991). In addition to the reduction in leaf area, it has been shown that CCC can induce stomatal closure which would reduce transpiration (Mishra and Pradhan 1972; Pill et al. 1979; De et al. 1982). Another possibility which has been suggested is that treatment with onium compounds causes an accumulation of solutes, such as amino acids and sugars, which allows plants to maintain turgor under reduced leaf water potentials (Bode and Wild 1984; Knapp et al. 1987). Onium-treated plants have also been shown to tolerate a variety of other abiotic stresses, such as salt and temperature stress (Gausman 1986); and biotic stresses such as insects (Zummo et al. 1984; Dreyer et al. 1983), diseases and nematodes (Erwin et al. 1976 1979); however, very little is known about how increased tolerance is achieved.

Pyrimidines.   The two most commonly used pyrimidine growth retardants are   ancymidol   [α-cyclopropyl-α-(4-methoxyphenyl)-5-pyrimidienemethanol]

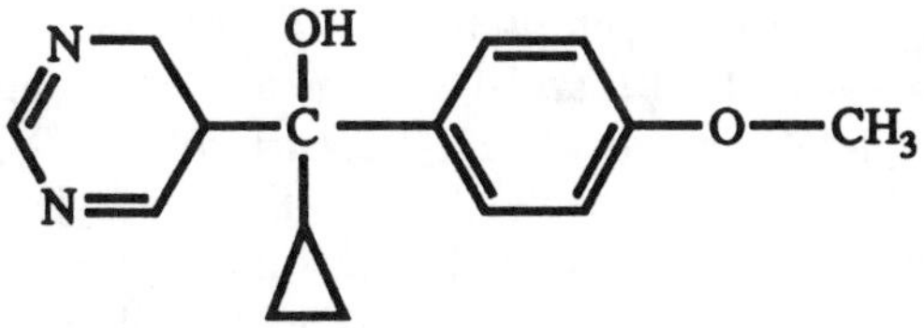

Ancymidol

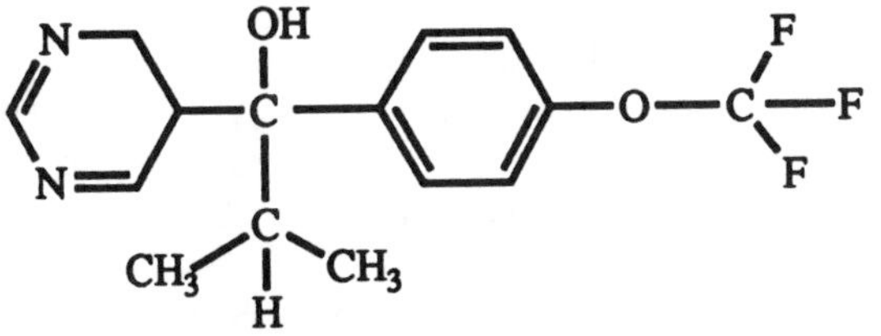

Flurprimidol

Figure 12.3.   Structural formulas for two pyrimidine growth retardants.

and flurprimidol [α-(1-methylethyl)-α-(4-trifluoromethoxyphenyl)-5-py-rimidine-methanol], which are shown in Figure 12.3. The primary mode of action of pyrimidine growth retardants is thought to be the inhibition of the cytochrome P-450, which controls the oxidation of kaurene to kaurenoic acid (Coolbaugh and Hamilton 1976). Although the major effect of pyrimidine growth retardants appears to be due to the inhibition of gibberellin biosynthesis it has also been shown to interfere with sterol (Shive and Sisler 1976) and abscisic acid biosynthesis (Cowan and Railton 1987). There have been reports showing that pyrimidine growth retardants either have no effect (Davis et al. 1986) or cause a slight reduction in photosynthesis (Sterrett et al. 1989). Although pyrimidine growth retardants have little effect on photosynthesis they have been shown to cause a reduction in water use (Johnson 1974; Barrett and Nell 1981 1982).

Triazoles.   The triazoles are a highly active group of plant-growth-retard-ing chemicals including paclobutrazol ([2RS, 3RS]-1-[4-chlorophenyl]-4,4-

dimethyl-2-[1,2,4-triazol-1-yl]pentan-3-ol); uniconazole ([E-1-[4-chloro-phenyl]-4,4-dimethyl-2-[1.2.4-triazol-1-yl]penten-3-ol); triapenthenol [(E)-(RS)-1-cyclohexyl-4,4-dimethyl-2-(1H-1,2,4-triazole-1-yl)-pent-1-en-3-ol]; BAS 111 ([1-phenoxy-5,5-dimethyl-3-(1,2,4-triazole-1-yl)hexan-5-ol]; and LAB 150 978 ([1-(4-trifluor-methyl)-2-(1,2,4-triazolyl(1)]-3-(5-methyl-1,3-dioxan-5-yl)-propen-3-ol and are shown in Figure 12.4. Triazole compounds reduce plant growth by inhibiting the microsomal oxidation of kaurene, kaurenol, and kaurenal, which is catalyzed by kaurene oxidase, a cytochrome P-450 oxidase (Davis et al. 1988; Izumi et al. 1985). In addition to blocking gibberellin biosynthesis, triazole compounds have been shown to inhibit sterol biosynthesis (Haughan et al. 1989); reduce ABA (Wang et al. 1987), ethylene (Sauerbrey et al. 1988), and indole-3-acetic acid (Law and Hamilton 1989); and increase cytokinin content (Izumi et al. 1988).

Although an increase in chlorophyll content is observed in triazole-treated plants, there is little direct effect on photosynthesis; however, it has indirectly been shown to influence photosynthetic activity (Davis and Curry 1991). Triazole-treated plants have been reported to protect plants against abiotic stresses due to water (Asore-Boamah et al. 1986; Atkinson and Chauhan 1987) and sulfur dioxide (Upadhyaya et al. 1991). It has been suggested that the ability of triazole compounds to induce tolerance to abiotic stress is due to increased antioxidant content or activity in treated plants (Upadhyaya et al. 1989, 1990, 1991; Senaratna et al. 1988). Triazole compounds have also been shown to reduce insect population densities; however, it is unclear how this is achieved (Campbell et al. 1989).

Others. Tetcyclacis (5-(4-chlorophenyl)-3,4,5,9,10-pentaza-tetra-cyclo-4,4,10^{2,6} O^{8,11}-dodeca-3,9-diene) is a norbornenodiazetine derivative shown in Figure 12.5. This compound reduces gibberellin biosynthesis by blocking microsomal oxidation of kaurene to kaurenoic acid (Rademacher and Jung 1986). Tetcyclacis also inhibits sterol biosynthesis, and in general appears to act like triazole plant growth retardants (Davis and Curry 1991). Prohexadione calcium (calcium 3,5-dioxo-4-propionyl-cyclohexanecarboxylate), shown in Figure 12.5, has recently been found to have growth-retardant activity. Its mode of action appears to be through the inhibition of the 3β hydroxylation of GA$_{20}$ to GA$_1$ and 2β hydroxylation of GA$_1$ to GA$_8$ (Nakayama et al. 1990). Since this compound is relatively new, little is known about its physiological and bio-chemical effects in plants. Inabenfide (4-chloro-2-(α-hydroxylbenzl)iso-nicotinanilide), an isonicotinic acid anilide derivative, is shown in Figure 12.5; it also inhibits gibberellin biosynthesis by blocking the oxidative conversion of kaurene to kaurenoic acid (Miki et al. 1990).

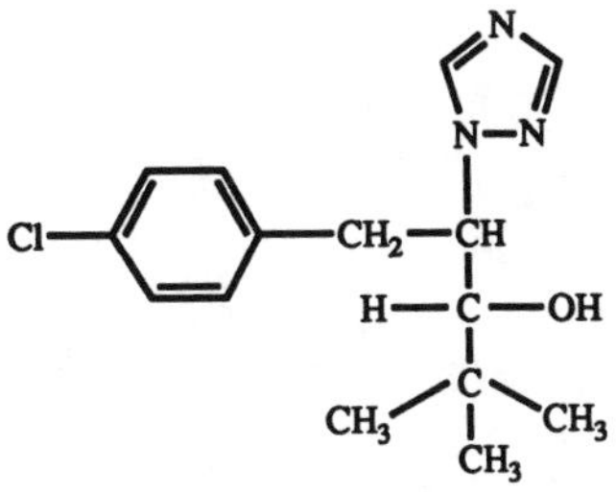

Paclobutrazol

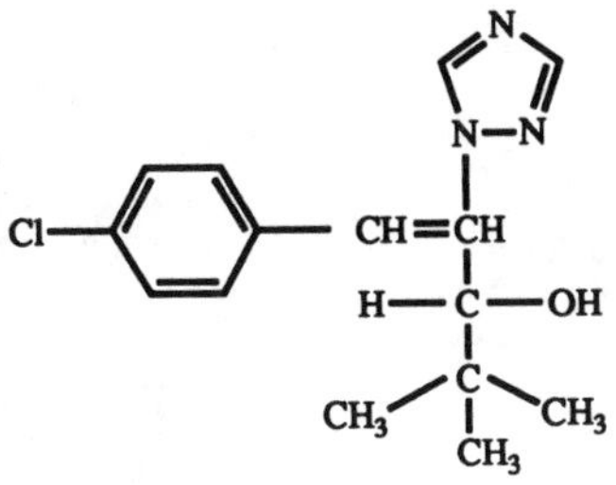

Uniconazole

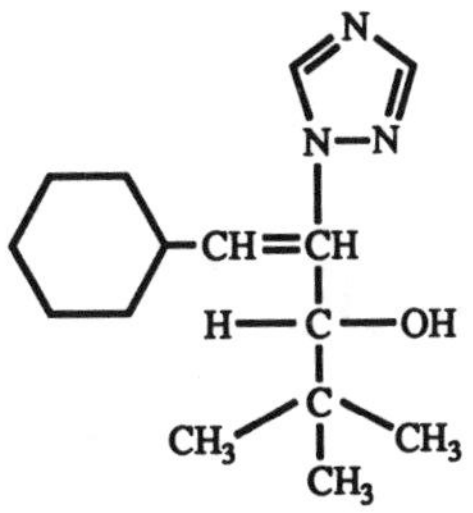

Tripenthenol

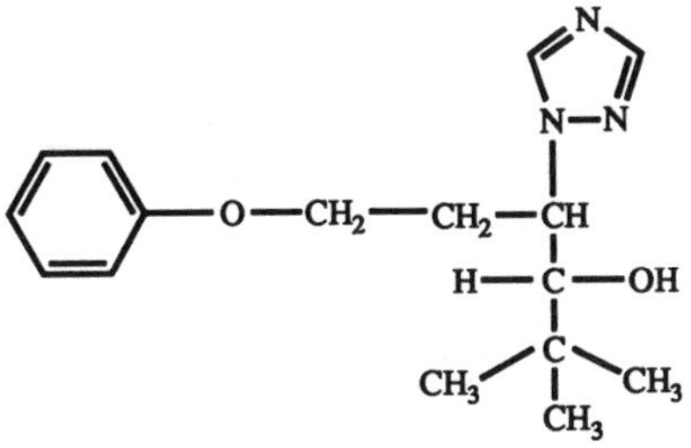

BAS 111

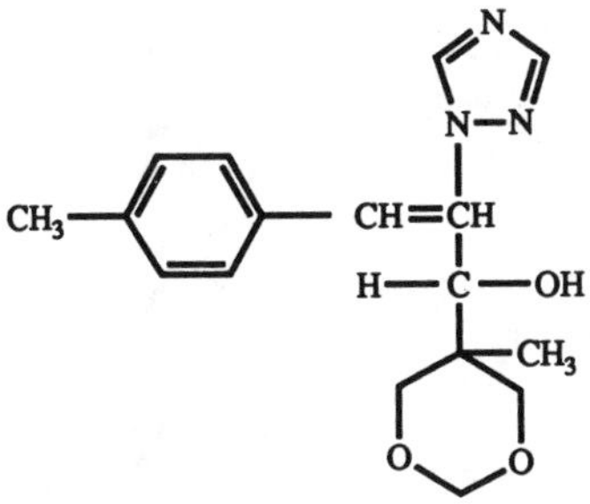

LAB 150 978

Figure 12.4. Structural formulas for five triazole growth retardants.

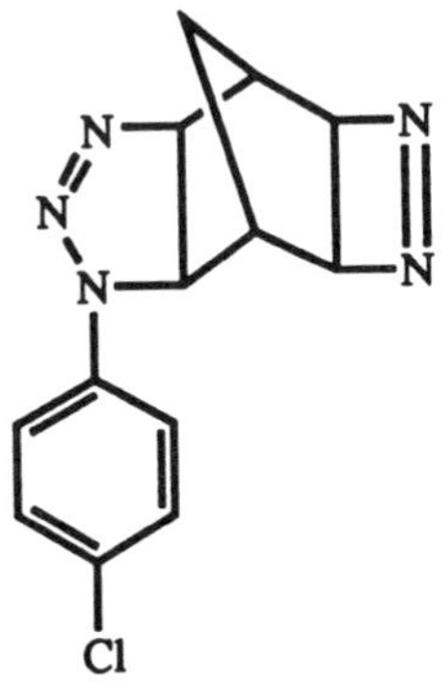

Tetcyclacis

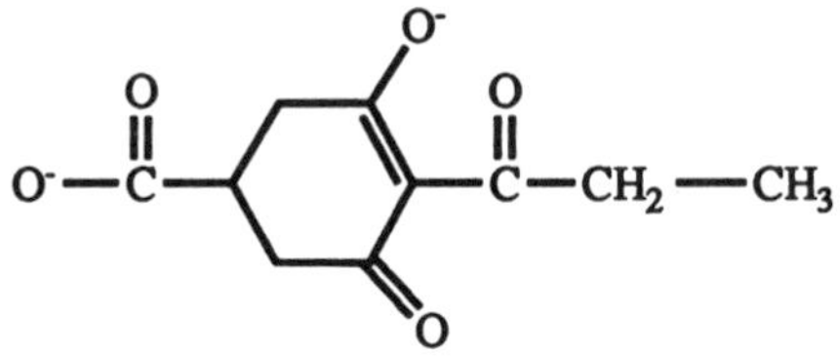

Prohexadione calcium

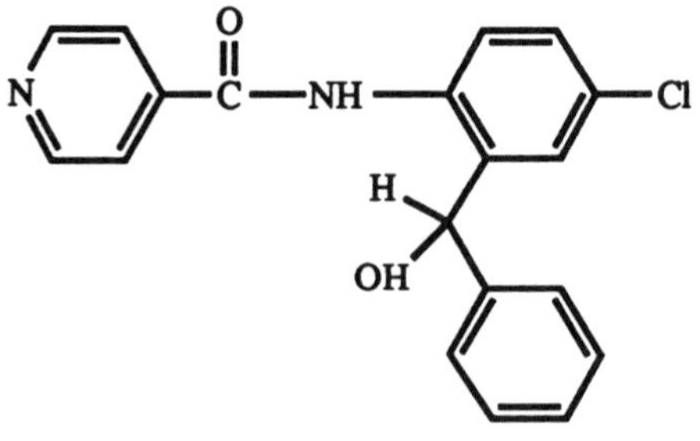

Inabenfide

Figure 12.5. Structural formulas for three other growth retardants which inhibit gibberellin biosynthesis.

Growth-Retarding Compounds Not Inhibiting Gibberellin Biosynthesis

Morphactins are a class of growth retarding compounds including fluorene, fluorene-9-carboxylic acid, and chlorflurenol (methyl[methyl-2-chloro-9-hydroxy fluorene-(9)-carboxylate]) shown in Figure 12.6. They received their name because they have the ability to affect plant morphogenesis, (morphologically active substances). As a general rule morphactins inhibit plant growth, while gibberellins have a promotive effect. There have been studies suggesting that morphactins do not block gibberellin biosynthesis, rather they act as competitive antagonists. Although morphactins have a wide range of physiological effects (Schneider 1970), in plants they appear to act in a nonspecific manner (Davis and Curry 1991).

The structure of dikegulac (2,3:4,6, bis-O-(1-methylethylidene)-x-L-xylo-2-hexulofuranosonic acid) is shown in Figure 12.7. It is thought that the primary response for this compound is the retardation of apical dominance

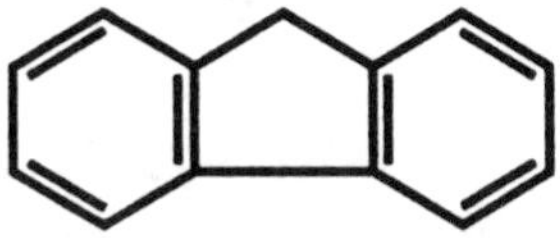

Fluorene 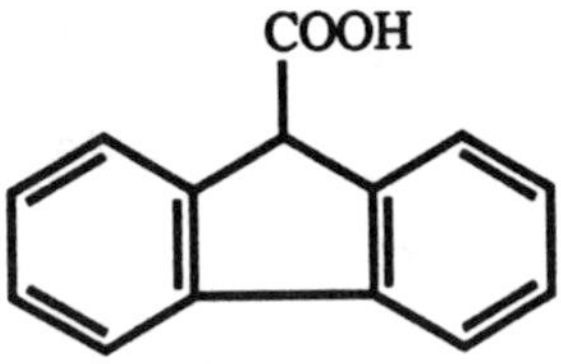   **Fluorene-9-carboxylic acid**

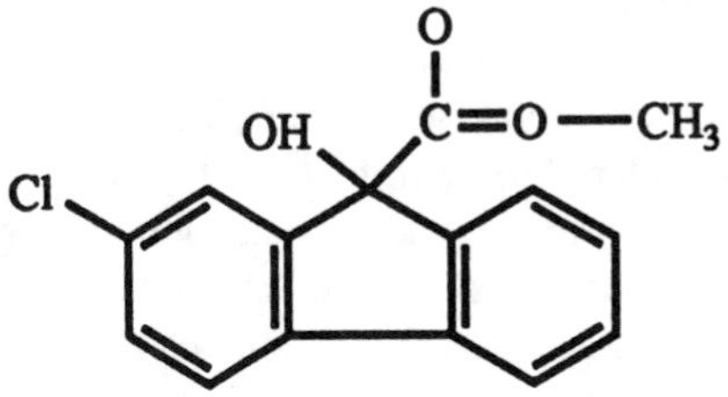

Chlorflurenol methyl

Figure 12.6. Structural formulas for three morphactin growth retardants.

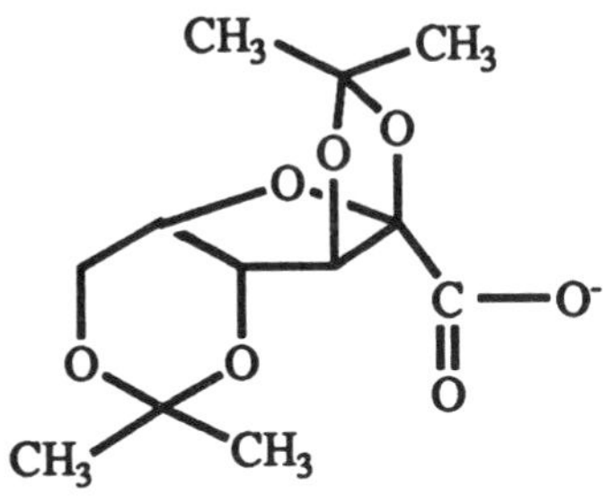

Dikegulac sodium

Figure 12.7.   Structural formula for the growth retardant dikegulac.

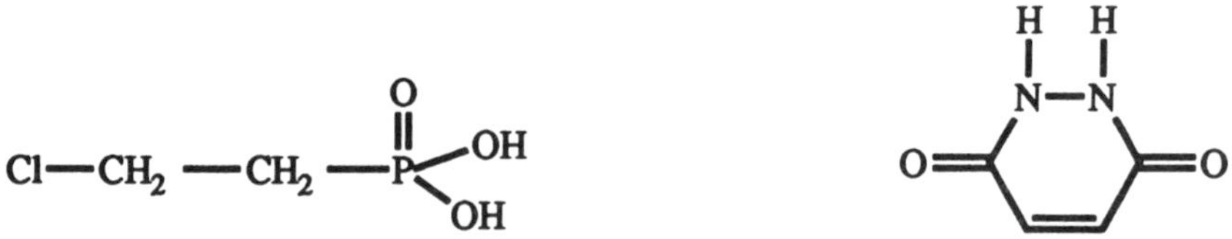

Ethephon

Maleic hydrazide

Figure 12.8.   Structural formulas for ethephon and maleic hydrazide.

leading to lateral bud break. This is supported by research showing that dividing cells are very susceptable to it, whereas stationary cells are less affected (Zillkah and Gressel 1978). Although the mode of action of dikegulac is unknown, it has been shown to reduce gibberellinlike substances while stimulating ABA-like substances and ethylene (Bhattacharjee 1984).

The ethylene-releasing compound ethephon (2-chloroethyl)phosphonic acid (Figure 12.8) can be considered to be a growth retardant since ethylene causes shorter, thicker stems. The primary use of ethephon as a growth retardant is for controlling lodging of cereal and grain crops (Davis and Curry 1991). Maleic hydrazide (1,2-dihydro 3,6-pyridazinedione) (Figure 12.8) is a growth retardant which blocks cell division by interfering with the production of uracil (Schoene and Hoffman 1949; Nooden 1969). The acetamide derivatives mefluidide [N-(2,4-dimethyl-5-[trifluoromethyl]sulfonyl)amino-phenyl)acetamide] and amidochlor N-[acetylamino)methyl]-2-chloro-N-(2,6-diethylphenyl)acetamide (Figure 12.9) have been used primarily for inhibiting turf grass growth. At the present time, the mechanism of action for both of these compounds is

Acetamide Derivatives

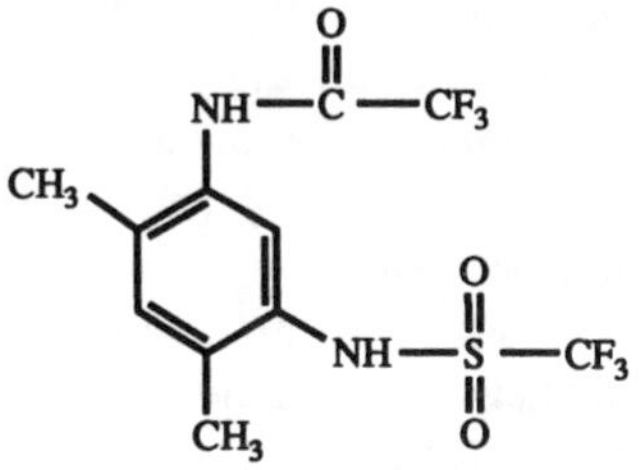

Mefluidide

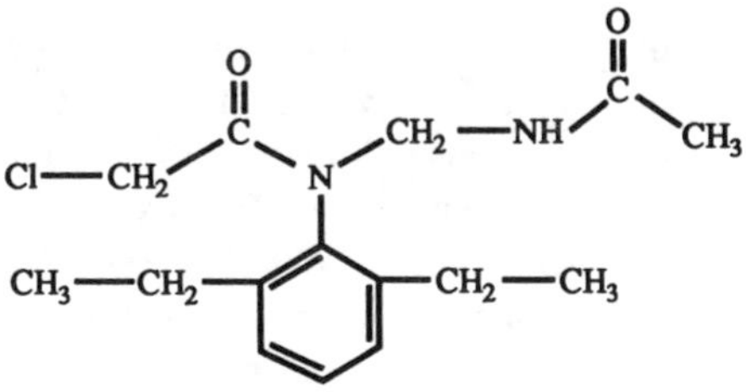

Amidochlor

Figure 12.9. Structural formulas for two growth retardants which are acetamide derivatives.

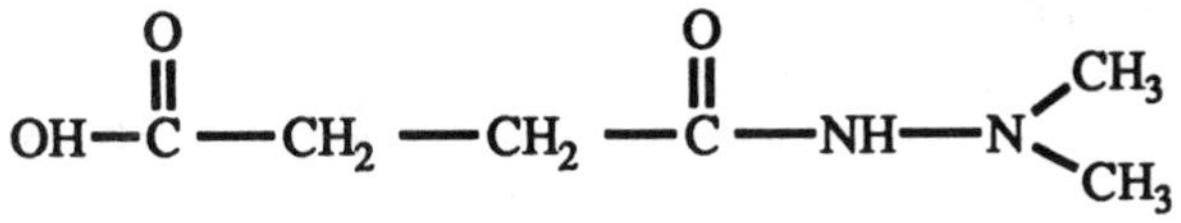

Daminozide

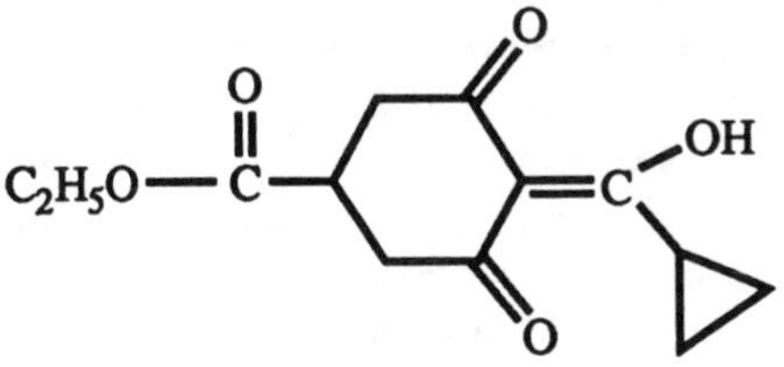

Cimectacarb

Figure 12.10. Structural formulas for the growth retardants daminozide and cimectacarb.

unknown (Davis and Curry 1991). The primary commercial application of daminozide (butanedioic acid mono-(2,2-dimethylhydrazide) (Figure 12.10) is to control the height of bedding plants. Its precise mechanism of action is unknown, although it has been suggested that it might affect gibberellin biosynthesis (Riddel et al. 1962). Cimetacarb [4(cyclopropyl-α-hydroxy-methylene)-3,5-dioxo-cyclohexanecarboxylic acid ethyl ester (Figure 12.10) has potential as a turf growth retardant (Cutler and Schneider 1990), but its mode of action remains unknown. Fatty acid derivatives including fatty alcohols (chain length 8–10) and methyl esters (chain length 8–12) have been shown to reduce plant height (Davis and Curry 1991) by an unknown mechanism. The commercial formulations of mixed methyl esters of fatty acids have been reported to increase branching and reduce shoot growth (Byers and Barden 1976).

USES OF PLANT GROWTH RETARDANTS

The control of vegetative growth is very important in agriculture. A summary of the major applications for each of the plant growth retardants previously discussed, broken down by commodities, will be given in this section.

In many floricultural crops it is important to reduce plant size to increase saleability; however, this must be done without adversely affecting aesthetic quality. The optimum height for most potted plants is 20–25 cm (Adriansen 1985); however, this varies depending upon container size, market preference and species. When using a growth retardant to reduce size it is important to make sure there are no adverse effects on the overall quality of the plant and its postharvest performance. Plant growth retardants have been used for many years to manipulate the size, shape, and overall quality of floricultural crops. At the present time, CCC, daminozide, ancymidol, and paclobutrazol are the primary compounds used for poinsettias and chrysanthemums, although uniconazol and tetcyclacis have been shown to have potential.

In many cases bedding plants (e.g., zinnia, geranium) are treated with growth retardants to promote compactness, maintain quality prior to sale, and also promote longer-shelf life because they do not rapidly outgrow their container. Prior to its removal from the market, daminozide was the primary growth retardant used to control the height of bedding plants for many years. Paclobutrazol has been shown to be very effective in reducing the height of a variety of bedding plants (Cox and Keever 1988; Keever and Cox 1989; Tayama and Carver 1990). Since paclobutrazol is very active and highly persistent as compared to other growth retardants, it has generally been shown that only a single application is necessary. In addition, paclobutrazol has been reported to be effective in bedding plant species where other retardants were not (e.g., impatiens, dianthus) (Barrett and Nell 1987). Uniconazol has been shown to have promise for retarding growth of bedding plants (Davis et al. 1988) and is even more active and persistent then paclobutrazol. The use of growth retardants is not a common practice in nursery crops or for the regulation of tree growth at the present time, although they have potential (Davis and Curry 1991). For many years there has been a considerable amount of interest in the use of growth retardants as chemical mowing agents on turf grass. The reason for this interest is that growth retardants have potential for reducing labor, fuel, and equipment costs for turf management. Although a wide range of growth-retarding compounds have been tested for regulating turf growth, it has been shown that paclobutrazol, flurprimidol, mefluidide, and amidochlor have the most potential. Problems associated with the use of growth retardants as chemical mowing agents, such as inconsistent results, phytotoxicity, and reduced recuperative potential have been encountered and must be solved before growth retardants can be used commercially for regulating turf grass growth (Breuninger and Watschke 1989).

One of the major uses of plant growth retardants in agriculture is to control lodging in cereals and grain crops such as wheat, rice, rye, and barley. Lodging is a serious problem causing reduced yields. By using plant growth retardants, lodging is reduced, thereby facilitating harvest. The two retardants most commonly used to control lodging are CCC and ethephon, although others have also been shown to be effective. Maleic hydrazide has been used as a sprout inhibitor in onions and potatoes during storage for many years. Prior to being removed from the market, daminozide was used to produce stocky tomato transplants. Both paclobutrazol (Pombo et al. 1985) and uniconazole (Hickman et al. 1989) have since been shown to facilitate tomato transplant production. Growth retardants have a number of potential uses in vegetable crops, however, at the present time their use is limited (Davis and Curry 1991).

The size and shape of decidous fruit and nut trees are very important commercially. In apples, size control can be achieved using clonal rootstocks and scions (Ferre et al. 1982); however, this has not been successful in other fruit or nut crops (Hansche and Beres 1980; Rom 1983; Graselly 1987; Rom and Carlson 1987). The use of cultural methods for restricting growth of deciduous fruit and nut trees has met with limited success due to the large number of variables encountered in the production of perennial crops. In the early 1960s, interest in the use of chemicals for the regulation of fruit tree growth began with daminozide (Batjer et al. 1964). This compound was effective in reducing vegetative growth and also stimulated floral bud induction, resulting in increased bloom the following year in apple, pear and cherry trees (Edgerton and Hoffman 1965; Rogers and Thompson 1968; Erez 1985). In 1989, the use of daminozide was banned in the United States because of controversial health and environmental concerns; therefore, alternative growth-retarding agents had to be evaluated. Although other plant growth retardants (Davis and Curry 1991) have been tested, paclobutrazol appears to have the most promise for retarding plant growth in deciduous fruit trees. Presently, paclobutrazol is registered in seven countries, excluding the United States, to control vegetative growth in apple and pear trees by foliar application. In addition, paclobutrazol is also registered for use on peach, nectarine, cherry, apricot, plum, and olive trees in 11 countries outside the United States. Other benefits gained by using paclobutrazol are mite control (Raese and Burts 1983; Dheim and Browning 1988a) and increases in floral bud initiation (Richardson et al. 1986; Dheim and Browning 1988a). Paclobutrazol can also be used to increase fruitlet abscission when applied at full bloom or shortly before (Dheim and Browning 1988b). The potential to control vegetative growth in certain nut species such as pecan has been demonstrated with triazole analogs. They have been shown to effectively control growth in young seedlings (Marquard 1985) and commercial plantings (Wood 1988); however, at the present time they are not registered for use in nut crops.

Regulation of vegetative growth in grapes is very important. Paclobutrazol has been shown to be an effective vegetative growth inhibitor in both table and wine grapes without affecting yield, berry quality, or cold-hardiness of dormant buds (Ahmedullah et al. 1986). However, there have also been reports showing that paclobutrazol promotes excessive vegetative growth rates, resulting in reduced berry size and brix (Shaltout et al. 1988). Therefore, the effects of paclobutrazol on berry quality appears to be closely linked to the species as well as the stage of growth. The use of high-density planting in the citrus industry is increasing in popularity due to many benefits associated with it such as earlier returns. However, at the present time there are only a few methods available which are acceptable for controlling growth. Paclobutrazol has been shown to reduce vegetative growth in tangelo seedlings (Aron et alo 1985), sour orange seedlings (Swietlik 1986), and lemon trees (Smeirat and Qrunfleh 1989); however, more research is necessary before it can be used on a commercial scale.

INVOLVEMENT OF PLANT GROWTH SUBSTANCES IN PHOTOSYNTHETIC PROCESSES AND PARTITIONING OF ASSIMILATES

Gibberellins and Photosynthesis

Foliar Applications.   It is well known that gibberellins have the ability to stimulate plant growth and development in a variety of test systems. A question which might be raised is whether gibberellins increase plant size as a result of increased photosynthetic rates or due to more efficient utilization of photosynthetic products? There are many different reports in the literature on the involvement of gibberellins in photosynthetic processes, some showing increases (Wareing et al. 1968; Borzenkova 1976; Chatterjee et al. 1976; Coulombe and Paquin 1959; Erkan and Bangerth 1980; Gale et al. 1974; Lester et al. 1972; Marcelle and Oben 1972; Marcelle et al. 1974) and others no effect (Haber and Tolbert 1957; Hayashi 1961; Little and Loach 1975) or a decrease (Sanhla and Huber 1974). Coulombe and Paquin (1959) demonstrated that tomato foliage entirely or partially sprayed with GA_3 exhibited an increase in respiration, photosynthesis, and transpiration within several hours. Similar results were found by Hayashi (1961), who demonstrated that foliar applications of GA_3 increased photosynthetic activity in tomato plants over a seven-day period. However, his conclusion was that the increase in photosynthetic rate was not due to increased photosynthetic activity, but to an increase in leaf area. Foliar sprays of GA_3 have been reported to stimulate photosynthetic rates of fully expanded primary leaves of bean plants (Marcelle et al. 1974; Treharne et al. 1970). However, Bidwell and Turner (1966) reported that GA_3 had no effect on photosynthetic rates of bean leaves. Foliar applications of gibberellins have also been shown to stimulate photosynthetic rates in maize (Wareing et al. 1968) and potato leaves (Borzenkova 1976).

When applied as a foliar spray gibberellins have been shown to increase rates of CO_2 fixation, transpiration, and stomatal aperture in barley leaves (Livine and Vaadia 1965); whereas, foliar applications of GA_3 were shown to have no effect on CO_2 fixation or the distribution among products of photosynthesis in chlorella and detached leaves of barley, oats, and peas (Haber and Tolbert 1957). In *Pennisetum* there was a decrease in CO_2 fixation and photosynthetic enzymes following GA_3 treatment (Wellburn et al. 1973). In plants where GA_3 stimulates photosynthetic rates or enhances ultrastructural morphogenesis of plastids it also enhances the in vitro activity of ribulose bisphosphate carboxylase/oxygenase (rubisco) a major photosynthetic enzyme in plants (Wareing et al. 1968; Arteca and Dong 1981; Hoad et al. 1977). The activity of cytoplasmic sucrose phosphate synthase (SPS), a key enzyme regulating the pool size of sucrose in the leaf has been shown to be stimulated by foliar applications of GA_3 or GA_4 and GA_7 in soybean plants. Source tissues have the capacity to load photoassimilates into phloem tissue for long-distance transport and foliar applications of GA_3 have been shown to promote phloem loading (Baker 1985).

Root Applications. As is evident from a brief review of the literature on the subject of plant growth substance involvement in photosynthetic processes there is a serious state of confusion with some reporting increase, decrease, or no effect. When positive results were obtained from GA_3 treatments, they were at higher concentrations than the known low endogenous levels of gibberellins found in plants (Treharne and Stoddart 1968). Until the work of Arteca and Dong (1981) there had been no detailed studies on root applications of any of the known classes of plant growth substances on photosynthetic processes. By applying plant growth substances to the roots, a number of benefits over foliar applications can be achieved. First, two of the eight classes of plant growth substances discussed in this text are thought to be produced in the roots and transported to the shoots where they elicit a response. Second, there may be greater uptake of these compounds through the roots and a more even distribution throughout the plant, since root applications avoid problems associated with the cuticle layer, leaf hairs, and other factors which make penetration of the leaf more difficult. Third, applying plant growth substances to the roots also has decided practical applications. It is feasible to supply plant growth substances to the roots to reduce transplanting shock, as a slow-release fertilizer, increasing availability over a long period of time. Even though there are a number of benefits which can be derived from using root applications there are also problems associated with this approach. The root environment has a profound effect on the effiency of root applications of plant growth substances. For example, when plant growth substances are applied to the roots of plants grown hydroponically there is maximal uptake and low concentrations are required to have an effect on photosynthesis (Arteca and Dong 1981; Dong and Arteca 1982; Arteca 1982; Tsai and Arteca 1985; Arteca et al. 1985a, 1985b; Arteca and Tsai 1988; Arteca et al. 1991). By contrast, high amounts of organic matter in the media can bind the growth substance, making it unavailable to the plant. Another disadvantage is the potential for degradation of plant growth substances by soil microorganisms. The effects of root applications of GA_3 on photosynthesis in tomato plants grown hydroponically was first shown by Arteca and Dong (1981). They showed that photosynthetic rates (mg CO_2/dm$_2$/ hr) determined using an open CO_2 gas exchange system were increased 40–50% within five hours of treatment with a 1.4 µM GA_3 treatment to the roots. If GA_3 was left in continual contact with the roots, photosynthetic rates remained elevated for the duration of the experiment (nine days). The dramatic effect of root applications of GA_3 over a range of very low concentrations on tomato plant growth is shown one week following treatment (Figure 12.11). Interestingly, at lower light levels the percent stimulation of photosynthesis by GA_3 was more dramatic. There was approximately a 90% increase in photosynthetic rate at 80 µE m^{-2} s^{-1}, while at saturating light conditions there was approximately a 40% increase over the control rate (Figure 12.12). In a subsequent paper with hydroponically grown tomato plants it was shown that root applications of GA_3 stimulated photosynthesis, relative growth rate (RGR), and total dry matter accumulation (Dong and Arteca 1982). When plant roots were

Figure 12.11.   Effects varying concentrations of gibberellic acid (GA_3) on tomato plant growth at one week after treatments; from left to right are 1.4 µM GA_3, 0.14 µM GA_3, 0.014 µM GA_3 and control (from Arteca and Dong, (1981)).

dipped in a gel containing GA_3 and planted in a medium containing 1 part sphagnum peat moss: 1 part vermiculite: 1 part sand at the same concentration as previously reported in studies using hydroponics, there was no significant effect on plant growth showing one of the potential problems associated with root applications (Arteca 1982). The effects of root applications of GA_3 on photosynthesis, transpiration, and growth of *Pelargonium* plants has also been studied (Arteca et al. 1985a, 1991). It was shown that root applications of GA_3 dramatically stimulated relative growth rates in a genetically diverse group of hydroponically grown *Pelargoniums* at concentrations as low as 0.014 µM with maximum stimulation at 1.4 µM GA_3 (Arteca et al. 1985a, 1991). As shown in Figure 12.13 root applications with GA_3 have a profound visual effect on the growth of *Pelargoniums* within one week of treatment. GA_3 caused a reduction in transpiration rate, whereas net photosynthetic rates and total chlorophyll content were unaffected (Arteca et al. 1985a). Therefore, the increase in RGR found with *Pelargoniums* was not due to increased photosynthesis; rather it may have been due to a modification in normal partitioning of photosynthates, which is similar to observations reported by others (Little and Loach 1975).

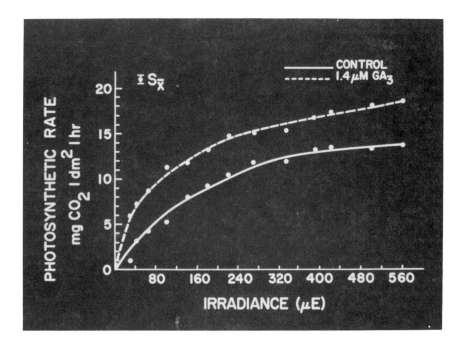

Figure 12.12.   Effects of varying light levels on the photosynthetic rates of GA_3 treated (1.4 μM GA_3) and control plants. Each point represents the mean of three replications plus or minus the standard error of the mean two days following treatment (from Arteca and Dong (1981)).

The response to GA_3 was also evaluated for *Pelargoniums* grown in a soil mix. It was shown that root applications of GA_3 increased the number of cuttings, length of the cuttings, as well as fresh and dry weights. There was a difference between GA_3 concentrations required to elicit a response in plants grown hydroponically and in a soil mix, with plants grown in soil requiring higher concentrations to be responsive. The difference may have been due to microbial breakdown of GA_3 in the soil or its retention by materials contained within the soil mix. The concentration of GA_3 required to stimulate growth appears to be cultivar-dependent. Therefore, the proper concentration required to elicit an optimal response should be tested experimentally prior to large-scale application. The results from studies previously mentioned suggested that root applications of GA_3 may have great potential for increasing cutting production in *Pelargoniums* (Arteca et al. 1985a, 1991). The effects of root applications of GA_3 on photosynthesis and growth have also been evaluated in many other species, including both C_3 and C_4 plants. The response of root applications of GA_3 appears to have no correlation with monocots or dicots, C_3 or C_4 photosyn-

Figure 12.13. Effects of root applications of 1.4 μM gibberellic acid on geranium (*Pelargonium x hortorum* Sincerity) growth one week after treatment. Untreated on left and treated plant on right (from Arteca et al. (1985a)).

thesis, but only depends on the species (Tsai and Arteca 1985; Latimer 1992). Although root applications are better than foliar sprays, providing more consistent results, it is still unclear as to how GA stimulates plant growth. In summary, the enhancement of growth rate by GA$_3$ might result from an increase in effective leaf area, a stimulation of photosynthetic rate, a modification in partitioning of photosynthate, or a cooperative effect.

Cytokinins and Photosynthesis

Foliar Applications. Researchers have shown that foliar sprays with kinetin can stimulate photosynthetic rates (Borzenkova 1976; Meidner 1969; Treharne et al. 1970). Meidner (1969) reported that the treatment of the primary leaves of barley with 3 μM solutions of kinetin promoted an increase in net CO_2 assimilation rates. The resulting reduction in CO_2 concentration inside the leaves was considered to be one of the factors causing decreases in stomatal resistance, but it also appeared that kinetin was affecting the stomatal mechanism directly (Meidner 1967). The increase in net CO_2 assimilation was later shown to be primarily due to a decrease in stomatal resistance (Meidner 1969). Others have reported that kinetin or benzyladenine can stimulate photosynthesis and rubisco in fully expanded bean leaves with no significant effects on specific radioactivity incorporated into plastid rRNA components (Feierabend 1969; Treharne et al. 1970). It has also been shown that foliar applications of cytokinins can promote chloroplast development (Parthier 1979). Both chloroplasts and amyloplasts are derived from proplastids, therefore, it is possible that amyloplast development might also be altered by cytokinin applications (Jones et al. 1986). Cytokinins have also been reported to enhance phloem loading (Daie 1986) and unloading (Clifford et al. 1986), suggesting that their ability to regulate transport of photoassimilates may be responsible for the observed effects on photosynthesis.

Root Applications. Dong and Arteca (1982) showed that there was a 30–35% increase in photosynthetic rate (mg $CO_2/dm^2/hr$) of attached tomato leaves within eight hours of root treatments with 0.47 μM kinetin. For up to two days following the initial treatment with 0.47 μM kinetin there was a stimulation of photosynthesis, RGR, and total dry matter accumulation. At low light levels there was approximately a 100% increase in photosynthetic rate two days following treatment with 0.47 μM kinetin, while at saturating irradiance there was a 30–35% increase (Figure 12.14). When roots were left in contact with 0.47 μM kinetin for longer than two days there was extensive branching of the root system and growth was dramatically reduced. Plants treated with kinetin for up to seven days exhibited optimal stimulatory effects on photosynthesis at considerably lower concentrations (0.0047 μM). This increase in photosynthesis is consistent with the findings of others who have shown that foliar sprays of kinetin enhanced photosynthetic rates (Borzenkova 1976; Meidner 1967 1969; Treharne et al. 1970) and is further supported by the dry weight gains in barley tiller buds resulting from short-term kinetin root dips (Sharif and Dale 1980).

In addition to the stimulatory effect on photosynthesis, root applications of 0.0047 μM kinetin also caused a decrease in stomatal resistance, thus allowing more CO_2 into the leaf and potentially increasing growth. This was consistent with the findings of Meidner (1967, 1969) and would partially explain the increased photosynthetic rates at low light levels. Since kinetin treatment decreases leaf resistance, thereby, increasing CO_2 uptake by the leaf, rubisco may be promoted and the wasteful effects of photorespiration due to the oxygenase form of this enzyme would be suppressed. It has been shown that root applications of kinetin can reduce transplanting shock in tomato plants (Arteca 1982). In this study kinetin was incorporated into a starch-acrylate polymer gel, roots were dipped in this mixture and then planted in a potting mixture containing 1 sphagnum peat moss: 1 vermiculite: 1 sand, and grown under green-

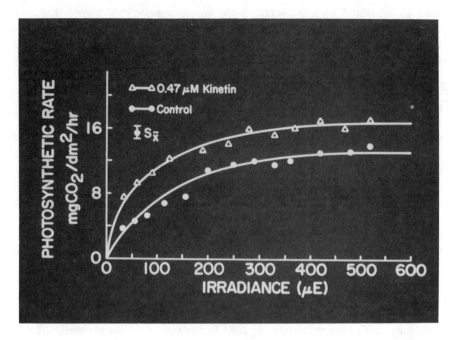

Figure 12.14. Effects of varying light levels on the photosynthetic rates of kinetin-treated (0.47 μM) and control plants. Each point represents the mean two days following treatment of three replications. Bar in the top left-hand corner indicates the maximum standard error (from Dong and Arteca (1982)).

house conditions. All kinetin treatments increased RGR, total leaf area, and plant dry weight, but the highest concentration tested was supraoptimum. This response to kinetin supports the findings of Dong and Arteca (1982), which showed that root applications of kinetin at low concentrations increased photosynthetic rates in tomato plants. Therefore, the use of root applications of kinetin to decrease tomato transplanting shock has the ability to increase transplanting efficiency; however, kinetin treatments must be field-tested before they can be incorporated into commercial practices.

Molecular Manipulation of Endogenous Levels of Cytokinins and Photosynthesis. Transgenic tobacco plants with T-DNA carrying gene 4 for cytokinin synthesis were shown to contain 10–20% higher levels of endogenous zeatin, zeatin riboside, and isopentenyladenosine. These changes were associated with changes in photosynthetic processes at the chloroplast and leaf level. Transgenic plants had a higher proportion of photosystem 2 light harvesting complex in the thylakoids as compared to control plants. Photosynthetic rates as measured by CO_2 uptake were approximately 25% higher than in nontransformed plants. In addition, the rates of dark and light respiration, and the CO_2 compensation point were higher in transgenic plants, while transpiration rates were not significantly effected (Catsky et al. 1993a). Research by the same group has also shown that high levels of endogenous cytokinins in transgenic plantlets limits photosynthesis (Catsky et al. 1993b). Transgenic potato plants containing the gene for cytokinin synthesis grown in vitro exhibited a dramatic increase in zeatin, zeatin riboside, isopentenyladenine, and isopentenyladenosine over control plants. The high levels of cytokinins resulted in decreased photosynthetic rates, while increased dark respiration, light respiration, and CO_2 compensation points were observed. The higher levels of endogenous cytokinins were also associated with lower activities of both photosystems in isolated chloroplasts (Catsky et al. 1993b). These studies, along with future studies utilizing molecular technology to modify endogenous levels of plant growth substances, may someday lead to a better understanding of how plant growth substances are involved in the enhancement of photosynthesis.

Abscisic Acid and Photosynthesis

There is a large amount of evidence indicating that ABA has an important regulatory role in the resistance of plants to environmental stress (Markhart et al. 1979; Riken et al. 1976; Davies and Jones 1991). It is also known to cause stomatal closure in a number of plants (Raschke and Hendrick 1985; Raschke 1975; Walton 1980; Davies and Jones 1991) and has been reported to modify the permeability of roots to water (Fiscus 1981; Glinka 1977; Glinka and Reinhold 1972; Markhart et al. 1979). In general, foliar applications of ABA have been shown to decrease photosynthetic rates and rubisco levels (Fisher et al. 1985; Mittelheuser and van Steveninck 1971; Poskuta et al. 1972; Wellburn et al. 1973; Makeev et al. 1992; Davies and Jones 1991) and have a potential role in photosynthate partitioning and seed filling (Brenner 1987; Makeev et al. 1992). Markhart (1982) reported that ABA applied to the roots of soybean plants was transported to the shoot via the transpiration stream. Since ABA has an important regulatory role in the response of plants to environmental stress and root applications of plant growth substances are more efficient than foliar applications (Arteca and Dong 1981; Dong and Arteca 1982), studies utilizing root applications of ABA to address practical problems were undertaken. The effects of root applications of ABA on reducing photosynthesis and transpiration in *Pelargoniums* were evaluated in order to reduce damage to cuttings which occurs due to active transpiration and growth during their shipment (Arteca et al. 1985b). In this study they showed that 30 minutes after ABA treatments (76 µM) to the roots, there was a reduction in both photosynthesis and transpiration (Arteca et al. 1985b). This response was faster than previously reported (Markhart et al. 1979) and is probably due to more efficient uptake of ABA by *Pelargonium* plants. When the roots of *Pelargonium* plants were kept in continual contact with ABA (76 μM), photosynthesis and transpiration remained low. Although this is an interesting finding it is not practical to keep ABA in contact with the roots for the duration of the shipping process; therefore pulse/chase experiments were performed. Plants treated with 76 μM ABA for six or nine hours followed by rinsing the roots and returning to full-strength Hoagland solution minus ABA showed an inhibition in photosynthesis and transpiration four days following treatment, however, after six days there was no difference between treated and control plants. The reduction in photosynthesis and transpiration reported in this paper may have practical value in the shipment and storage of *Pelargonium* cuttings. It is also possible that this technology could be applied to other genus and species as well (Arteca et al. 1985b). Another example of the practical application of ABA to reduce photosynthesis, transpiration, and growth is with tomato plants. When tomato transplants are set in the field there is a delay in growth of about 10 days following transplanting due to shipping injury (Arteca 1982). Part of this transplant shock is the result of damage arising from active transpiration and growth causing desiccation during transport. It had been suggested that root applications of ABA may have practical value in reducing photosynthesis and transpiration in other genus and species, thereby reducing shipping injury. Therefore, the information from the *Pelargonium* study was applied to tomato transplants. It was shown that short-term pulses of ABA to tomato transplants could reduce photosynthesis and transpiration for four days, but by the sixth day there was no difference from the untreated control (Arteca and Tsai 1988). This work supported the previous findings of Arteca et al. (1985b) and provides additional evidence that ABA treatments may have practical value in a wide range of genus and species; however, further research is necessary in order to fully evaluate the commercial or applied potential of ABA treatments.

Indole-3-Acetic Acid and Photosynthesis

IAA has been shown to have a stimulatory effect on photosynthetic rates in a number of different species when applied as a foliar spray (Borzenkova 1976; Turner and Bidwell 1965; Chatterjee et al. 1976). In a study by Turner and Bidwell (1965) they showed that when bean leaves were sprayed with a solution of IAA, the rate of leaf CO_2 assimilation increased shortly thereafter, then returned to the original rate. They also demonstrated that spraying the leaves of green beans swiss chard, strawberries, sunflowers, soybeans, wheat, geraniums and ferns with IAA caused increases in the rate of CO_2 assimilation 30–100% during the first half-hour to one hour following spraying. In subsequent experiments with isolated chloroplasts it was shown that IAA caused a stimulation in photosynthesis by increasing photophosphorylation and CO_2 fixation (Tamas et al. 1973). However, when similar experiments were performed (Robinson et al. 1978) there was no significant effect on the rate of CO_2 fixation; instead IAA treatments were shown to reduce aging of the isolated chloroplasts. Therefore, it was concluded that the increase in CO_2 uptake following IAA application was not due to a direct interaction with the chloroplasts. Erkan and Bangerth (1980) reported that foliar applications of IAA reduced photosynthesis of tomato and pepper plants. In a subsequent study with hydroponically grown tomato plants it was shown that root applications of IAA have either no effect on photosynthetic rate, RGR, or total plant dry weight, or an inhibitory effect was observed (Dong and Arteca 1982). It has also been shown that photosynthesis was inhibited in blade meristems of *Alaria esculenta* treated with IAA (Bugelin and Bal 1976). In addition, Zerbe and Wild (1981) reported that dry weight, soluble reducing sugars, soluble protein, chlorophyll, chlorophyll a/b ratio, and the rate of CO_2 fixation were lowered by foliar applications of IAA during the development of the primary leaves of *Sinapsis alba*. A review by Brenner (1987) outlined the possible control points at which plant growth substances may affect photosynthesis and partitioning of sucrose in developing seeds. In this scheme IAA was identified as a promoter of stomatal opening, rubisco, sucrose phosphate synthase, phloem loading, phloem transport, and activity of sinks. Considering these effects, the possibility exists that increases in photosynthesis were due to an increase in stomatal aperture and an increase in the activity of sinks, but further research is necessary to more clearly understand the direct involvement of IAA in photosynthesis.

Other Plant Growth Substances and Photosynthesis

There are a number of reports in the literature on the effects of ethylene and ethylene-releasing compounds on photosynthesis in a variety of species with some showing no effect to partial inhibition (Kays and Pallas 1980; Pallas and Kays 1982; Taylor and Gunderson 1986; Erkan and Bangerth 1980; Gunderson and Taylor 1988; Squier et al. 1985) or a promotive effect (Cliquet and Morot-Gaudry 1989). In most cases when ethylene promoted a decrease or increase in photosynthesis there was a partial effect on stomatal aperture. Recently, it was shown that when tomato plants were treated with high CO_2 levels for short or long periods of time there was a decrease in photosynthesis, which is typical of C_3 species. In conjunction with the decrease in photosynthesis there was an increase in ACC and ethylene, implicating ethylene in the reduction of photosynthesis. Plants grown at high levels of CO_2 for long periods of time maintained their capacity for increased levels of ethylene production (Woodrow and Grodzinski 1993). When epinasty was induced by foliar applications of ethephon in tomato plants there was a 60% decrease in the ability of the leaf to intercept light, thereby, resulting in a 35% reduction in photosynthesis. When the leaves from epinastic plants were returned to their original position normal rates of photosynthesis were observed (Woodrow et al. 1989). Although ethylene has been shown to reduce photosynthesis in a number of cases, it is difficult to determine if this is a direct effect, since it has the ability to modify stomate size, leaf angle, chloroplast integrity, and sink activity in tissues such as apical meristem (Abeles et al. 1992; Cliquet et al. 1991) and other complicating factors. In general, most workers conclude that ethylene does not have a direct effect on photosynthesis (Bradford 1983; Woodrow and Grodzinski 1989; Woodrow et al. 1989).

Brassinosteroids have been shown to have profound effects on plant growth and development. Initial studies by workers at the USDA in Beltsville, Maryland, provided evidence that brassinosteroids have the ability to increase crop yields (Maugh 1981) and suggested that these compounds could have potential economic benefits in agriculture. Due to variability in results, testing was discontinued in the United States, while extensive testing has been initiated in Japan and China using 24-epibrassinolide, a synthetic brassinosteroid. These studies showed that 24-epibrassinolide has the ability to increase yields in a variety of plant species, but depending upon cultural conditions, method of application, and other factors, there were variable effects. Although brassinosteroids have a profound effect on crop yields, little is known about their effects on photosynthesis. It is possible that future research in this area will lead to a better understanding of the involvement of BR in photosynthetic processes and provide an explanation for the variable effects on crop performance (Cutler et al. 1991).

Jasmonates have been shown to inhibit plant growth and speed the senescence process. Therefore, even though their involvement in photosynthesis has not been studied directly, it appears that the effects would be negative. While there have been no studies on the effects of salicylic acid on photosynthesis, it has been shown to stimulate flowering and other processes associated with a reduction in photosynthesis.

REFERENCES

Abeles, F. B., Morgan, P. W., and Saltveit Jr., M. E. (1992). *Ethylene in Plant Biology. Second Edition*, Academic Press, San Diego.

Adriansen, E. (1985). "Height control of *Beloperone guttata* by paclobutrazol". *Acta Hortic.* 167:395.

Ahmedullah, M., Kawakami, A., Sandidge, C. R., and Wample, R. L. (1986). "Effect of paclobutrazol on the vegetative growth, yield, quality and winter hardiness of buds of Concord grape". *HortScience* 21:273.

Aron, Y., Monselise, S. P., Goren, R., and Costo, J. (1985). "Chemical control of vegetative growth in citrus trees by paclobutrazol". *HortScience* 20:96.

Arteca, R. N. (1982). "Effect of root applications of kinetin and gibberellic acid on transplanting shock in tomato plants". *HortScience* 17:633–634.

Arteca, R. N. and Dong, C. N. (1981). "Stimulation of photosynthesis by application of phytohormones to the root systems of tomato plants". *Photosynthesis Research* 2:243–249.

Arteca, R. N., Holcomb, E. J., Schlagnhaufer, C., and Tsai, D. S. (1985a). "Effect of root applications of gibberellic acid on photosynthesis and growth of geranium plants grown hydroponically". *HortScience* 20:925–927.

Arteca, R. N., Schlagnhaufer, C. D., and Arteca, J. M. (1991). "Effects of root applications of gibberellic acid on growth of seven different *Pelargonium* cultivars". *HortScience* 26:555–556.

Arteca, R. N. and Tsai, D. S. (1988). "Effects of abscisic acid on photosynthesis, transpiration and growth of tomato plants". *Crop Research* 27:91–96.

Arteca, R. N., Tsai, D. S., and Schlagnhaufer, C. (1985b). "Abscisic acid effects on photosynthesis and transpiration in geranium cuttings". *HortScience* 20:370–372.

Asore-Boamah, N. K., Hofstra, G., Fletcher, R. A., and Dumbroff, E. B. (1986). "Triadimedon protects bean plants from water stress through its effects on abscisic acid". *Plant Cell Physiol.* 27:383.

Atkinson, D. and Chauhan, J. S. (1987). "The effects of paclobutrazol on the water use of fruit plants at two temperatures". *J. Hortic. Sci.* 62:421.

Baker, D. A. (1985). "Regulation of phloem loading". *British Plant Growth Regulator Group* Monograph 12:163–176.

Barrett, J. E. and Nell, T. A. (1981). "Transpiration in growth retardant treated poinsettia, bean and tomato". *Proc. Fla. State Hortic. Soc.* 94:85.

Barrett, J. E. and Nell, T. A. (1982). "Irrigation interval and growth retardants affect poinsettia development". *Proc. Fla. State Hortic. Soc.* 95:167.

Barrett, J. E. and Nell, T. A. (1987). "Bonzi for bedding plants. At last, is there a way to control run away impatiens?". *Grower Talks* 50:52.

Batjer, L. P., Williams, M. W., and Martin, G. C. (1964). "Effects of N-dimethyl amino succinamic acid (B-Nine) on vegetative and fruit characteristics of apples, pears, and sweet cherries". *J. Am. Soc. Hortic. Sci.* 85:11.

Bhattacharjee, A. (1984). "Responses of sunflower plants toward growth retardation with special reference to growth, metabolism and yield". Ph.D. Thesis, Burdwan University

Bidwell, R. G. S. and Turner, W. B. (1966). "Effect of growth regulators on CO_2 assimilation in leaves and its correlation with the bud break response in photosynthesis". *Plant Physiol.* 41:267–270.

Bode, J. and Wild, A. (1984). "The influence of (2-chloroethyl)trimethylammonium chloride (CCC) on growth and photosynthetic metabolism of young wheat plants (*Triticum aestivum* L.)". *J. Plant Physiol.* 116:435.

Borzenkova, R. A. (1976). "Effect of phytohormones on the photosynthetic metabolism of potato leaves". *Mater. Ekol. Fiziol. Rast. Ural. Flory.* 104:110.

Bradford, K. J. (1983). "Involvement of plant growth substances in the alternation of leaf gas exchange of flooded tomato plants". *Plant Physiol.* 73:480–483.

Brenner, M. L. (1987). "The role of hormones in photosynthate partitioning and seed filling". In *Plant Hormones and Their Role in Plant Growth and Development*, ed., P. J. Davies, Martinus Nijhoff Publishers, Boston, pp. 474.

Breuninger, J. M. and Watschke, T. L. (1989). "Growth regulation of turfgrass". *Rev. Weed Sci.* 4:153.

Buggeln, R. G. and Bal, A. K. (1976). "Effects of auxins and chemically related nonauxins on photosynthesis and chloroplast ultrastructure in *Alaria esculenta* (Laminariales)". *Canadian J. Botany* 55:2098–2105.

Byers, R. E. and Barden, J. A. (1976). "Chemical control of vegetative growth and flowering of non-bearing Delicious apple trees". *HortScience* 11:306.

Campbell, C. A. M., Easterbrook, M. A. and Fisher, A. J. (1989). "Effects of plant growth regulators paclobutrazol and chlormequat chloride on pear psyllid (*Cacopsylla pyricola* [Folster]) and pear rust mite (*Epitrimerus piri* [Nal.])". *J. Hortic. Sci.* 64:561.

Catsky, J., Pospisilova, J., Machackova, I., Synkova, H., Wilhelmova, N., and Sestak, Z. (1993a). "High-level of endogenous cytokinins in transgenic potato plantlets limits photosynthesis". *Biologia Plant.* 35:191–198.

Catsky, J., Pospisilova, J., Machackova, I., Wilhelmova, N. and Sestak, Z. (1993b). "Photosynthesis and water relations in transgenic tobacco plants with T-DNA carrying gene 4 for cytokinin synthesis". *Biologia Plant.* 35:393–399.

Chatterjee, A., Mandal, R. K., and Sircar, S. M. (1976). "Effects of growth substances on productivity, photosynthesis and translocation of rice varieties". *Indian J. Plant Physiol.* 19:121–138.

Clifford, P. E., Offler, C. E., and Patrick, J. W. (1986). "Growth regulators have rapid effects on photosynthate unloading from seed coats of *Phaseolus vulgaris* L.". *Plant Physiol.* 80:635–637.

Cliquet, J. B., Boutin, J. P., Deleens, E., and Morot-Gaudry, J. F. (1991). "Ethephon effects on translocation and partitioning of assimilates in *Zea mays*". *Plant Physiol. Biochem.* 29:623–630.

Cliquet, J. B. and Morot-Gaudry, J. F. (1989). "Ethephon and photosynthesis control in maize". *C. R. Acad. Sci. Paris* 309:317–322.

Coolbaugh, R. C. and Hamilton, R. (1976). "Inhibition of ent-kaurene oxidation and growth by alpha-cyclopropyl-alpha(p-methoxyphenyl)-5-pyrimidine methylalcohol". *Plant Physiol.* 57:245.

Coulombe, L. J. and Paquin, R. (1959). "Effects de lacide gibberellique wurle metabolisme des plantes". *Canadian J. Botany* 37:897–901.

Cowan, A. K. and Railton, I. D. (1987). "Cytokinins and ancymidol inhibit abscisic acid biosynthesis in *Persea gratissima*". *J. Plant Physiol.* 130:273.

Cox, D. A. and Keever, G. J. (1988). "Paclobutrazol inhibits growth of zinnia and geranium". *HortScience* 23:1029.

Cutler, H. G. and Schneider, B. A. (1990). *Plant Growth Regulator Handbook*, Plant Growth Regulator Society of America, Ithaca, NY.

Cutler, H. G., Yokota, T., and Adam, G. (1991). *Brassinosteroids: Chemistry, Bioactivity and Aplications*, American Chemical Society, Washington, DC.

Daie, J. (1986). "Turgor-mediated transport of sugars". *Plant Physiol.* 80S:98.

Davies, W. J.and Jones, H. G. (1991). *Abscisic Acid: Physiology and Biochemistry*, Bios Scientific Publishers, Oxford, U.K.

Davis, T. D. and Curry, E. A. (1991). "Chemical regulation of vegetative growth". *Critical Reviews in Plant Science* 10:151–188.

Davis, T. D., Steffens, G. L., and Sankhla, N. (1988). "Triazole plant growth regulators". In *Horticultural Reviews. Volume 10*, ed., J. Janick, Timber Press, Portland, OR, pp. 63.

Davis, T. D., Walser, R. H., and Sankhla, N. (1986). "Growth and photosynthesis of poinsettias as affected by plant growth regulators". *J. Current Biosci.* 3:121.

De, R. Giri, G., Saran, G., Singh, R. K. and Chaiturvedi, G. S. (1982). "Modification of water balance of dryland wheat through the use of chlormequat chloride". *J. Agric. Sci.* 98:593.

Dheim, M. A. and Browning, G. (1988a). "The mechanism of the effect of (2RS, 3RS)-paclobutrazol on flower initiiation of pear cvs Doyenne du Comice and Conference". *J. Hortic. Sci.* 63:393.

Dheim, M. A. and Browning, G. (1988b). "Preliminary studies on the use of (2RS, 3RS)-paclobutrazol for fruitlet thinning and growth control of 'Conference' pear". *J. Hortic. Sci.* 63:407.

Dong, C. N. and Arteca, R. N. (1982). "Changes in photosynthetic rates resulting from phytohormone treatments to the roots of tomato plants". *Photosynthesis Research* 3:45–52.

Dreyer, D. L., Campbell, B. C., and Jones, K. C. (1983). "Effect of bioregulator-treated sorghum on greenbug fecundity and feeding behavior: Implication to host plant resistance". *Phytochemistry* 23:1593.

Edgerton, L. J. and Hoffman, M. B. (1965). "Some physiological responses of apple to N-dimethyl amino succinamic acid and other growth regulators". *Proc. Am. Soc. Hortic. Sci.* 86:28.

Erez, A. (1985). "Growth control with paclobutrazol of peaches grown in a meadow orchard system". *Acta Hortic.* 160:26.

Erkan, Z. and Bangerth, F. (1980). "Investigations on the effect of phytohormones and growth regulators on the transpiration, stomata aperture and photosynthesis of pepper (*Capsicum annuum* L.) and tomato (*Lycopersicon esculetum* Mill.) plants". *Botany* 54:207–220.

Erwin, D. C., Tsai, S. D., and Khan, R. A. (1979). "Growth retardants mitigate Verticillium wilt and influence yield of cotton". *Phytopathology* 69:283.

Erwin, D. C., Tsai, S. D., and Khan, R. A. (1976). "Reduction of severity of Verticillium wilt of cotton by the growth retardant, tributyl[(5-chloro-2-thienyl)methyl]phosphonium chloride". *Phytopathology* 66:106.

Feiocrabend, J. (1969). "Influence of cytokinins on the formation of photosynthetic enzymes in rye seedlings". *Planta* 84:11–29.

Ferre, D. C., Schmid, J. C. and Morrison, C. A. (1982). "An evaluation over 16 years of Delicious strains and other cultivars on several rootstocks and hardy interstems". *Fruit Var. J.* 36:37.

Fiscus, E. L. (1981). "Effects of abscisic acid on the hydraulic conductance of and the total ion transport through *Phaseolus* root systems". *Plant Physiol.* 68:169–174.

Fisher, E., Still, M., and Raschke, K. (1985). "Effects of abscisic acid on photosynthesis in whole leaves: Changes in CO_2 assimilation, levels of carbon reduction cycle intermediates and activity of ribulose-1,5-bisphophate carboxylase". *International Plant Growth Substance 12th Conference 28.*

Gale, M. D., Edrich, J., and Lupton, F. G. H. (1974). "Photosynthetic rates and the effects of applied gibberellin in some dwarf, semi-dwarf and tall wheat varieties (*Triticum aestivum*)". *J. Agric. Sci. Camb.* 83:43–46.

Gausman, H. W. (1986). *Onium Bioregulators, Including Pix and Cycocel and Their Biorelevancy*, West Printing, Lubbock, TX.

Glinka, Z. (1977). "Effects of ABA and of hydrostatic pressure gradient on water movement through excised sunflower roots". *Plant Physiol.* 59:933–935.

Glinka, Z. and Reinhold, L. (1972). "Induced changes in permeability of plant cell membranes to water". *Plant Physiol.* 49:602–606.

Graselly, C. (1987). "New French stone fruit rootstocks". *Fruit Var. J.* 41:65.

Gunderson, C. A. and Taylor Jr., G. E. (1988). "Kinetics of inhibition of foliar exchange by exogenous ethylene: an ultrasensitive response". *New Phytologist* 110:517–524.

Haber, A. H. and Tolbert, M.(1957). "Photosynthesis in gibberellin treated leaves". *Plant Physiol.* 32:152–153.

Hansche, P. E. and Beres, W. (1980). "Genetic remodeling of fruit and nut trees to facilitate cultivar improvement". *HortScience* 15:710.

Haughan, P. A., Burden, R. S., Lenton, J. R., and Goad, J. L. (1989). "Inhibition of celery cell growth and sterol biosynthesis by the enantiomers of paclobutrazol". *Phytochemistry* 28:781.

Hayashi, T. (1961). "The effect of gibberellin treatment on the photosynthetic activity of plants". *Sixth International Conf. Plant Growth Regulation* 579–587.

Hickman, G. W., Perry, E. J., Mullen, R. J. and Smith, R. (1989). "Growth regulator controls tomato transplant height". *Calif. Agric.* 43:19.

Hoad, G. V., Loveys, B. R. and Skenek, G. M. (1977). "The effect of fruit removal on cytokinins and gibberellin-like substance". *Planta* 136:25–30.

Izumi, K., Kamiya, Y., Sakurai, A., Oshio, H., and Takahashi, N. (1985). "Studies of sites of action of a new plant growth retardant (E)-1-(4-chlorophenyl)-4,4-dimethyl-

2-(1,2,4-triazol-1-pentaen-3-ol (S-3307) and comparative effects of its stereoisomers in a cell-free system from *Cucurbita maxima*". *Plant Cell Physiol.* 26:821.

Izumi, K., Nakagawa, S., Kobayashi, M., Oshio, H., Sakurai, A., and Takahashi, N. (1988). "Levels of IAA, cytokinins, ABA and ethylene in rice plants as affected by a gibberellin biosynthesis inhibitor, uniconazole-P". *Plant Cell Physiol.* 29:97.

Johnson, C. R. (1974). "Response of chrysanthemums grown in clay and plastic pots to soil application of ancymidol". *HortScience* 9:58.

Jones, R. J., Griffith, S. M., and Brenner, M. L. (1986). "Sink regulation of source activity: Regulation by hormonal control". In *Regulation of Carbon and Nitrogen Reduction and Utilization in Maize*, ed., J. Shannon, Martinus Nijhoff, Hague, The Netherlands.

Kays, S. J. and Pallas Jr, J. E. (1980). "Inhibition of photosynthesis by ethylene". *Nature* 285:51–52.

Keever, G. J. and Cox, D. A. (1989). "Growth inhibition of marigold following drench and foliar-applied paclobutrazol". *HortScience* 24:390.

Knapp, J. S., Harms, C. L., and Volenec, J. J. (1987). "Growth regulator effects on wheat culm nonstructural and structural carbohydrates and lignin". *Crop Sci.* 27:1201.

Latimer, J. G. (1992). "Drought, Paclobutrazol, abscisic acid and gibberellic acid as alternatives to daminozide in tomato transplant production". *J. Amer. Soc. Hort. Sci.* 117:243–247.

Law, D. M. and Hamilton, R. H. (1989). "Reduction in the free indole-3-acetic acid levels in Alaska pea by the gibberellin biosynthesis inhibitor uniconazol". *Physiol. Plant.* 76:535.

Lester, D. C., Carter, O. G., Kelleher, F. M. and Laing, D. R. (1972). "The effect of gibberellic acid on apparent photosynthesis and dark respiration of simulated swards of *Pennisetum clandestinum* Hochst". *Australian J. Agric. Research* 23:205–213.

Little, C. H. A. and Loach, K. (1975). "Effect of gibberellic acid on growth and photosynthesis in *Abies basamea*". *Cana. J. Botany* 53:1805–1810.

Livine, A. and Vaadia, Y. (1965). "Stimulation of transpiration rate in barley leaves by kinetin and gibberellic acid". *Physiol. Plant.* 18:658–664.

Makeev, A. V., Krendeleva, T. E., and Mokronosov, A. T. (1992). "Photosynthesis and abscisic acid". *Soviet Plant Physiol.* 39:118–126.

Marcelle, R. H., Clijsters, H., Oben, G., Bronchart, R. and Micheal, J. M. (1974). "Effects of CCC and GA3 on photosynthesis of primary bean leaves". *Proc. Eighth International Conference of Plant Growth Substances* 1169–1174.

Marcelle, T. and Oben, G. (1972). "Effects of some growth regulators on the CO_2 exchanges of leaves". *Acta Horticulturae* 34:55–58.

Markhart, A. H., Fiscur, E. L., Naylor, A. W., and Kramer, P. J. (1979). "Effect of abscisic acid on root hydraulic conductivity". *Plant Physiol.* 64:611–614.

Markhart. A. H. (1982). "Penetration of soybean roots by abscisic acid isomers". *Plant Physiol.* 69:1350–2.

Marquard, R. D. (1985). "Chemical growth regulation of pecan seedlings". *HortScience* 20:119.

Maugh II, T. H. (1981). "New chemicals promise lager crops". *Science* 212:33–34.

Meidner, H. (1967). "The effect of kinetin on stomatal opening and the rate of intake of carbon dioxide in mature primary leaves of barley". *J. Exp. Botany* 18:556–561.

Meidner, H. (1969). "Rate limiting resistances and photosynthesis". *Nature* 222:876–877.

Miki, T., Kamiya, Y., Fukazawa, M., Ichikawa, T., and Sakurai, A. (1990). "Sites of inhibition by a plant-growth regulator, 4′-chloro-2′-(alpha-hydroxybenzyl) isonicotinanilide (inabenfide), and its related compounds in the biosynthesis of gibberellins". *Plant Cell Physiol.* 31:201.

Mishra, D. and Pradhan, G. C. (1972). "Effect of transpiration reducing chemicals on growth, flowering, and stomatal opening of tomato plants". *Plant Physiol.* 50:271.

Mittelheuser, C. J. and Van Steveninck, R. F. M. (1971). "Rapid action of abscisic acid on photosynthesis and stomatal resistance". *Planta* 97:83-86.

Nakayama, I., Miyazawa, T., Kobayashi, M., Kamiya, Y., Abe, H., and Sakurai, A. (1990). "Effects of a new plant growth regulator prohexadione calcium (BX-112) on shoot elongation caused by exogenously applied gibberellins in rice (*Oryza sativa* L.) seedlings". *Plant Cell Physiol.* 31:195.

Nooden, L. (1969). "The mode of action of maleic hydrazide: inhibition of growth". *Physiol. Plant.* 22:260.

Pallas Jr. J. E. and Kays, S. J. (1982). "Inhibition of photosynthesis by ethylene – a stomatal effect". *Plant Physiol.* 70:598–601.

Parthier, B. (1979). "The role of phytohormones (cytokinins) in chloroplast development". *Biochem. Physiol. Pflanzen* 174:173–214.

Pill, N. G., Lambeth, V. N., and Hinchley, T. M. (1979). "Effects of nitrogen forms and level on ion concentrations, water stress and blossom-end rot incidence in tomato". *J. Am. Soc. Hortic. Sci.* 103:265.

Pombo, G., Orzolek, M. D., Tukey, L. D., and Pyzik, T. P. (1985). "The effect of paclobutrazol, daminozide, glyphosate and 2,4-D in gel on the emergence and growth of germinated tomato seeds". *J. Hortic. Sci.* 60:353.

Poskuta, J., Antoszewski, R., and Faltynowicz, M. (1972). "Photosynthesis, photoresiration and respiration of strawberry and maize leaves as influenced by abscisic acid". *Photosynthetica* 6:370–374.

Rademacher, W. and Jung, J. (1986). "GA biosynthesis inhibitors — An update". *Proc. Plant Growth Reg. Soc. Am.* 13:102.

Raese, J. T. and Burts, E. C. (1983). "Increased yield and suppression of shoot growth and mite populations of d'Anjou pear trees with nitrogen and paclobutrazol". *HortScience* 18:212.

Raschke, K. (1975). "Stomatal action". *Annu. Rev. Plant Physiol.* 26:309–340.

Raschke, K. and Hendrick, R. (1985). "Simultaneous and independent effects of abscisic acid on stomata and the photosynthetic apparatus in whole leaves". *Planta* 163:105–118.

Richardson, P. J., Webster, A. D., and Quinlan, J. D. (1986). "The effect of paclobutrazol sprays with or without the addition of surfactants on the shoot growth, yield, and fruit quality of the apple cultivars Cox and Suntan". *J. Hortic. Sci.* 61:439.

Riddell, J. A., Hageman, H. A., J'Anthony, C. M., and Hubbard, W. L. (1962). "Retardation of plant growth by a new group of chemicals". *Science* 136:391.

Riken, A., Blemenfeld, A., and Richmond, A. E. (1976). "Chilling resistance as affected by stressing environments and ABA". *Bot. Gaz.* 137:307–312.

Robinson, S. P., Wiskich, J. T., and Paleg, L. G. (1978). "Effects of indoleacetic acid on CO_2 fixation, electron transport and phosphorylation in isolated chloroplasts". *Aust. J. Plant Physiol.* 5:425–431.

Rogers, B. L. and Thompson, A. H. (1968). "Growth and fruiting response of young apple and pear trees to annual applications of succinic acid 2,2-dimethylhydrazide on fruit shape of Delicious apples". *HortScience* 93:16.

Rom, R. C. (1983). "The peach rootstock situation: An international perspective". *Fruit Var. J.* 41:65.

Rom, R. C. and Carlson, R. F. (1987). *Rootstocks for Fruit Crops*, John Wiley and Sons, New York.

Sanhla, N. and Huber, W. (1974). "Eco-physiological studies on India arid zone plants. IV. Effect of salinity and gibberellin on the activities of photsynthetic enzymes and CO_2 fixation products in leaves of *Pennisetum typhoides* seedlings". *Biochem. Physiol. Pflanzen* 166:181–187.

Sauerbrey, E., Grossman, K., and Jung, J. (1988). "Ethylene production by sunflower cell suspensions effects of plant growth retardants". *Plant Physiol.* 87:510.

Schneider, G. (1970). "Morphactins: Physiology and performance". *Annu. Rev. Plant Physiol.* 21:499.

Schoene, D. L. and Hoffman, D. L. (1949). "Maleic hydrazide, a unique growth regulant". *Science* 109:588.

Senaratna, T., Mackay, C. E., McKersie, B. D., and Fletcher, R. A. (1988). "Uniconazole-induced chilling tolerance in tomato and its relationship to antioxidant content". *J. Plant Physiol.* 133:56.

Shaltout, A. D., Salem, A. T., and Kilany, A. S. (1988). "Effect of pre-bloom sprays and soil drenches of paclobutrazol on growth, yield, and fruit composition of Roumi Red grapes". *J. Am. Soc. Hortic. Sci.* 113:13.

Sharif, R. and Dale, J. E. (1980). "Growth regulating substances and the growth of tiller buds in barley; effects of cytokinins". *J. Exp. Botany* 31:921–930.

Shive, J. B. and Sisler, H. D. (1976). "Effects of ancymidol (a growth retardant) and triarimol (a fungicide) on the growth, sterols and gibberellins of *Phaseolus vulgaris* (L.)". *Plant Physiol.* 57:640.

Smeirat, N. and Qrunfleh, M. (1989). "Effect of paclobutrazol on vegetative and reproductive growth of Lisbon lemon". *Acta Hortic.* 239:261.

Squier, S. A., Taylor, G. E., Selvidge, W. J., and Gunderson, C. A. (1985). "Effect of ethylene and related hydrocarbons on carbon assimilation and transpiration in herbaceous and woody species". *Environ. Sci. Technol.* 19:432–437.

Sterrett, J. P., Tworkoski, T. J., and Kujawshi, P. T. (1989). "Physiological responses of deciduous tree root collar drenched with flurprimidol". *J. Arboricult.* 15:120.

Swietlik, D. (1986). "Effect of gibberellin inhibitors on growth and mineral nutrition of sour orange seedlings". *Sci. Hortic.* 29:325.

Tamas, I. A., Schwartz, J. W., Breithaupt, B. J., Hagin, J. M., and Arnold, P. H. (1973). "Effect of indoleacetic acid on photosynthetic reactions in isolated chloroplasts". In *Proc. Eighth International Conf. Plant Growth Substances*, pp. 1159–1168.

Tayama, H. K. and Carver, S. A. (1990). "Zonal geranium growth and flowering responses to six growth regulators". *HortScience* 25:82.

Taylor Jr., G. E. and Gunderson, C. A. (1986). "The response of foliar gas exchage to exogenously applied ethylene". *Plant Physiol.* 82:653–657.

Treharne, K. J. and Stoddart, J. L. (1968). "Effects of gibberellin on photosynthesis in red clover *(Trifolium pratense* L.)". *Nature* 220:457–458.

Treharne, K. J., Stoddart, J. L., Pughe, J., Paranjothy, K., and Wareing, P. F. (1970). "Effects of gibberellins and cytokinins on the activity pf photosynthetic enzymes and plastid ribosomal RNA synthesis in *Phaselous vulgaris* L.". *Nature* 228:129–131.

Tsai, D. S. and Arteca, R. N. (1985). "Effects of root applications of gibberellic acid on photosynthesis and growth in C_3 and C_4 plants". *Photosynthesis Res.* 6:147–157.

Turner, W. B. and Bidwell, R. G. S. (1965). "Rates of photosynthesis in attached and detached bean leaves and the effect of spraying with indole-3-acetic acid". *Plant Physiol.* 40:446–451.

Upadhyaya, A., Davis, T. D., Larsen, M. H., Walser, R. H., and Sankhla, N. (1990). "Uniconazole-induced thermotolerance in soybean seedling root tissue". *Physiol. Plant.* 79:78.

Upadhyaya, A., Davis, T. D., Walser, R. H., Galbraith, A. B. and Sankhla, N. (1989). "Uniconazole-induced alleviation of low-temperature damage in relation to anti-oxidant activity". *HortScience* 24:955.

Upadhyaya, A., Davis, T. D., and Walser, R. H. (1991). "Alleviation of sulfur dioxide-induced phytotoxicity in cucumber plants by uniconazole". *Biochem. Physiol. Pflanzen* 187:59.

Walton, D. C. (1980). "Biochemistry and physiology of abscisic acid". *Annu. Rev. Plant Physiol.* 31:453–490.

Wang, S. A. Y., Sun, T., Zuo, L. J., and Faust, M. (1987). "Effect of paclobutrazol on water stress-induced abscisic acid in apple seedling leaves". *Plant Physiol.* 84:1051.

Wareing, P. F., Khalifa, M. M., and Treharne, K. J. (1968). "Rate-limiting processes in photosynthesis at saturating light intensities". *Nature* 222:453–457.

Wellburn, F. A. M., Wellburn, A. R., Stoddart, J. L., and Treharne, K. L. (1973). "Influence of gibberellic acid and abscisic acid and the growth retardant, CCC upon plastid development". *Planta* 111:337–346.

Wood, B. W. (1988). "Paclobutrazol supresses shoot growth and influences nut quality and yield of young pecan trees". *J. Am. Soc. Hortic. Sci.* 113:374.

Woodrow, L. and Grodzinski, B. (1989). "An evaluation of the effects of ethylene on carbon assimilation in *Lycopersicon esculentum* Mill.". *J. Exp. Botany* 40:361–368.

Woodrow, L. and Grodzinski, B. (1993). "Ethylene exchange in *Lycopersicon esculentum* Mill. leaves during short-term and long-term exposures to CO_2". *J. Exp. Botany* 44:471–480.

Woodrow, L., Jiao, J., Tsujita, M. J., and Grodzinski, B. (1989). "Whole plant and leaf steady state gas exchange during ethylene exposure in *Xanthium strumarium* L.". *J. Amer. Soc. Hort. Sci.* 90:85–90.

Zerbe, R. and Wild, A. (1981). "The effect of indole-3-acetic acid on the photosynthetic apparatus of *Sinapis alba*". *Photosynthetic Research* 1:71–81.

Zillkah, S. and Gressel, J. (1978). "Differential inhibition by dikegulac of dividing and stationary cells in vitro". *Planta* 147:274.

Zummo, G. R., Benedict, H. J. H., and Segers, J. C. (1984). "Effect of the plant growth regulator mepiquat chloride on host plant resistance in cottton to bollworm (Lepidoptera: Noctuidae)". *J. Econ. Entomol.* 77:922.

CHAPTER 13

Weed Control

Most if not all people working with plants understand what is meant by the term *weed*. The most common definition of a weed is a plant growing where it is not desired (Buchholtz 1967). Now even though this definition makes a lot of sense, it is not one which is agreed upon by all. In fact, there are a number of definitions which have been used to describe weeds (Harlan and deWet 1965), and a considerable amount of debate about the pros and cons of each (Zimdahl 1993). Weeds are very costly because they compete with crop plants for water, nutrients, and light, while harboring diseases and insects which attack crop plants, thereby increasing production costs. In addition, weeds:

1. Reduce quality of farm products.
2. Reduce plant and animal yields.
3. Increase production costs.
4. Interfere with water management.
5. Pose problems with human health.

6. Limit human efficiency.
7. Decrease land value.
8. Reduce crop options which can be planted on a given piece of land.
9. May be a fire hazard.
10. Can be unsightly.

Weeds are classified as pests in the same manner as insects, plant diseases, nematodes, and rodents. A chemical used to control a pest is called a pesticide, and one which is used specifically for weed control is known as an herbicide. The cost of weeds to man is much higher then generally recognized. Weeds are widespread in their occurrence and cost billions of dollars annually due to crop losses and control costs on farms in the United States (Chandler et al. 1984). Weeds also have a major economic impact in home site maintenance, lowered human efficiency, and recreational areas. At the present time there are no accurate dollar amounts available; however, weeds result in major losses annually in these areas.

WEED CONTROL METHODS

There are many reasons why weed control is necessary from successful crop production to aesthetics in the urban environment. The following sections outline some of the factors and practices which should be considered when developing a weed management program.

Prevention, Control, and Eradication

Prevention is a practical way to stop a given species of weeds from contaminating an area. In fact, a good weed manager emphasizes prevention rather than control. This is accomplished by:

1. Making sure no new weed seeds are carried into a given area by contaminated crop seeds, feed, digestive tracts of animals, or on machinery. In addition, one should be careful to inspect nursery stock and any materials such as sand, soil, etc. brought onto the property and to pay special attention to the perimeter of the property line as a source of new weeds.
2. Preventing weeds in a given area from going to seed.
3. Preventing the spread of perennial weeds which reproduce vegetatively.

Prevention is a very important management practice because it is a practical and cost-effective way to control weed problems, especially preventing outbreaks of new weed problems; however, it is very difficult to fully implement.

Control is an important form of weed management which decreases the population of weeds to levels which do not reduce yields, interfere with harvest operation, or create other problems typically associated with weeds.

Eradication is the complete elimination of a weed species from an area including live plants and reproductive parts such as seeds and vegetative structures. On a large scale, eradication is not generally economically feasible, however, on a smaller scale where high-value horticultural or ornamental plants are grown, this approach may be a viable option. Eradication generally uses soil fumigation to get rid of weed seeds and vegetative parts found in the soil, although chemicals can also be used successfully. An added benefit of soil fumigation is insect and disease control.

Weed Management

As mentioned earlier it is always better to prevent than to control; however, this is not always possible because in many cases weed seeds are airborne and/or animalborne and infiltration cannot be prevented. Weed management is both a science and an art combining prevention, eradication, and control to successfully reduce the weed problem to an acceptable level. A good weed manager takes into account many factors such as the field cropping history, objectives of the control, technology available, financial resources, and many other factors in order to successfully control weeds. Once good management practices have been established to prevent weeds there are four general areas of weed control:

1. Mechanical practices include tillage, hand weeding, mowing, mulching, electric weeding, burning, sound weeding and flooding.
2. Cultural practices such as crop selection (competition), rotation, variety selection, planting date, plant population, and spacing, plus fertility and irrigation all affect weed management.
3. Biological control uses insects and diseases that naturally attack weeds and the competitive ability of crops and herbivores to reduce or eliminate the detrimental effects of weed populations.
4. Chemical control is used to nonselectively kill weeds prior to planting or to selectively kill weeds in crops.

The best reduction in weed population incorporates mechanical, cultural, and chemical control. In specific cases biological controls can be added to this combination, but its use is limited. Although all groups and their integration are important, further discussion will be limited to chemical control, since it directly relates to the general theme of this text (Ashton and Monaco 1991; Zimdahl 1993; *Herbicide Handbook* 1989).

INTRODUCTION TO CHEMICAL CONTROL

The use of chemicals to selectively kill weeds is a key part of any weed management system. The word *herbicide* comes from the Latin *herba*, or plant, and *caedere*, to kill. Since Theophrastus reported that trees, especially young trees, could be killed by pouring olive oil over their roots, many nonselective materials such as salt, sodium arsenate, carbon bisulfide, petroleum oils and others have been shown to kill weeds (Zimdahl 1993). It was not until 1932 that the first synthetic organic chemical, 2-methyl-4,6-dinitrophenol was introduced for selective weed control in crops. It was used for many years thereafter for selective control of some broad-leaf weeds and grasses in large-seeded crops such as beans. Today, there are over 130 different selective herbicides used in the world. In the United States one billion pounds of pesticides are sold annually and over 65% are herbicides. It must be stressed that selectivity is the key to the widespread use of herbicides. Herbicides, like any technology, have advantages and disadvantages which must be carefully evaluated prior to use.

There are several methods for classifying herbicides, each having a problem associated with it. For discussion purposes, herbicides can be grouped into the following categories:

1. Crop use.
2. Visual effects.
3. Site of uptake (root versus shoot absorption).
4. Contact versus systemic activity.
5. Selectivity.

Although it is very important to know whether an herbicide is selective, it is a poor way to classify herbicides. Selectivity is primarily a function of rate and other factors such as:

1. Plant age and stage of growth.
2. Morphology.
3. Adsorption characteristics.
4. Translocation.
5. Type of treatment.
6. Time and method of application.
7. Chemical formulation.
8. Environmental conditions.
9. Time of application.
10. Chemical structure.
11. Mode of action.

The pros and cons of each method of classification are presented by Zimdahl (1993) and Ashton and Monaco (1991).

The three major mechanisms of herbicide action which will be discussed are:

1. Inhibition of plant growth.
2. Inhibition of respiration or photosynthesis.
3. Inhibition of plant biosynthetic processes.

Of the eight classes of plant growth substances discussed in this text there are herbicides which mimic auxin action and those which interfere with gibberellin biosynthesis, while there are no herbicides which directly interfere or promote any of the other plant growth substances.

Herbicides Which Mimic IAA Action

Phenoxy Acids. The introduction of 2,4-dichlorophenoxyacetic acid (2,4-D) and 4-chloro-2-methylphenoxyacetic acid (MCPA) in the mid-1940s, immediately following World War II, had a profound effect on weed control. It was demonstrated that synthetic organic compounds could be developed economically for selective control of weeds in crops. After these compounds were introduced, the chemical industry initiated a major program on research and development of a wide variety of herbicides which are available and commonly used today.

The discovery that phenoxy herbicides are phytotoxic was a direct result of basic plant growth regulator research. They are often referred to as auxinlike herbicides because they induce epinasty in broad-leaf plants in the same way as IAA, the naturally occurring auxin found in plants. Today there are six phenoxy herbicides used in the United States. All have the basic phenoxy structure with a chlorine at position 4 of the ring and an aliphatic acid (acetic, butyric, or propanoic acid) attached to the oxygen atom with either a chlorine or methyl group at position 2 on the ring. The common name, chemical name, and structure are given in Table 13.1. As is evident from Table 13.1, each of the six phenoxy herbicides have similar characteristics; however, they each have unique uses. 2,4,5-T, which had been used as a phenoxy herbicide, has been withdrawn from the market because of the presence of a carcinogenic contaminant, 2,3,8,9-tetrachlorodibenzo-p-dioxin (TCDD), occurring during its synthesis. Current methods for the synthesis of the six phenoxy herbicides used today do not contain TCDD.

2,4-D is a white crystalline solid which is slightly soluble in water. Common forms of 2,4-D are acid, amine salts, inorganic salts, and esters. The different forms resulting from substitution at the terminal hydrogen atom on the acetic acid side chain of the parent molecule are shown in Table 13.1. These substi-

Table 13.1.   The common name, chemical name, and structural formula for six phenoxy herbicides.

Common name	Chemical name	R_1	R_2
2,4-D	(2,4-Dichlorophenoxy)acetic acid	$-CH_2-COOH$	$-Cl$
MCPA	(4-Chloro-2-methylphenoxy)acetic acid	$-CH_2-COOH$	$-CH_3$
2,4-DB	4-(2,4-Dichlorophenoxy)butanoic acid	$-(CH_2)_3-COOH$	$-Cl$
MCPB	4-(4-Chloro-2-methylphenoxy)butanoic acid	$-(CH_2)_3-COOH$	$-CH_3$
Dichloroprop (2,4-DP)	($\pm$)-2-(2,4-Dichlorophenoxy)propanoic acid	CH_3 $\vert$ $-CH-COOH$	$-Cl$
Mecoprop (MCPP)	($\pm$)-2-(4-Chloro-2-methylphenoxy) propanoic acid	CH_3 $\vert$ $-CH-COOH$	$-CH_3$

tutions modify physical and biological characteristics of the parent molecule, facilitating their use and/or increasing their effectiveness. The efficiency of a particular form of 2,4-D is associated with increased absorption, however, volatility of the compound generally increases with greater absorption. 2,4-D in its acid form is not commonly used because it is slightly soluble in water, has a low volatility, and is fairly expensive to formulate. There are many formulations of 2,4-D currently available with various combinations of amines, salts, esters, or acids. One formulation of 2,4-D in its acid form together with an ester has been shown to be very effective for the control of field bindweed, Russian knapweed, Canada thistle, leafy spurge, and cattails (Ashton and Monaco 1991). There are several different amine forms of 2,4-D which can be broken down into two types, both having very different properties:

1. Water-soluble (dimethylamine, isopropylamine, and triethanolamine).
2. Oil-soluble amines (dodecylamine) (Table 13.2).

The water-soluble amine forms are commonly used because they are very soluble in water, have a low volatility, are easy to handle under field conditions, and are inexpensive. They are typically less effective than the highly

Table 13.2. The structure of 2,4-dichlorophenoxyacetic acid along with different formulations, their chemical names, solubility, and volatility

$$O-CH_2-\overset{\overset{\displaystyle O}{\|}}{C}-O-R'$$

(attached to a 2,4-dichlorophenyl ring, substituents Cl)

Form	Chemical name	R'	Solubility	Volatility
Salts	Ammonium	NH_4^+	water-soluble	nonvolatile
	Sodium	Na^+	water-soluble	nonvolatile
	Potassium	K^+	water-soluble	nonvolatile
	Lithium	Li^+	water-soluble	nonvolatile
Amines	Dimethylamine	CH_3 $-N-CH3$	water-soluble	low
	Isopropylamine	CH_3 $-NH-CH-CH_3$	water-soluble	low
	Triethanolamine	$-NH-(CH_2-CH_2OH)_3$	water-soluble	low
	Dodecylamine	$-NH-CH_2-(CH_2)_{10}-CH_3$	oil-soluble	nonvolatile
Esters	Ethylester	$-CH_2CH_3$	insoluble	volatile
	Isopropylester	CH_3 $-CH_2-CH_3$	insoluble	volatile
	Butoxyethylester	$-CH_2-CH_2-O-(CH_2)_3$ CH_3	insoluble	low
	Propylene glycol butyl etherester	$(-CH-CH_2-O)_x-(CH_2)_3-CH_3$ CH_3	insoluble	low

volatile esters, however, they provide good weed control at a reasonable cost. Oil-soluble amines must be used as an emulsifiable concentrate and have the advantage of being as effective as the low-volatility esters of 2,4-D with minimal volatility even at high temperatures. There are a number of different 2,4-D esters which are used, each varying in their degree of volatility, all of which are insoluble in water and used as emulsifiable concentrates. When the length of the alcohol side chain is increased there is a reduction in the volatility of the compound, and this typically reduces its absorption by the plant (Table 13.2). In general, esters are absorbed much more readily than other forms of 2,4-D. When a salt such as ammonium, potassium, sodium, or lithium is attached to the 2,4-D parent molecule it becomes water-soluble and nonvolatile. Unfortunately, it also makes it the least effective type of 2,4-D. In fact, of the four salts mentioned, the lithium salt is fairly effective and still available, while the other three are limited in their use.

2,4-D is used as a postemergence treatment to control annual and perennial broad-leaf weeds in grass crops and noncrop areas. It can also be used for woody plant and aquatic weed control. In order to obtain maximal herbicidal activity 2,4-D should be applied as a foliar spray since it can readily be leached from the soil, adsorbed onto soil colloids, and can be readily degraded by microorganisms in the soil. The most obvious visual effect of 2,4-D and other phenoxy herbicides on broad-leaf plants is epinasty (Figure 13.1). Leaves readily absorb 2,4-D, which is translocated both acropetally and basipetally. It accumulates in areas where there is high photosynthate utilization, such as developing organs and meristems. Although applied as a foliar spray, 2,4-D is translocated to underground roots and rhizomes, making it very useful in the control of perennial weeds. 2,4-D acts in a similar manner to IAA, however, IAA has endogenous control mechanisms which regulate its levels within a physiological range, whereas the plant has limited ability to control 2,4-D levels once it enters the plant. When applied to plants over a range of concentrations from low to high, 2,4-D exhibits a typical bell-shaped curve, initially showing a stimulation of growth at low concentrations followed by an inhibition of plant growth and related metabolic processes. When applied as a foliar spray, the level of 2,4-D in meristems and developing organs increases with time after application, at first being low and later high. Therefore, initially 2,4-D promotes a dramatic stimulation of metabolic processes and growth resulting in epinasty followed by a suppression in these processes and growth, which results in death of the plant (Penner and Ashton 1966; van Overbeek 1964). 2,4-D has been shown to promote a dramatic stimulation in ethylene production when applied to vegetative tissues, therefore, it is possible that the uncontrolled growth promoted by 2,4-D may be due to ethylene (Abeles et al. 1992).

As shown in Table 13.1, MCPA is very similar in structure to 2,4-D,

Figure 13.1.   Epinasty induced by 2,4-D in flowering pear (courtesy of L. J. Kuhns).

differing at the number 2 position on the parent ring where it has a methyl group instead of a chlorine. It can be formulated as a free acid, amine, inorganic salt, or ester which affects both its physical and biological properties in the same manner as 2,4-D. Although its herbicidal properties are similar to 2,4-D, it has been shown to be more selective on cereals, legumes, and flax at equal rates (Ashton and Monaco 1991). Since the structures are almost identical, the soil influence and mode of action of MCPA is similar to 2,4-D.

2,4-DB is similar in structure to 2,4-D; however, it has a butanoic acid attached to the oxygen on the parent ring instead of acetic acid (Table 13.1). Although 2,4-DB can be formulated in the same way as 2,4-D, it is commonly used as a water-soluble dimethylamine or a low-volatility butoxyethanol ester. It has been shown to be very selective on small-seeded and other legume crops as a postemergence treatment for controlling broad-leaf weeds. The amine form is used in alfalfa, bird's-foot trefoil, peanuts, and soybean, whereas, the ester is used in alfalfa and bird's-foot trefoil. It is affected by soil in the same way as 2,4-D, in fact, for it to be effective it must undergo β-oxidation converting it to the 2,4-D acid form. This reaction is more rapid in susceptible rather than tolerant plants, thereby permitting its use as a selective herbicide (*Herbicide Handbook* 1989).

MCPB is identical in structure to 2,4-DB except it has a methyl group at the number 2 position of the ring instead of a chlorine atom (Table 13.1). It is formulated as a sodium salt and used as a postemergence herbicide to control Canada thistle and other broad-leaf weeds in peas. The soil influence of MCPB is similar to 2,4-D, and the mode of action is similar to 2,4-DB (Ashton and Monaco 1991).

Dichlorprop is similar in structure to 2,4-DB except it has a propanoic acid group attached to the oxygen on the parent ring as opposed to a butanoic acid (Table 13.1). It is commonly used as a low-volatility butoxyethyl ester for woody plant control on roadsides and utility right-of-way. Mecoprop is identical to dichlorprop except it has a methyl group at the number 2 position on the parent ring as opposed to a chlorine (Table 13.1). It is usually formulated as a soluble amine salt and in some cases it is used together with 2,4-D to broaden its effect. Mecoprop is used to control 2,4-D-tolerant weeds; however, there are some restrictions for certain grass species, environmental conditions, and time of mowing. For more details see product label and consult with local authorities. The soil influence and mode of action for both dichlorprop and mecoprop are similar to 2,4-D.

Benzoic Acids. Dicamba is the common name for 3,6-dichloro-2-methoxybenzoic acid (Figure 13.2). This structure should look familiar because it is similar to salicylates, however, there is a methoxy group in the number 2 position in place of a hydroxyl group and chlorines are added at the number 3

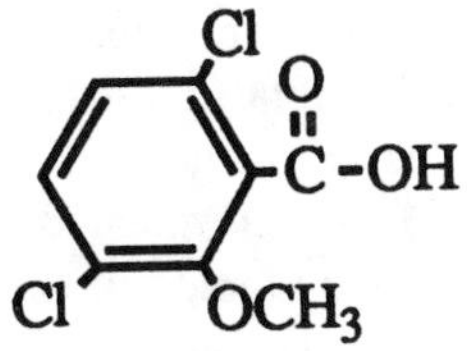

Cl   O
       ‖
       C-OH

Cl   OCH₃

Dicamba
(3,6-dichloro-2-methoxybenzoic acid)

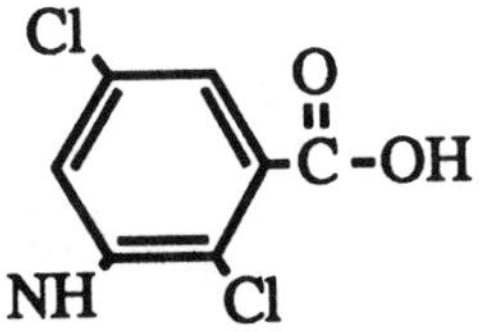

Cl   O
       ‖
       C-OH

NH   Cl

Chloramben
(3-amino-2,5-dichlorobenzoic acid)

Figure 13.2. Structural formulas for two benzoic acid herbicides.

and 6 positions. Dicamba is a growth-regulator herbicide, which controls a similar spectrum of weeds as 2,4-D. However, it is more effective on many perennial weeds at lower rates than 2,4-D. It is used as a preemergence or postemergence herbicide to control a wide range of annual and perennial broad-leaf weeds, primarily in grass crops. It is typically formulated as a dimethylamine salt and is also available together with 2,4-D, atrazine, or several other herbicides to broaden the spectrum of weed control. Dicamba is fairly mobile within the soil and subject to leaching. Under warm-moist, slightly acidic soil conditions dicamba is degraded by soil microorganisms and goes through one half-life in 14 days; however, this is under ideal conditions. When soil conditions are cool and dry, dicamba can persist in the soil for up to several months. Dicamba has a similar mode of action as 2,4-D, promoting epinasty and other physiological effects (*Herbicide Handbook* 1989).

Chloramben is the common name for 3-amino-2,5-dichlorobenzoic acid (Figure 13.2). Its structure is similar to dicamba, however, it has a chlorine in the number 2 position in place of the methoxy and an amine group in the number 3 position. Chloramben is formulated as an ammonium or sodium salt and is used as a preemergence herbicide in vegetable and field crops. In order to increase the spectrum of weeds controlled by chloramben it is also used in combination with other herbicides. In sandy soils chloramben is readily leached from the soil, however, as the level of organic matter increases, the degree of leaching is also reduced. Microorganisms in the soil will degrade chloramben, but it has been shown to persist in most soils for six to eight weeks. Chloramben is readily absorbed by roots, seeds, or leaves and is transported depending upon the species (Ashton and Crafts 1981). Seedlings treated with chloramben show epinastic symptoms which are characteristic of auxinlike compounds.

Pyridines

The basic chemical structure of a pyridine consists of one nitrogen atom and five carbon atoms in a six-membered ring (Figure 13.3). There are three herbicides which fall into this class: picloram, triclopyr, and clopyralid. Picloram is the common name for 4-amino-3,5,6-trichloro-2-pyridinecarboxylic acid (Figure 13.3). It is a highly potent, persistent, and relatively nonselective herbicide which can be used selectively in small grains giving excellent control of woody plants and many perennial broad-leaf species including field bind-weed and Canada thistle. Picloram can be applied to either the foliage or soil where it is readily adsorbed by organic matter and certain clays (*Herbicide Handbook* 1989), while it is readily leached through sandy soils. Its use as an herbicide has been restricted for a number of reasons, one of the main ones being that it is very persistent in soils because it is slowly degraded by soil microorganisms and has been shown to have phytotoxic effects on plants more than one year after application. Picloram is readily absorbed by all parts of the plant and is readily transported to areas of rapid growth, where it accumulates and persists because it cannot be metabolized by plants. Its mode of action is thought to be similar to 2,4-D because it induces epinasty and other effects in a similar manner (Ashton and Crafts 1981; Fedtke 1982).

Clopyralid is the common name for 3,6-dichloro-2-pyridine carboxylic acid (Figure 13.3). Its structure differs from picloram because it does not have an amino group in the number 1 position nor does it have a chlorine in the number 6 position. It is used as a postemergence herbicide and is typically applied as a foliar spray; however, in certain species it has been shown to be taken up by the roots. Clopyralid is effective in controlling many annual and perennial broad-leaf weeds and certain woody species. It has been shown to be very effective in controlling weeds from polygonaceae, compositae, and legumin-

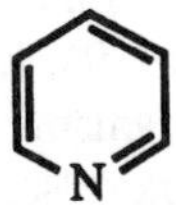

Pyridine

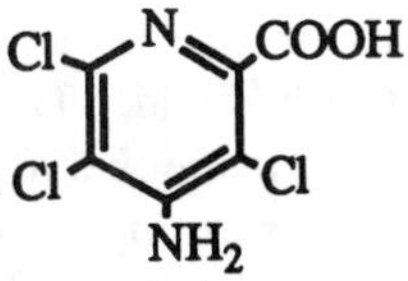

Pichloram
(4-amino-3,5,6-trichloropicolinic acid)

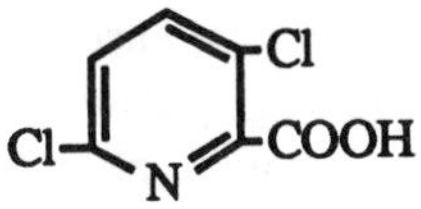

Clopyralid
(3,6-dichloro-2-pyridinecarboxylic acid)

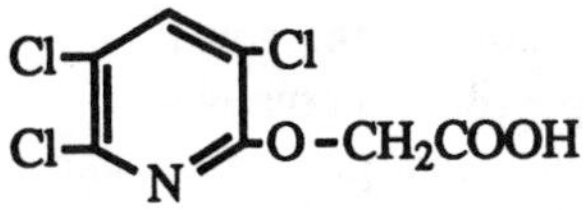

Triclopyr
[(3,5,6-trichloro-2-pyridinyl)oxy]acetic acid

Figure 13.3. Structural formulas for three pyridine herbicides.

osae families (Ashton and Monaco 1991). Clopyralid is not readily adsorbed to soil colloids (*Herbicide Handbook* 1989), in sandy soils it is readily leached and can be degraded by soil microorganisms fairly rapidly. The mode of action is similar to that of 2,4-D and it is more selective than picloram.

Triclopyr is the common name for [(3,5,6-trichloro-2-pyridinyl)oxy] acetic acid (Figure 13.3), which differs from clopyralid because it has an oxyacetic acid at the number 3 position and a chlorine in the number 6 position. It is effective on woody plants and a variety of broad-leaf weeds. It is more effective than other auxinlike herbicides in controlling ash, oaks, and other root-sprouting species. Most grasses are tolerant and while not used extensively in crops, it is widely used as a turf herbicide. Triclopyr is not strongly adsorbed onto organic matter (*Herbicide Handbook* 1989), is readily leached in sandy soils and degraded fairly rapidly in soils, by microorganisms. Its mode of action is similar to 2,4-D and other auxinlike herbicides.

Gibberellin Synthesis Inhibitors

At the present time there are no herbicides which specifically interfere with gibberellin synthesis or action. However, it has recently been suggested that gibberellin synthesis inhibitors may have practical use in the regulation of weed species. It is well established that aquatic plants play a key role in providing oxygen, habitate, and sediment stabilization in fresh water systems. However, when aquatic plant growth becomes too dense it poses many problems, such as restricting fishing, boating, swimming, and other uses (Riemer 1984; Ross and Lembi 1985). At the present time aquatic plant management strategies cause a major reduction or even the elimination of plant populations due herbicides (Ashton and Monaco 1991), many of which are not selective resulting in adverse effects such as oxygen depletion and others. There are some aquatic plants which elongate rapidly, making their way above the water surface and forming a canopy which shades plants with prostrate growth forms, thereby restricting their growth (Barko and Smart 1981; Barko et al. 1986). These rapidly elongating plants are considered to be aquatic weeds. However, if these plants were kept short they would no longer be considered weeds and could still serve useful functions in the aquatic environment. Netherland and Lembi (1992) explored a possible alternative to current weed control strategies by interfering with normal plant growth substance synthesis or action to alter the morphology of the plant.

Gibberellins are known to induce elongation (Takahashi et al. 1991), while pyrimidine and triazole plant growth retardants inhibit gibberellin synthesis (Graebe 1987; Davis and Curry 1991). These compounds reduce stem length in a wide variety of plants (Sterret 1988; Sterret and Tworkoski 1987) including deep-water rice (Raskin and Kende 1984) and the aquatic plant *Callitriche* spp. (Muskgrave et al. 1972), suggesting that aquatic weeds capable of rapid elongation may also be inhibited by these kinds of compounds. The potential of the gibberellin synthesis inhibitors flurprimidol (a substituted pyrimidine), paclobutrazol, and uniconazole (triazole derivatives) to inhibit stem elongation of the weedy submersed plants Eurasian watermilfoil and hydrilla under laboratory conditions without adversely affecting selected physiological parameters were evaluated by Netherland and Lembi (1992). They showed that these inhibitors reduced plant height but did not affect photosynthesis, respiration, or chlorophyll content in the two weedy submersed aquatic plants. Both plants required only a 24-hour exposure to the growth retardant to maintain stem length reduction for a six week period, suggesting that these compounds have potential for reducing plant height in aquatic systems. It should be noted that the sensitivity and morphological responses to plant growth retardants differed between the two species. Although the use of plant growth retardants to reduce aquatic plant growth is promising under laboratory conditions, field studies must be done in order to more completely evaluate their usefulness as a new strategy in aquatic plant management on a large scale.

Inhibition of Respiration, Photosynthesis, and Biosynthetic Processes

Herbicides which act by inhibiting respiration, photosynthesis, and biosynthetic processes are very important in agriculture, probably more so than those which inhibit plant growth. There are many herbicides involved in the inhibition of each of these processes with a great deal of information about each one. In keeping with the focus of this text, only examples of the different classes of herbicides and the processes which they affect will be discussed. For more details on these herbicides see the *Herbicide Handbook* (1989).

Respiration Inhibitors. Herbicides can interfere with respiration in two ways, by uncoupling oxidative phosphorylation or by blocking electron transport. Examples of compounds which uncouple oxidative phosphorylation are methylarsonic acid (MAA) and phenols such as dinoseb, while hydroxybenzonitriles such as bromoxynil inhibit electron transport.

Chemicals based on methylarsonic acid (Figure 13.4), such as monosodium or disodium methanearsonate, are uncouplers of oxidative phosphorylation. They are currently used as selective postemergence herbicides in cotton and noncrop areas to control johnsongrass, nutsedge, water grass, foxtail, cocklebur, pigweed, and others. In cotton it is applied as a directed postemergence spray when cotton is 3 in. tall and prior to full bloom, while in noncrop vegetation all foliage should be sprayed to obtain maximum coverage. All formulations of methylarsonic acid-based compounds are highly water-soluble. These compounds are almost completely inactivated in the soil by surface adsorption and ion exchange (*Herbicide Handbook* 1989).

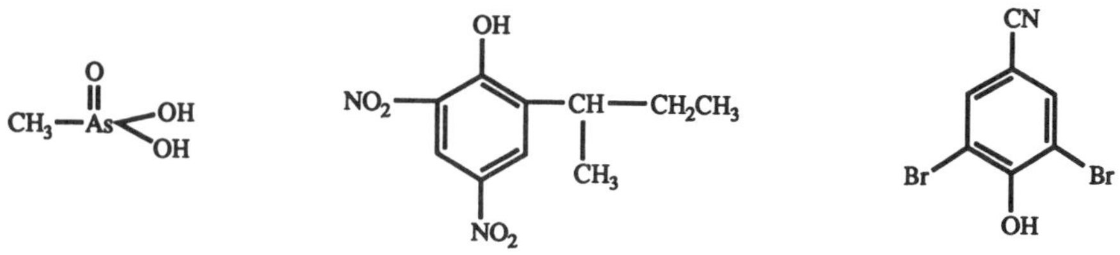

Methylarsonic acid (MAA)   **Dinoseb**   **Bromoxynil**

Figure 13.4.   Structural formulas for herbicides which inhibit respiration.

The substituted phenol dinoseb (2-1(1-methylpropyl)-4,6-dinitrophenol) (Figure 13.4) also uncouples oxidative phosphorylation. Dinoseb is effective in controlling seedling weeds and grasses in crops such as small grains, soybeans, peanuts, beans, strawberries, some forage crops, and in grape vineyards, and fruit and nut orchards. It is applied as a preplant, preemergence, postemergence, and directed postemergence application depending on the weed to be controlled, crop, and formulation used. The usual carrier for dinoseb is either an oil-based emulsion or oil solution. As a foliar spray it is absorbed without subsequent translocation and causes cell necrosis. Salt formulations are readily washed from the foliage, while oil solutions are more resistant. Dinoseb is loosely adsorbed in most soils and in sandy soils some leaching can occur. It has been shown that dinoseb can be broken down by microorganisms and some volatilization from soils can occur. The persistence of dinoseb in the soil is from two to four weeks under normal conditions when used at the recommended rates (*Herbicide Handbook* 1989).

The hydroxybenzonitrite bromoxynil (3,5-dibromo-4-hydroxybenxonitrile) (Figure 13.4) is an inhibitor of electron transport. It is registered for use on wheat, barley, oats, rye, corn, grain sorghum, garlic, onions, seedling alfalfa, flax, mint, annual canary grass, newly seeded turf to control specific broad-leaf weeds, and on noncrop sites. Bromoxynil is applied as a postemergence foliar spray and water is its usual carrier. It is readily absorbed by the foliage and is not generally translocated. When applied as a sodium salt formulation it is water-soluble and readily washed from the leaf; however, when applied as an oil-soluble amine or ester it can resist removal from the foliage. In plants, bromoxynil can be hydroxylated to form benzoic acid, while little is known about how it is affected in the soil (*Herbicide Handbook* 1989).

Photosynthetic Inhibitors.   More classes of herbicides act on photosynthesis by inhibiting the conversion of light energy to chemical energy than on any other physiological processes. Herbicides from different chemical classes such as ureas, uracils, triazines, triazinones, acylanilids, and pyridazinones interfere with the reduction of plastoquinone on the acceptor site of PS II (Schulz et al. 1990). These substances with very different structures bind to the thylakoid membrane of chloroplasts, thereby blocking photosynthetic electron transport.

Urea is commonly used as a nitrogen fertilizer. It can be made into an effective herbicide by substituting three of the hydrogen atoms of urea (Figure 13.5) with a phenyl, methyl, and/or methoxy group. Diuron (N'-(3,4-dichlorophenyl)-N,N-dimethyl urea) (Figure 13.5), also called DCMU, is used by scientists as an inhibitor for studies in photosynthesis. At low application rates, diuron is used to selectively control germinating broad-leaf and grass weeds in crops such as cotton, sugarcane, pineapple, grapes, apples, pears, citrus, and alfalfa, whereas at higher rates it is used as a nonselective weed killer. For selective weed control, diuron is sprayed on the soil as a preemergence and/or direct postemergence treatment. As a nonselective weed killer, diuron can be applied either by spraying the soil or as dry granules at any time except when the ground is frozen. When used as a selective herbicide, diuron is applied as a suspension in water; however, when used as a nonselective herbicide it can be applied in an oil suspension or as a dry granular formulation. Diuron is most readily absorbed via the root system and is translocated upward by the xylem, where it is a strong inhibitor of the Hill reaction. In soils diuron is adsorbed greatest by soils with a high organic matter content and to a lesser extent in sandy soils where it can be leached. Diuron is readily broken down by microbes found in soil and aquatic environments. When applied at low rates diuron will persist in the soil for one season, while at higher concentrations it may persist for more than one season (*Herbicide Handbook* 1989).

Bromacil (5-bromo-6-methyl-3-(1-methylpropyl)-2,4(1*H*, 3*H*)pyrimidine-dione) is a uracil-type herbicide (Figure 13.5) which can be used for selective weed control of annual and perennial weeds in orange, grapefruit, and lemon orchards and for seedling weeds in pineapple. It is also commonly used on noncropland for the control of a wide range of annual and perennial grasses and broad-leaf weeds. Bromacil is applied as a spray or spread dry on the soil surface either prior to or shortly after a period of active weed growth. Carriers used are water for suspensions, oil solutions or suspensions, or granular formulations. Bromacil is readily absorbed through the root system and is a specific inhibitor of photosynthesis. In soil there is little adsorption of bromacil to soil colloids, however, it is highly susceptible to microbial degradation. When used as a selective herbicide it can persist in the soil for one year, whereas at the higher concentrations it can persist for more than one year (*Herbicide Handbook* 1989).

Simazine (6-chloro-N,N'-diethyl-1,3,5-triazine-2,4-diamine) is an example of a triazine herbicide shown in Figure 13.5. It is widely used as a selective herbicide for control of broad-leaf and grass weeds in corn, citrus, fruits, nuts,

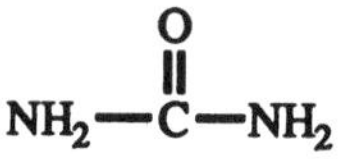

Urea

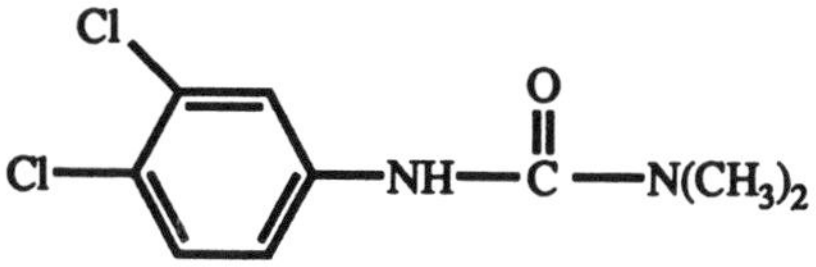

Diuron

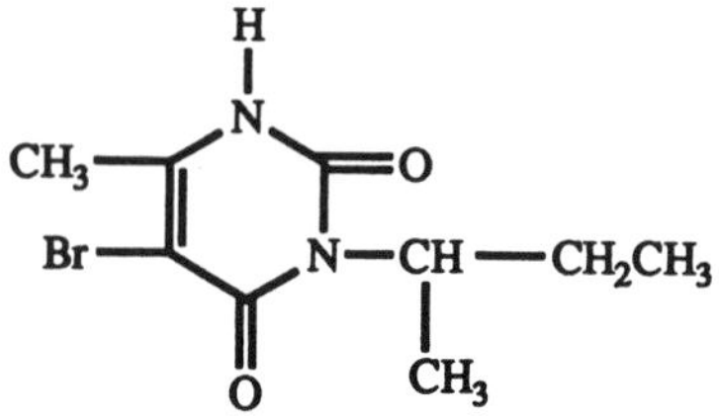

Bromacil

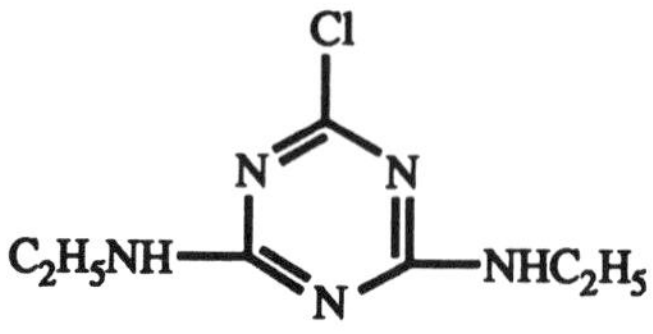

Simazine

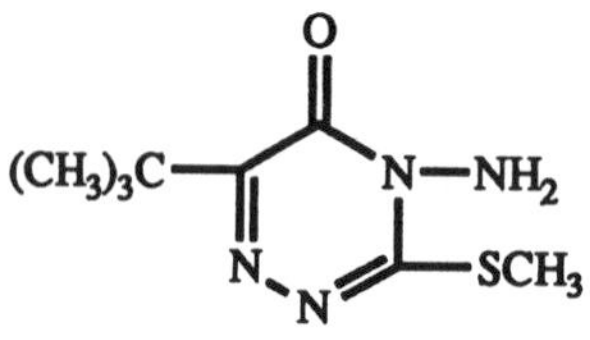

Metribuzin

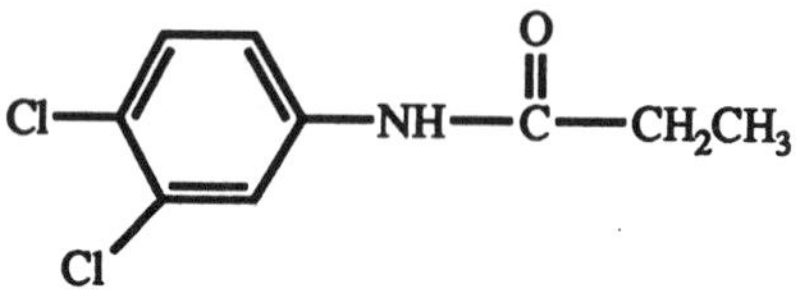

Propanil

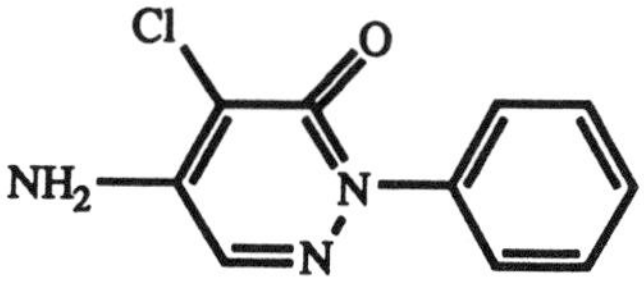

Pyrazon

Figure 13.5. Structural formulas for herbicides which inhibit photosynthesis.

olives, established alfalfa, and perennial grasses grown for seed or pasture, some turf grasses used for sod, ornamentals, nursery plantings, Christmas tree plantations, sugarcane, asparagus, and artichokes. It can be used to selectively control algae and submerged weeds in ponds. Simazine is also used as a nonselective herbicide for vegetation control in noncropland. It is applied as a spray in the form of a suspension or as a granular to bare soil. It is absorbed mostly by the roots and transported via the xylem to apical meristems and leaves where it accumulates with little or no foliar absorption. The mechanism of action of simazine and triazine herbicides is due to their ability to bind a pigment protein of photosystem II complex in the thylakoid membrane of the chloroplast, interfering with normal electron transport into the plastoquinone pool. In tolerant plants, simazine is readily metabolized to hydroxysimazine and amino acid conjugates, whereas in sensitive plants it accumulates, causing chlorosis and eventually death (Figure 13.6). Simazine is more readily adsorbed on soils with high organic matter than in sandy soils. Due to its low water solubility and adsorption to soil colloids there is little problem with leaching. Microbial degradation accounts for the major breakdown of simazine in soils (*Herbicide Handbook* 1989).

Figure 13.6.   Injury in barberry leaves caused with increasing dosage of simazine (courtesy of L. J. Kuhns).

Metribuzin   (4-amino-6-(1,1-dimethylethyl)-3-(methylthio)-1,2,4-triazin-5(4*H*)-one) (Figure 13.5) is an example of a triazinone herbicide which is effective against annual grasses and many broad-leaf weeds, including cocklebur, velvet leaf, jimson weed, coffee weed, tea weed, and sickle pod. Potatoes, sugarcane, established asparagus, tomatoes, soybeans, citrus, corn, established cereals, peas, and some turf, range, and pasture grasses are tolerant to metribuzin. It is typically applied as a preemergence or postemergence spray to the soil as a water suspension. Although metribuzin can be adsorbed in some cases by the foliage, its major route of uptake is via the roots by osmotic diffusion and is subsequently transported to the top of the plant by the xylem. In tolerant plants metribuzim is detoxified by oxidation and conversion to inactive conjugated forms. In soils with high organic matter metribuzin is moderately adsorbed, however, as the amount of organic matter decreases the degree of leaching becomes greater. The major means of detoxification is by soil microorganisms. Phytotoxic levels in the soil do not persist from one season to another (*Herbicide Handbook* 1989).

Propanil   (N-(3,4-dichlorophenyl)propanamide)   is   an   example   of   an acylanilide herbicide (Figure 13.5) which is used as a selective postemergence herbicide to control grasses and broad-leaf weeds in cultivated rice in the southern United States. It is applied as a ground or aerial application in water and is translocated from the leaf to the growing point, then back to other leaves. In plants such as rice it is completely metabolized and in the soil it is degraded one to three days following application due to microbes found in the soil (*Herbicide Handbook* 1989).

Pyrazon (5-amino-4-chloro-2-phenyl-3-(2*H*)-pyridazinone) (Figure 13.5) is an example of a pyridazinone herbicide which is typically applied preemergence, broadcast, or early postemergence, banded for control of annual broadleaf weeds in sugar beets and red beets. It is rapidly absorbed by the roots and translocated via the xylem throughout the plant. It is not recommended on soils classified as sand or loamy sands because of leaching and possible crop injury; however adsorption on soils containing greater than 5% organic matter dramatically reduces herbicide activity. Pyrazon is broken down by microbial organisms at a moderate rate, the degradation product being dephenylated pyrazon, which has no significant herbicidal activity and does not persist in the soil for longer than four to eight weeks depending upon soil moisture and temperature (*Herbicide Handbook* 1989).

Inhibitors of Biosynthetic Processes.   Herbicides have been shown to block a variety of biosynthetic processes in plants such as cell division—carbamates (propham), dinitroanilines (oryzalin), and difenzoquat (difenzoquat methyl sulphate); nucleic acid and protein synthesis—alphachloroacetamides (alachlor); amino acid synthesis—sulfonylureas (bensulfuron), imidazolinones (imazapyr), glyphosate; carotinoid synthesis inhibitors—amitrole, py-ridazinones (norflurazon), fluridone; inhibition of lipid biosynthesis—car-bamothioates (EPTC), cyclohexanediones (sethoxydim), aryloxyphenoxyp-ropionates (diclofop); and cell membrane disruption—p-nitro substituted diphenylethers (oxyfluorfen), bipyridylliums (paraquat) (Zimdahl 1993; Ash-ton and Monaco 1991; *Herbicide Handbook* 1989).

CELL DIVISION INHIBITORS.   Propham (1-methylethyl phenylcarbamate) (Fig-ure 13.7) is a carbamate-type herbicide. It effectively controls many annual grasses such as annual ryegrass, downy brome, annual bluegrass and certain broad-leaf weeds such as dodder, curly dock, and purslane. Due to its highly volatile nature, propham is mainly used in western states where winter and cool weather weeds are of economic importance. Propham's herbicidal action is mainly through the roots. Three application methods have been used: preemer-gence, preplant incorporated, and postemergence. It is applied as a flowable suspension formulation in water or as a granular. Propham is absorbed through the coleoptiles of emerging grass seedlings and to a lesser extent by the roots and leaves. When absorbed through the roots propham is readily transported to the top of the plant. The only major metabolite of propham is isopropyl-N-2-hydroxycarbanilate, which has been isolated and characterized from soybean plants. Propham is weakly adsorbed to organic matter and is readily leached from the soil. Soil microorganisms readily degrade propham to aniline and aniline is further degraded. Propham does not persist in the soil from one season to the next (*Herbicide Handbook* 1989).

Oryzalin (4-(dipropylamino)-3,5-dinitrobenzenesulfonamide) (Figure 13.7) is a dinitroaniline compound which is used as a selective preemergence herbi-cide on soybeans, cotton, potatoes, tobacco, fruit crops, nut crops, vineyards, ornamental trees, turf, shrubs, flowers, and noncropland areas. At the recom-mended dose it is effective in many annual grasses (e.g., barnyard grass) and dicot weeds (e.g., velvet leaf). It is applied as a preplant or postplant emerg-ence spray to the soil surface; its usual carrier is water. Oryzalin does not directly inhibit seed germination, rather it affects physiological growth pro-cesses associated with it. Under natural conditions oryzalin is not readily leached. The organic matter and clay content in the soil has an influence on the rate at which oryzalin is applied. Microorganisms are thought to play a key role in the rapid degradation of oryzalin from the soil. It has also been shown under laboratory conditions that oryzalin can be photodecomposed. When ap-plied at the recommended rate, oryzalin has been shown to have no adverse effect on succeeding crops (*Herbicide Handbook* 1989).

Difenzoquat methyl sulphate (1,2-dimethyl-3,5-diphenyl-1*H*-pyrazolium methyl sulphat) (Figure 13.7) is an example of a difenzoquat herbicide. It is used for the postemergence control of wild oats in spring and winter cereals

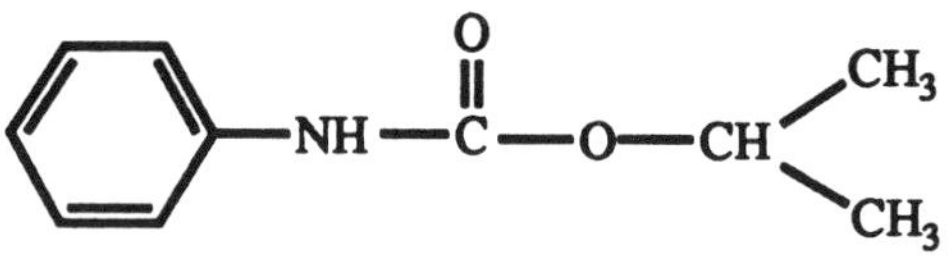

Propham

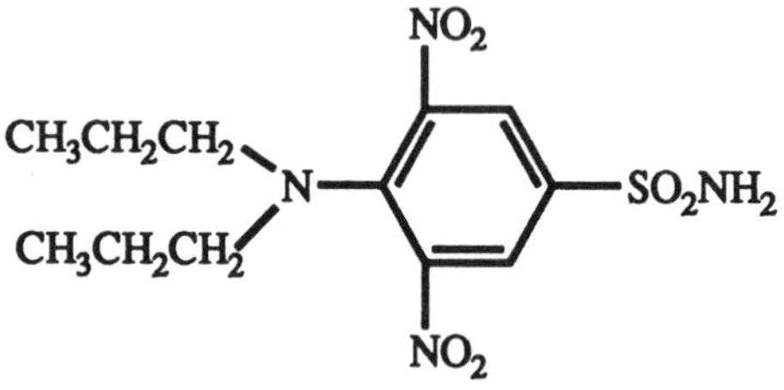

Oryzalin

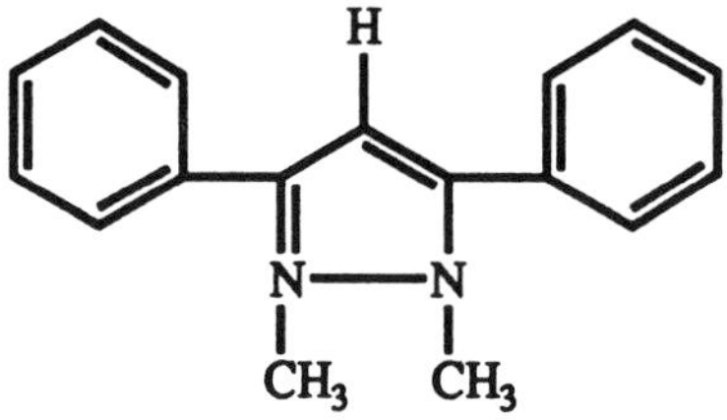

**Difenzoquat Methyl
Sulfate**

Figure 13.7. Structural formulas for herbicides which inhibit cell division.

(e.g., barley, wheat) and other crops such as maize and grass seed crops as a foliar spray using water as a carrier. Difenzoquat methyl sulphate penetrates the foliage very rapidly and accumulates in the treated area, although there is some acropetal transport. It is not significantly metabolized in plants, and in soils it is strongly adsorbed by soil particles, which negates problems with leaching. It is readily demethylated photochemically to a fairly volatile mono-methylated pyrazol, whereas microorganisms have little affect on this com-pound. In the soil difenzoquat methyl sulphate has a half-life of 16 weeks; therefore, rotation to other crops can be made the following season (*Herbicide Handbook* 1989).

NUCLEIC ACID AND PROTEIN SYNTHESIS INHIBITORS. Alachlor (2-chloro-N-(2,6-diethylphenyl)-N-(methoxymethyl)acetamide) (Figure 13.8) is an example of an alphachloroacetamide compound which is applied as a preemer-gence, early postemergence, or preplant incorporated as a spray formulation in water or liquid fertilizer or as a granular. It is effective for the control of most annual grasses and certain broad-leaf weeds and yellow nutsedge. Tolerant crops include corn, soybeans, grain sorghum, dry beans, peanuts, cotton, sun-flowers and some other ornamental and established turf species. Alachlor is mainly absorbed through the germinating plant shoots and to a lesser extent by the roots. It is readily translocated throughout the plant and accumulates in vegetative parts to a greater extent than reproductive parts. In plants alachlor is rapidly metabolized. It is readily adsorbed by soil colloids and the major form of degradation in soil is due to microorganisms. Alachlor persists in soil for six to eight weeks depending on soil type and climactic conditions (*Herbicide Handbook* 1989).

INHIBITORS OF AMINO ACIDS. Both sulfonylureas (e.g., bensulfuron methyl) and imidazolinones (e.g., imazapyr) interfere with the biosynthesis of essential amino acids in plants. The sulfonylureas and imidazolinones achieve their effect by inhibition of the same enzyme acetolactate synthase (ALS). ALS is the first enzyme in the biosynthetic chain resulting in the synthesis of branched-chain amino acids valine, leucine, and isoleucine (Schulz et al. 1990).

Bensulfuron methyl (methyl 2 [[[[(4,6-dimethoxy-2-pyrimidinyl)amino] carbonyl]amino]sulfonyl]methyl]benzoate) (Figure 13.8) is an example of a sulfonylurea, it is applied as a suspension in water as a preemergence or early postemergence treatment to weeds in standing water. It is effective on most annual and perennial broad-leaf weeds and nutsedge weeds. In an aqueous medium it is rapidly absorbed by the foliage, and tolerant plants have the ability to metabolize it to inactive metabolites. In high organic matter soils it is readily adsorbed and the rate of leaching is minimal. Bensulfuron methyl is broken down by aerobic and anaerobic soil and aquatic microorganisms and can also be inactivated by chemical hydrolysis. Under field conditions its half-life in

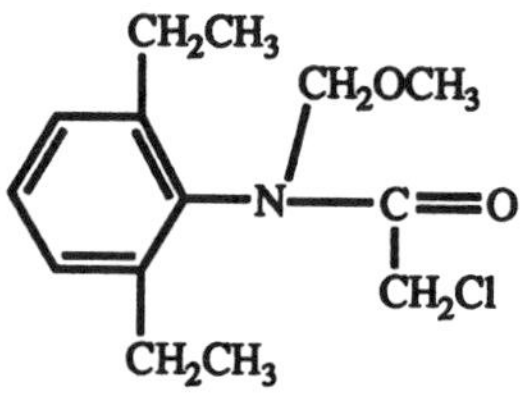

Alachlor

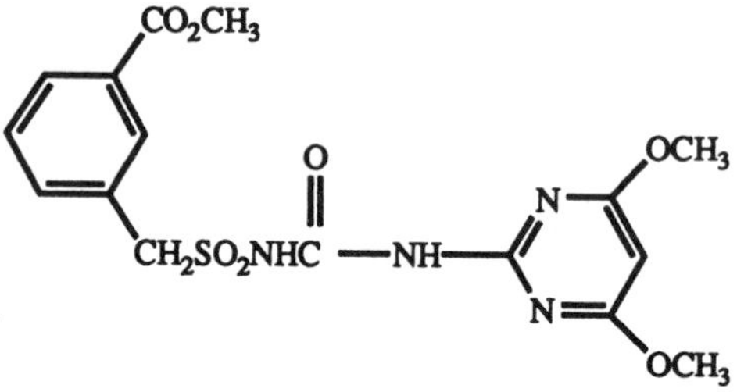

Bensulfuron methyl

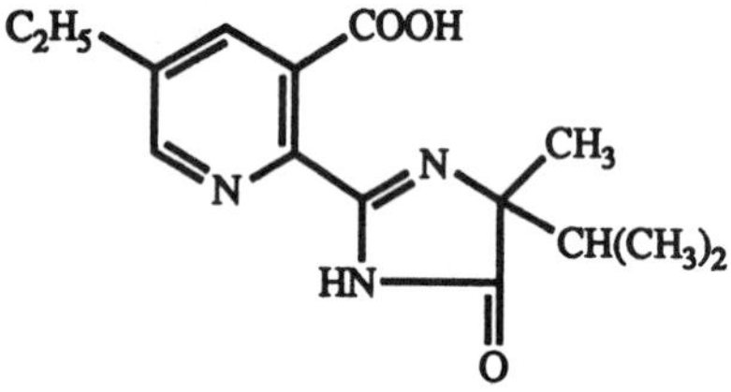

Imazethapyr

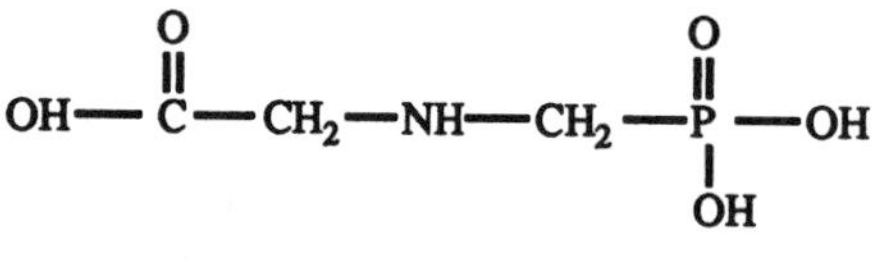

Glyphosate

Figure 13.8. Structural formulas for herbicides which inhibit nucleic acid, protein, and amino acid biosynthesis.

water is five to ten days, whereas in soil it is four to eight weeks (*Herbicide Handbook* 1989). Imazethapyr ((+ or −)2-[4,5-dihydro-4-methyl-4-(1-methylethyl)-5-oxo-1*H*-imidazol-2-yl]-5-ethyl-3-pyridinecarboxylic acid) (Figure 13.8) is an example of a imidazolinone compound; it is applied as a preplant incorporated, preemergence at cracking, and postemergence with water as a carrier for broad-spectrum weed control in soybeans, peanuts, edible beans, and alfalfa. It is readily absorbed through the roots and foliage then translocated via the xylem and phloem to growing points where it accumulates. Imazethapyr is readily metabolized by tolerant plants such as soybeans. In the soil imazethapyr is adsorbed on soil colloids and leaching is minimal. Its persistence in the soil at phytotoxic levels is from several weeks to several months depending on environmental conditions. The loss of imazethapyr from the soil is attributed to microbial activity and uptake by the plant (*Herbicide Handbook* 1989; Ashton and Monaco 1991).

Glyphosate (N-(phosphonomethyl)glycine) (Figure 13.8) is a broad-spectrum herbicide useful in crop, noncrop, and aquatic weed control as a postemergence spray. It is a very effective nonselective herbicide on deep-rooted perennial species and on annual and biennial species of grasses, sedges, and broad-leaf weeds. Directional applications may be made with glyphosate to obtain selectivity. Glyphosate is absorbed through the foliage and translocated throughout the plant. Since it is readily transported through the plant it is effective in killing underground propagules of perennial species preventing regrowth from these sites. The specific site of action of glyphosate is an enzyme in the shikimic acid pathway, 3-phosphoshikimate-1-carboxyvinyltransferase (PSCV) formally known as 5-enolpyruvoyl shikimate phosphate synthase (EPSP synthase). The dramatic inhibition of this enzyme results in a decrease in the level of aromatic amino acids, ultimately leading to a slow cessation of growth and other symptoms. Through the use of labeling studies it has been shown that plants do not metabolize glyphosate. In the soil glyphosate is strongly adsorbed by soil colloids and leaching is very low. Glyphosate is degraded in the soil very rapidly by microorganisms, therefore, crops can be planted or seeded directly into treatment areas following application (*Herbicide Handbook* 1989; Ashton and Monaco 1991).

INHIBITORS OF CAROTINOID SYNTHESIS. Amitrole (1*H*-1,2,4-triazol-3-amine) (Figure 13.9) is applied as a foliar spray for the control of perennial broad-leaf weeds and grasses in noncropland and some aquatic weeds. It is absorbed by the plant slowly; however, once in the plant it is readily translocated. Amitrole inhibits pigment formation and regrowth of buds. It is metabolized by plants to form β-(3-amino-s-triazolyl-1-)α-alanine. In the soil amitrole is broken down by microorganisms and persists in the soil for approximately two to four weeks (*Herbicide Handbook* 1989).

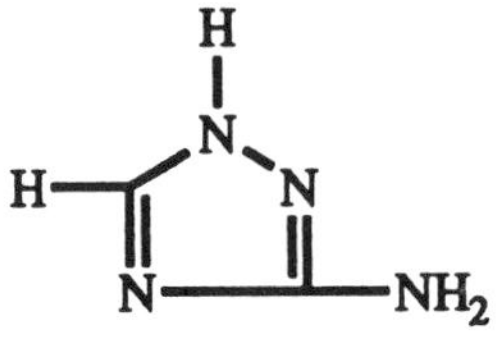

Amitrole

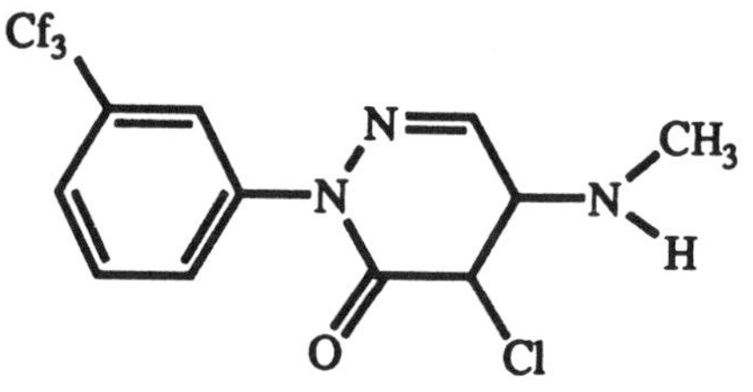

Norflurazon

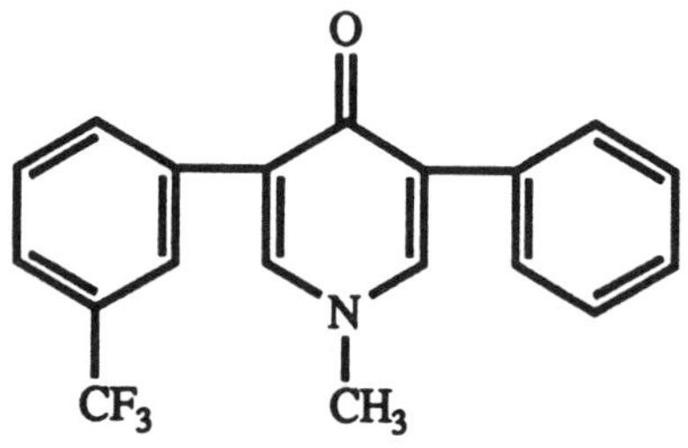

Fluridone

Figure 13.9. Structural formulas for herbicides which inhibit carotinoid synthesis.

Norflurazon (4-chloro-5-(methamino)-2-(3-(trifluoromethyl)phenyl-3-(2*H*-pyridazinone) (Figure 13.9) is an example of a pyridazinone compound which is applied preemergence or preplant incorporated to weeds, as a band or broadcast treatment. It controls many grasses (e.g., annual bluegrass, sedges) and broad-leaf weeds (e.g., ragweed, puncture vine) in tolerant crops such as alfalfa, asparagus, blueberries, cranberries, cotton, peanuts, sugarcane, soybeans, and tree and vine crops. Norflurazon is absorbed through the roots and readily translocated to growing points of susceptible plants. It acts by inhibiting the biosynthesis of carotinoid pigments, and without these pigments to filter light, photodegradation of chlorophyll occurs, leading to chlorosis in plants which are not tolerant. Norflurazon is adsorbed by organic matter and is not readily leached. It is degraded in the soil partially by microorganisms and by volatilization and photodecomposition. Norflurazon has a half-life in soils of between 45 and 180 days, depending on the organic matter content of the soil (*Herbicide Handbook* 1989).

Fluridone   (1-methyl-3-phenyl-5-[3-trifluoromethyl)phenyl]-4-($1H$)-py-ridinone) (Figure 13.9) is applied as a spray to ponds, lakes, reservoirs, or canals (where water movement is minimal) for control of submerged and emerged aquatic plants (e.g., fanwort, elodea). In susceptible plants fluridone is absorbed by the roots and readily translocated into the shoot where it inhibits carotenoid synthesis. There is little metabolism of fluridone in plants. Fluridone is strongly adsorbed by soil colloids and is not readily leached. Microorganisms are the major factors responsible for the degradation of fluridone in soils, whereas in an aquatic environment it is degraded principally by photolytic processes. Its half-life in pond water is approximately 21 days, whereas in hydrosoil it is 90 days (*Herbicide Handbook* 1989).

INHIBITORS OF LIPID BIOSYNTHESIS.   EPTC (5-ethyl dipropylcarbamothioate) (Figure 13.10) is an example of a carbamothioate compound which is applied mainly by incorporation into the soil as a selective preemergence herbicide for control of grasses (e.g., quack grass) and broad-leaf weeds (e.g., morning glory). It is registered for use in alfalfa, beans, bird's-foot trefoil, caster beans, citrus, clovers, corn, cotton, flax, potatoes, pineapples, pine seedlings, saf-flowers, strawberries, sugar beets, table beets, peas, almonds, walnuts, grapes, and ornamental plants. EPTC is readily absorbed by roots and is translocated upward to the leaves and stems where it inhibits growth in the meristem regions by affecting lipid biosynthesis. In plants it is rapidly metabolized to CO_2 and other naturally occurring plant constituents. In soil it is adsorbed to dry soil but can be leached. The major mode of degradation of EPTC is by microor-ganisms, although it can be lost from the soil by volatilization. At the recom-mended rate EPTC has a half-life of one week (*Herbicide Handbook* 1989).

Sethoxydim   (2-[1-(ethoxyimino)butyl]-5-2-(ethylthio)propyl]-3-hydroxy-2-cyclohexen-1-one) (Figure 13.10) is an example of a cyclohexanedione com-pound which is applied as a ground or aerial postemergence spray for selective control of essentially all annual and perennial grass weeds in nearly all broad-leaf crops for food and nonfood use. It is readily absorbed through the foliage, and once in the plant is translocated acropetally and basipetally to meristematic regions where it interferes with lipid metabolism. In plants, sethoxydim has been shown to be oxidized, structurally rearranged, and conjugated very rapid-ly. In soils, sethoxydim can be adsorbed depending on the organic matter content of the soil. It is broken down by microbes and photolysis in the soil with a half-life of five to 11 days (*Herbicide Handbook* 1989).

Diclofop (+ or −)-2-[4-(2,4-dichlorophenoxy)phenoxy]propanoic acid) (Figure 13.10) is an example of an aryloxphenoxypropionate compound which is generally applied as an early postemergence spray for the selective control

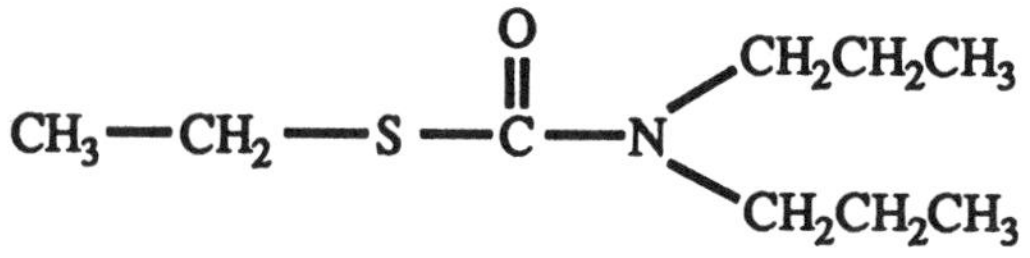

**S-ethyl depropylcarbamothioate
(EPTC)**

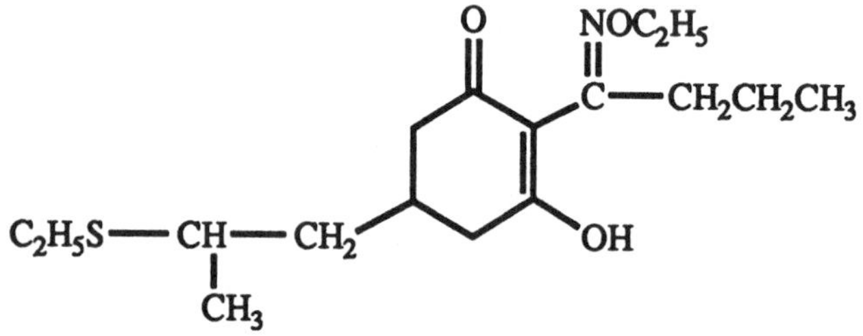

Sethoxydim

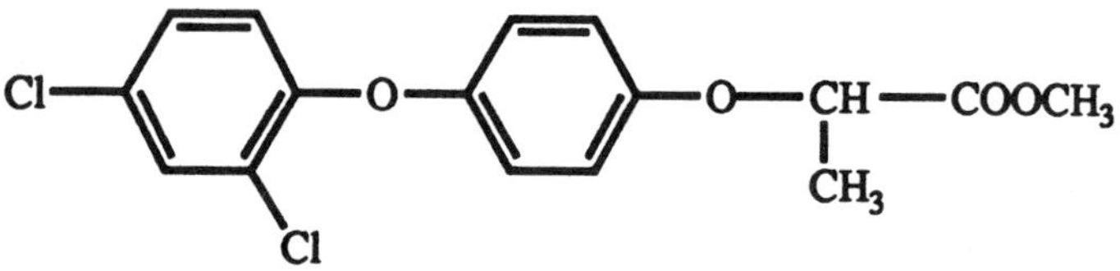

Diclofop

Figure 13.10. Structural formulas for herbicides which inhibit lipid biosynthesis.

of annual grassy weeds (e.g., annual ryegrass). Its primary use is on tolerant crops such as wheat, barley, soybeans, field peas, lentils, flax, sugar beets, and most broad-leaf agronomic and horticultural crops. Diclofop is absorbed by the foliage and is not readily translocated. In the plant it has been shown to be metabolized to 2-[4-(2′,4′-dichlorohydroxyphenoxy)phenoxy]propanoic acid in soybeans and 2-[4-(2′,4′-dichlorophenoxy)phenoxy]propanoic acid in wheat. In the soil, diclofop decomposes in a few days to 2-[4-(2′,4′-dichloro-phenoxy)phenoxy] propanoic acid, which in turn is metabolized with a half-life of 10–30 days depending on soil conditions (*Herbicide Handbook* 1989).

CELL MEMBRANE DISRUPTERS. Oxyfluorfen (2-chloro-1-(3-ethoxy-4-nit-rophenoxy)-4-trifluoromethyl)benzene) (Figure 13.11) is an example of a p-nitrosubstituted diphenyl ether compound which is effective as a preemergence and/or postemergence herbicide for the control of annual broad-leaf weeds in a variety of agronomic, horticultural, fruit tree, and tropical plantation crops. It can be applied to the roots or shoots but it is more effective in the shoots, once absorbed there is very little translocation. In plants, oxyfluorfen is not readily metabolized. It is strongly adsorbed by the soil, not readily leached or

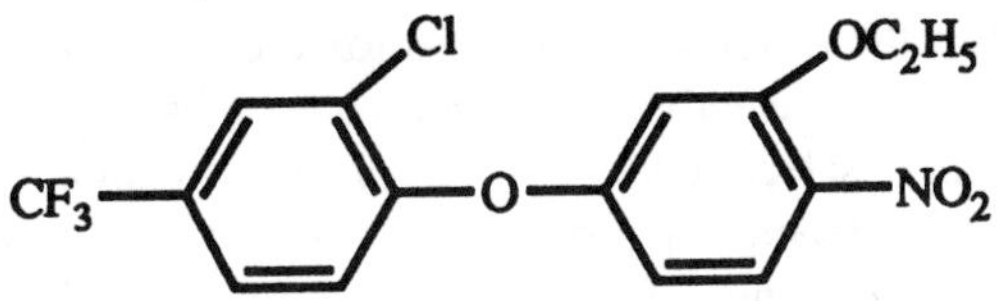

Oxyfluorfen

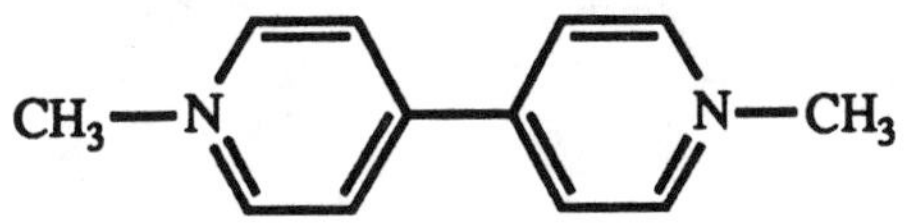

Paraquat

Figure 13.11. Structural formulas for herbicides which act as cell membrane disrupters.

broken down by microorganisms; however, photodecomposition of oxyfluorfen is rapid in water. When used at the recommended rate it has a half-life of 30–40 days (*Herbicide Handbook* 1989).

Paraquat (1,1'-dimethyl-4,4'-bibyridinium ion) (Figure 13.11) is a bipyridullium compound. It is applied as a foliar contact spray for the control of aquatic weeds and weeds in sugarcane and noncropland. It is absorbed very rapidly and can be translocated. In plants it has been demonstrated that there is no metabolic breakdown of paraquat, however, some photochemical breakdown has been shown to occur. Paraquat is rapidly and completely inactivated in the soil. There is little microbial breakdown of paraquat in the soil; however, it is strongly adsorbed by soils, and once bound is very persistent but biologically unavailable (*Herbicide Handbook* 1989).

GENETIC ENGINEERING FOR HERBICIDE RESISTANCE IN HIGHER PLANTS

The use of herbicides as part of an effective weed management program is an essential practice in modern agriculture. Recently, the demand for environmental safety has lead to the development of less toxic compounds which are safer and, in some cases, more selective. However, at the present time there are only a few herbicides which selectively control all plants without adversely affecting the cultivated crop. Safeners, also called antidotes or protectants, have been developed to broaden the range of crop selectivity for particular herbicides. The principle by which safeners work is simple, they interfere with the activity of an herbicide in a crop where the herbicide is not normally selective, thereby protecting the crop plant from injury. Flurazole (Figure 13.12) is a safener; when applied to sorghum it makes it possible to use alachlor as a selective herbicide. Another example of a safener is cyometronil (Figure 13.12), which

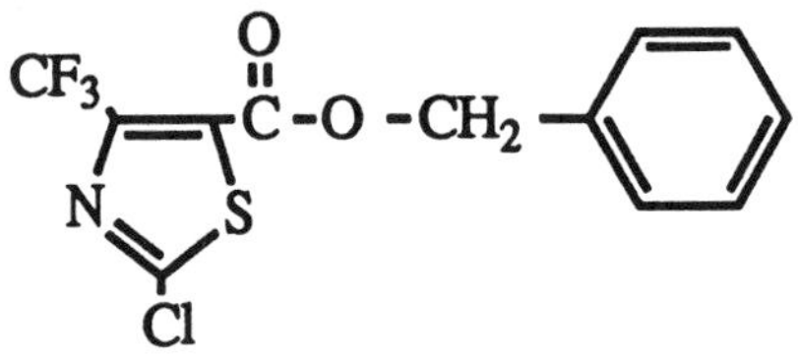

Flurazole 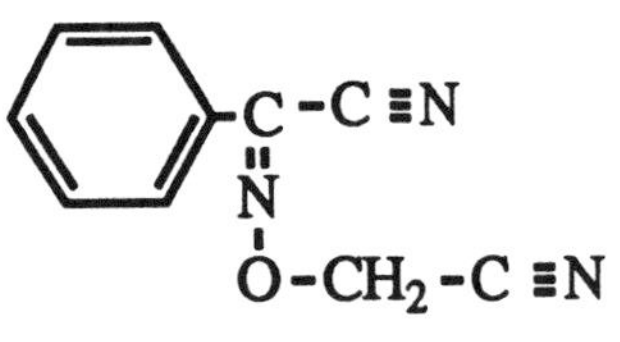   **Cyometrinil**

Figure 13.12. Examples of two safeners which are used to protect crops against herbicide damage.

is used to protect sorghum plants from metolachlor. These safeners and others have been used in sorghum and other crops such as rice, corn, wheat, and oats (Ashton and Monaco 1991; Zimdahl 1993; Schulz et al. 1990). Future research will lead to the development of new safeners for other herbicides to broaden their range of crop selectivity while maintaining their spectrum of weed control. One drawback of using safeners is that in the majority of cases where they are used the mode of action is unknown, which will make the development of new safeners more difficult.

The development of specific herbicide resistance is another more promising approach to safen the use of herbicides. Herbicide resistance can be achieved by reducing herbicide uptake, metabolism, modification, or conjugation of the herbicide and altering the target site where the herbicide binds. In recent years tissue culture and genetic engineering approaches have been used to obtain herbicide-resistant plants. Tissue culture is an effective way to produce herbicide-tolerant plants because of the ability of plants to be regenerated from plant parts. The successful selection of plants which are resistant to herbicides such as chlorate (Pental et al. 1982), picloram (Chaleff 1981), glyphosate (Singer and McDaniel 1985), chlorsulfuron and sulfometuron methyl (Chaleff and Ray 1984), and imazapyr (Shaner and Anderson 1985) has been accomplished using tissue culture methods. Resistant plants can be produced in several ways. One approach is by using the proper concentrations of auxins and cytokinins to produce undifferentiated tissues (calli) which generally form at the wound edges of an explant. Callus can then be maintained on an artificial medium for indefinite periods of time, and in a limited number of species the callus can be regenerated into plants. Herbicide-resistant plants can be obtained by subjecting callus to mutagenic treatments and placing it on herbicide-containing media. The calli which survive can then be regenerated into herbicide-tolerant plants. It has also been shown that resistant callus can be obtained without mutagenic treatment. This occurs due to variability known as somaclonal variation (Larkin et al. 1984; Larkin and Scowcroft 1981). Another approach is to use protoplasts, which are single cells without a cell wall, the cell wall having been degraded by hydrolytic enzymes, to select for herbicide-resistant plants (Kartha et al. 1974; Shepard and Totten 1977; Zapata et al. 1977; Dos Santos et al. 1980; Fujimura et al. 1985). Protoplasts have the added benefit of being able to directly take up DNA (Fowke 1985), cell organelles (Potrykus 1973), and chromosomes (Szabados et al. 1981), which can be used to develop herbicide-tolerant plants.

Mutations leading to resistance are random, therefore the molecular mechanisms of resistance in many cases are unknown. An example of this is glyphosate resistance in tobacco plants. These plants have been shown to exhibit cross-resistance with amitrole, an herbicide with a completely different chemical structure and mode of action. Some of the regenerated herbicide-resistant plants show slower growth rates than corresponding wild types, which has been attributed to cell culture conditions (Gressel 1985). However, these resistant plants can be used as donors of resistance in further crossing experiments.

Another approach to obtaining herbicide-resistant plants is by utilizing genetic engineering. Once the mode of action of a given herbicide is established, it is possible to produce herbicide-resistant plants by modifying the target site where it binds. At the present time there are a variety of approaches using genetic engineering to develop herbicide-resistant plants in connection with herbicides involved in the regulation of plant growth (2,4-D), photosynthetic inhibitors, and those which interfere with amino acid biosynthetic enzymes. In this section a brief explanation of work done with 2,4-D detoxification; PSII, s-triazines; amino acid biosynthesis, glyphosate, sulfonylurea, imidazolinone, and phoshinothricin herbicides will be given. However, for more details see the reviews by Schulz et al. (1990) and Llewellyn et al. (1990).

There are a variety of microorganisms found in soils including bacteria, yeast, and fungi from several taxonomic groups which have the ability to break down 2,4-D. The best characterized organisms are strains of *Alcaligenes eutrophus*, which is a simple gram-negative rod bacteria found in most aerated soils. Initial studies in this area showed that these strains had the ability to grow on synthetic media with 2,4-D as a sole source of carbon (Don and Pemberton 1981). All strains which were resistant contained a 75-kb plasmid which has since been shown to encode many of the enzymes necessary for the breakdown of 2,4-D. The overall pathway of 2,4-D degradation and associated genes in *A. eutrophus* are outlined in papers by Don et al. (1985) and Streber et al. (1987). A chimeric gene construct of the tfdA gene which is responsible for the conversion of the first step in the degradation of 2,4-D to 2,4-dichlorophenol was inserted into tobacco plants. When a group of control and transgenic plants were sprayed with a commercial preparation of 2,4-D isopropylester at concentrations from 0 to 1000 ppm, control plants showed injury at 30 ppm, whereas transgenic plants began to show the same symptoms at 1000 ppm (Llewellyn et al. 1990). Further work is in progress on the development of cotton and other crop plants which are resistant to 2,4-D.

There are many classes of chemicals which interfere with the reduction of plastoquinone on the acceptor site of PSII (Fedtke 1982). Early research showing that herbicides from different classes had the ability to dislodge one another from the thylakoid membrane lead to the idea that there were distinct but overlapping herbicide binding sites (Trebst and Draber 1978). Since this time an herbicide-binding protein which has several names (32-kDa protein, Q_b protein, D_1, herbicide-binding protein) has been studied in detail (Schulz et al. 1990). A comparison of the sequences of the genes coding for the herbicide-binding proteins from atrazine-resistant and wild-type plants made it possible to classify different amino acid exchanges which lead to herbicide resistance (Hirschberg and McIntosh 1983; Schulz et al. 1990); however, before this information can be used to engineer resistance in PSII herbicides further research is necessary.

Herbicides which act by inhibiting essential amino acid biosynthesis have a great deal of potential. The essential amino acids are phenylalanine, tyrosine, tryptophan, valine, leucine, isoleucine, histidine, lysine, threonine, and methionine make up about half of the amino acids used for the synthesis of proteins. Metabolic pathways for amino acid biosynthesis are essentially the same for plants and microorganisms, therefore, extensive information on microbial biosynthesis could be useful for future work in plant systems. In recent years three amino acid biosynthetic enzymes have been identified as targets of important classes of herbicides which include glyphosate, chlorsulfuron, sulfometuron methyl, imazapyr, and L-phosphinothricin.

Glyphosate (Round-up) is a widely used nonselective postemergence herbicide. Early in vitro studies with glyphosate have shown that it is a potent competitive inhibitor of the enzyme 5-enol-pyruvylshikimic acid 3-phosphate (EPSP) synthase (Steinrucker and Amrhein 1980), which is the sixth enzyme in the shikimic acid pathway leading to the biosynthesis of aromatic amino acids. Since this time EPSP synthase has been isolated from a number of plants and microbial organisms (Amrhein 1986) and the aroA gene which codes EPSP synthase has been isolated and sequenced (Duncan et al. 1984; Stalker et al. 1985; Klee et al. 1987). It has been shown that glyphosate tolerance in bacterial and plant cell cultures correlates with increases in EPSP synthase levels (Schultz et al. 1990). Two approaches have been taken to genetically engineer glyphosate-resistant plants: to overexpress a glyphosate-sensitive EPSP synthase gene into plants and to transfer a mutated gene which codes for a glyphosate-resistant EPSP synthase.

Both of these methods have been used to produce glyphosate-resistant plants. In a study with a petunia cell line which overproduces EPSP synthase due to gene amplification a cDNA of the EPSP synthase gene was isolated. A chimeric gene construct with a 35S cauliflower mosaic promoter was used to transform petunia cells, and the resulting transformants were shown to be tolerant to glyphosate at commercially applied rates (Shah et al. 1986). The other approach which involved the insertion and expression of a mutated bacterial aroA gene which codes for a glyphosate-resistant EPSP synthase has also been shown to be successful (Comai et al. 1985; Filatti et al. 1987). Although both methods can produce glyphosate-resistant plants, they still have some deleterious effects on plant growth in the absence of the herbicide; therefore, further research is necessary before transformants can be obtained for commercial application.

There are two other groups of herbicides, the sulfonylureas and the imidazolinones, which also interfere with the biosynthesis of essential amino acid synthesis. Both groups of herbicides inhibit the same enzyme, acetolactate synthase (ALS), which is the first enzyme in the biosynthetic chain resulting in the synthesis of the branched-chain amino acids valine, leucine, and isoleucine (Schulz et al. 1990). It was shown that there was a correlation between resistance to sulfonylureas and possession of a mutated herbicide-insensitive ALS (Chaleff and Mauvais 1984) in plants regenerated from resistant plant cell cultures (Chaleff and Ray 1984). Mutants resistant to the imidazolinone herbicides have also been reported; these tolerant cells contain 1000 times more mutated ALS than the wild type (Shaner and Anderson 1985). Both ALS genes and proteins have been extensively investigated (Schulz et al. 1990). It appears that the ease with which resistant ALS can be selected for use in the laboratory correlates with the rapid emergence of resistant weeds in the fields treated with sulfonulurea or imidazolinone herbicides.

L-Phosphinothricin is a naturally occurring amino acid possessing herbicidal activity by acting as a potent inhibitor of glutamine synthetase (Leason et al. 1982), which is the enzyme responsible for the assimilation of ammonia, nitrate reduction, direct uptake, or amino acid metabolism (Wallsgrove et al. 1987). When glutamine synthetase is inhibited by L-phosphinothricin there is a rapid increase in the ammonia concentration within the plant leading to death (Schulz et al. 1990). Selection for resistance to L-phosphinothricin has been successful in alfalfa tissue cultures (Donn et al. 1984). This resistance was found to be due to the amplification of a glutamine synthetase gene leading to the overproduction of this enzyme. However, it was not possible to regenerate plants from resistant tissue culture cells. Since this time several genes for glutamine synthetase have been isolated and characterized (Schulz et al. 1990). The alfalfa glutamine synthetase gene was overexpressed in tobacco, resulting in a three fold increase in specific glutamine synthetase activity and a 20-fold increase in resistance to L-phosphinothricin in plants growing on a medium containing phosphinothricin. Further work is in progress to engineer plants with a high level of resistance to L-phosphinothricin.

SUMMARY AND CONCLUSIONS

Until an ideal herbicide which can control all plant species without affecting the crop of interest and possessing a high degree of environmental safety with minimal persistence in the soil can be produced, alternative measures must be taken. One approach is to produce herbicide-resistant plants, a concept which is not a new practice among farmers. The development of herbicide-resistant crop plants has a number of advantages and disadvantages which are hotly debated. The principle advantages of herbicide-resistant crop plants follows:

1. Herbicides do not need to be applied as a preventative measure.

2. Soil erosion is reduced.

3. Older herbicides are replaced with modern products having a higher degree of environmental safety.

4. Herbicide mixtures are discontinued to simplify weed control programs.

The disadvantages associated with herbicide-resistant crops follows:

1. Resistant crops lead to a broader use of chemicals.

2. The use of herbicide-resistant varieties plus their companion herbicide may promote the appearance of herbicide-resistant weeds.

3. Large farms may be favored, to the detriment of small ones.

It is important to note that herbicide resistance is not a new concept in plant breeding or weed control and the use of selective herbicides has been a common agricultural practice for many years. An example of how classical breeding has lead to the development of herbicide resistance is the soybean variety Tracy M, which was bred for metribuzin resistance. The use of modern genetic engineering techniques to produce herbicide-resistant crop plants should be looked upon as a means of speeding up the process of obtaining these plants. It is possible that the economic and ecological advantages of herbicide-resistant plants outweigh the disadvantages and the use of herbicide-resistant crop plants along with the corresponding herbicides will become a routine practice in modern agriculture, however, only time will tell.

REFERENCES

Abeles, F. B., Morgan, P. W., and Saltveit Jr., M. E. (1992). *Ethylene in Plant Biology. Second Edition*, Academic Press, Inc., San Diego, CA.

Amrhein, N. (1986). "Specific inhibitors as probes into the biosynthesis and metabolism of aromatic amino acids". In *The Shikimic Acid Pathway*, eds., E. E. Conn, Plenum Press, New York, pp. 83.

Ashton, F. M. and Crafts, A. S. (1981). *Mode of Action of Herbicides. Second Edition*, Wiley Interscience, New York.

Ashton, F. M. and Monaco, T. J. (1991). *Weed Science. Principles and Practices. Third Edition*, Wiley Interscience, New York.

Barko, J. W., Adams, M. S., and Clesceri, N. L. (1986). "Environmental factors and their consideration in the management of submersed aquatic vegetation: A review". *J. Aquatic Plant Management* 24:1–10.

Barko, J. W. and Smart, R. M. (1981). "Comparative influences of light and temperature on the growth and metabolism of selected submersed freshwater macrophytes". *Ecol. Monogram* 51:219–235.

Buchholtz, K. P. (1967). "Report of the terminology committee of the Weed Science Society of America". *Weeds* 15:388–389.

Chaleff, R. S. (1981). "Variants and mutants". In *Genetics of Higher Plants. Application of Cell Culture*, eds., D. R. Newth and J. G. Torrey, Cambridge University Press, London, pp. 41.

Chaleff, R. A. and Mauvais, C. J. (1984). "Acetolactate synthase is the site of action of two sulfonylurea herbicides in higher plants". *Science* 224:1443.

Chaleff, R. S. and Ray, T. B. (1984). "Herbicide resistant mutants from tobacco cell cultures". *Science* 223:1148.

Chandler, J. M., Hamill, A. S., and Thomas, A. G. (1984). *Crop Losses due to Weeds in the United States and Canada*, Weed Science Society of America, Champaign, IL.

Comai, L., Facciotti, D., Hiatt, W. R., Thompson, G., Rose, R. E., and Stalker, P. M. (1985). "Expression in plants of a mutant aroA gene from *Salmonella typhimurium* confers tolerance to glyphosate". *Nature* 317:741.

Davis, T. D. and Curry, E. A. (1991). "Chemical regulation of vegetative growth". *Critical Reviews in Plant Science* 10:151–188.

Don, R. and Pemberton, J. (1981). "Properties of six pesticide degradation plasmids isolated from *Alcaligenes eutrophus* and *Alcaligenes paradoxus*". *J. Bacteriology* 145:681–686.

Don, R., Weightman, A., Knackmuss, H., and Timmis, K. (1985). "Transposon mutagenesis and cloning analysis of the pathways for degradation of 2,4-dichlorophenoxyacetic acid and 3-chlorobenzoate in *Alcaligenes eutrophus* JMP134(pJP4)". *J. Bacteriology* 161:85–90.

Donn, G., Tischer, E., Smith, J. A., and Goodman, H. M. (1984). "Herbicide-resistant alfalfa cells: an example of gene amplification in plants". *J. Mol. Appl. Genet.* 2:621.

Dos Santos, A. V. P., Outka, D. E., and Cocking, E. C. (1980). "Organogenesis and somatic embryogenesis in tissues derived from leaf protoplasts and leaf explants of *Medicago sativa*". *Z. Pflanzenphysiol.* 99:261.

Duncan, K., Lewendon, A., and Coggins, J. R. (1984). "The complete amino acid sequence of *Escherichia coli* 5-enolpyruvylshikimate 3-phosphate synthase". *FEBS Letters* 170:59.

Fedtke, C. (1982). *Biochemistry and Physiology of Herbicide Action*, Springer-Verlag, Berlin.

Filatti, J. J., Kiser, J., Rose, R., and Comai, L. (1987). "Efficient transfer of a glyphosate tolerance gene into tomato using a binary *Agrobacterium tumefaciens* vector". *Biotechnology* 5:726.

Fowke, C. C. (1985). "Plant protoplasts". In *Plant Protoplasts*, ed., F. Constable, CRC Press, Boca Raton, FL.

Fujimura, T., Sakurai, M., Akagi, H., Negishi, T., and Hirose, A. (1985). "Regeneration of rice plants from protoplasts". *Plant Tissue Culture Letters* 2:74.

Graebe, J. E. (1987). "Gibberellin biosynthesis and control". *Annu. Rev. Plant Physiol.* 38:419–465.

Gressel, J. (1985). "Biotechnologically conferring herbicide resistance in crops: The present realities". In *Molecular Form and Function of the Plant Genome*, ed., L. van Vloten-Doting, Plenum Press, New York, pp. 489.

Harlan, J. R. and de Wet, J. M. J. (1965). "Some thoughts about weeds". *Econ. Bot.* 19:16–24.

Herbicide Handbook (1989). Weed Science Society of America, Champaign, IL.

Hirschberg, J. and McIntosh, L. (1983). "Molecular basis of herbicide resistance in *Amaranthus hybridus*". *Science* 222:1346.

Kartha, K. K., Michayluk, M. R., Nao, K. N., Gamborg, O. L., and Constable, F. (1974). "Callus formation and regeneration from mesophyll protoplasts of rape plants *Brassica napus* cultivar zephyr". *Plant Science Letters* 3:265.

Klee, H. J., Muskopf, Y. M., and Gasser, C. S. (1987). "Cloning of an *Arabidopsis thaliana* gene encoding 5-enolpyruvylshikimate-3-phosphate synthase: Sequence analysis and manipulation to obtain glyphosate tolerant plants". *Mol. Gen. Genet.* 210:437.

Larkin, P. J., Ryan, S. A., Brettel, R. I. S., and Scowcroft, W. R. (1984). "Heritable somaclonal variation in wheat". *Theor. Appl. Genet.* 67:443.

Larkin, P. J. and Scowcroft, W. R. (1981). "Somaclonal variation—A novel source of variability from cell cultures for plant improvement". *Theor. Appl. Genet.* 60:197.

Leason, M., Cunliffe, D., Parkin, D., Lea, P. J., and Miflin, B. J. (1982). "Inhibition of *Pisum sativum* leaf glutamine synthetase EC-6.3.1.2 by methionine sulfoximine phosphinotricin and other glutamate analogues". *Phytochemistry* 21:855.

Llewellyn, D., Lyon, B. R., Cousins, Y., Huppatz, J., Dennis, E. S., and Peacock, W. J. (1990). "Genetic engineering of plants for resistance to the herbicide 2,4-D". In *Genetic Engineering of Crop Plants*, eds., G. W. Lycett and D. Grierson, Butterworths, London, pp. 67.

Muskgrave, A., Jackson, E., and Ling, E. (1972). *Callitriche* stem elongation is controlled by ethylene and gibberellin". *Nature New Biology* 238:93–96.

Netherland, M. D. and Lembi, C. A. (1992). "Gibberellin synthesis inhibitor effects of submersed aquatic weed species". *Weed Science* 40:29–36.

Penner, D. and Ashton, F. M. (1966). "Biochemical and metabolic changes in plants induced by chlorophenoxy herbicides". *Residue Review* 14:39–113.

Pental, D., Cooper-Bland, S., Harding, K., Cocking, E. C., and Muller, A. J. (1982). "Cultural studies on nitrate reductase deficient *Nicotiana tabaccum* mutant protoplasts". *Z. Pflanzenphysiol.* 105:219.

Potrykus, I. (1973). "Transplantation of chloroplasts into protoplasts of petunia". *Z. Pflanzenphysiol.* 70:364.

Raskin, I. and Kende, H. (1984). "The role of gibberellin in the growth response of submerged deep-water rice". *Plant Physiol.* 76:947–950.

Riemer, D. N. (1984). *Introduction to Freshwater Vegetation*, The AVI Publishing Co., Westport, CT.

Ross, M. A. and Lembi, C. A. (1985). *Applied Weed Science*, MacMillan Co., New York.

Schulz, A., Wengenmayer, F., and Goodman, H. M. (1990). "Genetic engineering of herbicide resistance in higher plants". *Critical Reviews in Plant Sciences* 9:1–15.

Shah, D. M., Horsch, R. B., Klee, H. J., Kishore, G. M., Winter, J. A., Tuner, N. E., Hironaka, C. M., Sanders, P. R., Gasser, C. S., Aykent, L., Siegel, N. R., Rogers, S. G., and Fraley, R. T. (1986). "Engineering herbicide tolerance in transgenic plants". *Science* 233:478.

Shaner, D. L. and Anderson, P. C. (1985). "Mechanism of action of the imidazolinones and cell culture selection of tolerant maize". In *Biotechnology in Plant Science, Relevance Agriculture Eighties*, eds., M. Zaitlin, P. R. Day, and A. Hollaender, Academic Press, Orlando, FL, pp. 287.

Shepard, J. F. and Totten, R. E. (1977). "Mesophyll cell protoplasts of potato: isolation, proliferation and plant regeneration". *Plant Physiol.* 60:313.

Singer, S. S. and McDaniel, C. N. (1985). "Selection of glyphosate-tolerant tobacco calli and the expression of this tolerance in regenerated plants". *Plant Physiol.* 78:411.

Stalker, D. M., Hiatz, W. R., and Comai, L. (1985). "A single amino acid substitution in the enzyme 5-enolpyruvulshikimate 3-phosphate synthase confers resistance to the herbicide glyphosate". *J. Biol. Chem.* 260:4725.

Steinrucker, H. C. and Amrhein, N. (1980). "The herbicide glyphosate is a potent inhibitor of 5-enol-pyruvylshikimic acid 3-phosphate-synthase". *Biochem. Biophys. Res. Comm.* 94:1207.

Sterret, J. P. (1988). "XE-1019: Plant response, translocation and metabolism". *J. Plant Growth Reg.* 7:19–26.

Sterret, J. P. and Tworkoski, T. (1987). "Flurprimidol: Plant response, translocation and metabolism". *J. Am. Soc. Hortic. Sci.* 112:341–345.

Streber, W., Timmis, K., and Zenk, M. (1987). "Analysis, cloning and high-level expression of 2,4-dichlorophenoxyacetate monooxygenase gene *tfdA* of *Alcaligenes eutrophus* JMP134". *J. Bacteriology* 169:2950–2955.

Szabados, L., Hadlaczky, G., and Dudits, D. (1981). "Uptake of isolated plant chromosomes by plant protoplasts". *Planta* 151:141.

Takahashi, N., Phinney, B. O., and MacMillan, J. (1991). *Gibberellins*, Springer-Verlag, Berlin.

Trebst, A. and Draber, W. (1978). "Structure activity correlation of recent herbicides in photosynthetic reactions". In *Advances in Pesticide Science, Volume 2*, eds., H. Geissbuhler, Pergamon Press, New York, pp. 223.

van Overbeek, J. (1964). "Survey of mechanism of herbicide action". In *The Physiology and Biochemistry of Herbicides*, ed., L. J. Audus, Academic Press, London, pp. 387–400.

Wallsgrove, R. M., Turner, J. C., Hall, N. P., Kendall, A. C., and Bright, S. W. J. (1987). "Barley mutants lacking chloroplast glutamine synthetase. Biochemical and genetic analysis". *Plant Physiol.* 83:155.

Zapata, F. J., Evans, P. K., Powers, J. B., and Cocking, E. C. (1977). "The effect of temperature on the division of protoplasts of *Lycopersicon esculentum* and *Lycopersicon peruvianum*". *Plant Science Letters* 8:119.

Zimdahl, R. L. (1993). *Fundamentals of Weed Science*, Academic Press, San Diego, CA.

Glossary

Abscisic acid—A 15-carbon sesquiterpenoid which is partially produced in the chloroplasts and other plastids via the mevalonic acid pathway. It is a naturally occurring compound which is involved in many aspects of plant growth and development, some of which are inhibitory (growth) and some which are promotive (normal embryogenesis, seed storage proteins).

Abscission—The separation of a plant part, such as a leaf, flower, seed, stem, or other from the parent plant.

Acropetal—Referring to movement from the base to the tip.

Adventitious rooting—The formation of roots at locations other then where roots occur under natural conditions.

Agent orange—A defoliant used in Vietnam which is a mixture of free 2,4-D and the n-butyl ester of 2,4,5-T.

Annual plants—Plants which complete a life cycle in one growing season.

Antibody—An immunoglobulin present in the serum of an animal and synthesized by plasma cells in response to invasion by an antigen, conferring immunity against later infection by the same antigen.

312

Anti-gibberellin—A compound which counteracts the action of gibberellins.

Autocatalytic ethylene production—Exogenous applications of ethylene which cause a burst in ethylene biosynthesis.

Autolysis—Self-digestion.

Auxins—The term *auxin* is derived from the Greek word *auxein* meaning to grow. Auxin is a generic term representing a class of compounds which are characterized by their capacity to induce elongation in shoot cells and resemble indole-3-acetic acid in physiological action. Auxins may, and generally do, affect other processes besides elongation, but elongation is considered critical. Auxins are generally acids with an unsaturated nucleus or their derivatives.

Bakanae disease—Foolish seedling.

Basipetal—Referring to movement from the tip to the base.

Biennial plants—Plants which complete their life cycle in two years.

Bioassay—A system used to test the activity of a substance with respect to a physiological response.

Bolting—Rapid stem elongation.

Bound auxins—Auxins not readily available and which can only be released from plant tissues after they are subjected to hydrolysis, enzymolysis, or autolysis.

Brassins—Crude lipoidal extracts from rape pollen causing swelling and splitting of bean second internodes.

Brassinolide—A steroid which is the active component found in brassins causing swelling and splitting of bean second internodes.

Brassinosteroids—A class of steroid compounds having activity similar to brassinolide in the bean second internode bioassay.

Climacteric—Refers to fruits which will ripen in response to ethylene.

Cytokinins—Substituted adenine compounds that promote cell division and other growth regulatory functions in the same manner as kinetin (6-furfurylaminopurine).

Day-neutral plants—Plants where flower initiation is determined solely by genotype and that have no specific light requirement.

Density labeling—Use of heavy water to show that an enzyme is produced via de novo synthesis.

Dioecious—Plants having male and female parts on separate plants, an example of this is spinach.

Distal stem cuttings—Shoots which form close to the stem tip.

Distally—Located on the leaf side of the abscission zone.

Dormancy—A temporary suspension of visible growth of any plant structure containing a meristem.

Ecdysteroid—Moulting hormones of insects and other arthropods.

Ecodormancy—Dormancy due to one or more unsuitable factors of the environment which are nonspecific in their effect. In seeds this term is equivalent to *quiescence*.

Endodormancy—Dormancy regulated by physiological factors inside the affected structure, such as, the rest period in buds. In seeds, this type is present if embryo excision fails to produce either more rapid germination or normal seedling growth.

Endogenous—Referring to something that occurs within the plant.

Enzyme-linked immunosorbent assay (ELISA)—An immunoassay which utilizes enzymes such as alkaline phosphatase or horseradish peroxidase as a tracer.

Enzymolysis—Enzymatic breakdown.

Epicotyl—Section of the stem above the cotyledons.

Epigeous germination—When the hypocotyl elongates and brings the cotyledons above the ground.

Epinasty—Downward bending of the petiole.

Ethylene—A simple unsaturated hydrocarbon which promotes fruit ripening and causes a triple response in etiolated pea seedlings including inhibition of elongation, increased radial expansion and horizontal growth of stems in response to gravity.

Exogenous—Referring to something that is applied externally.

Fertilization—Union of the male and female gametes to form the zygote.

Flower development—The differentiation of the flower structure including events from flower formation to anthesis (flowering).

Flower formation—The visible initiation of flower parts.

Flower initiation—An internal physiological change in the meristem which precedes any morphological change.

Free auxins—Auxins which can readily diffuse out of the tissue, are easily extracted with various solvents, and can be immediately used to regulate physiological processes in plants. Examples are indole-3-acetaldhyde, indole-3-acetonitrile, indole-3-pyruvic acid or indole-3-ethanol.

Fruit—The structure which results from the development of tissues which support the ovules of the plant.

Fruit set—The rapid growth of the ovary which usually follows pollination and fertilization.

Fusicoccin—A fungal toxin.

Geotropism—The movement of an organ in response to gravity. The are two forms of geotropism: negative, movement against the force of gravity; and positive, movement with the force of gravity.

Gibberellins—A class of plant growth substances having an ent-gibberellane skeleton that stimulate cell division and/or cell elongation and other regulatory functions in the same manner as gibberellic acid (GA_3).

Gummosis—Promotion of a gummy exudate from the stem.

Herbicide—Derived from the Latin *herba* or plant and *caedere*, to kill. It is a pesticide which is specifically used for weed control.

Heteroauxin—Other auxin. Today it is known as indole-3-acetic acid.

Hormone—Initially defined as a chemical arousing activity. The current definition in animal systems is a compound synthesized at a localized site and transported via the bloodstream to a target tissue where it regulates a physiological response based on concentration in that tissue.

Hypocotyl—Section of the stem below the cotyledons.

Hydrolysis—Chemical splitting of a bond and adding a water, can be by acid or base.

Hypersensitive reaction—Following infection by a pathogen a small area around the initial point of penetration dies, acting as a protective cell suicide which prevents to spread of disease.

Hypogeous germination—Occurs when the epicotyl emerges and the cotyledons remain below the soil surface.

Imperfect flower—Containing either pistils or stamens.

Jasmonates—A specific class of cyclopentanone compounds with activity similar to ($-$)-jasmonic acid and its methyl ester.

Juvenile phase—An initial period of growth when apical meristems will not respond to internal or external conditions to initiate flowers characterized by exponential increases in size; absence of the ability to shift from vegetative growth to reproductive maturity leading to the formation of flowers; specific morphological and physiological traits, including leaf shape, thorniness, vigor, or disease resistance; and a greater ability to regenerate adventitious roots and shoots.

Long-day plants—Plants which flower only when the dark period is shorter than a certain critical length.

Long-short-day plants—Plants which require long days first followed by short days.

Matric potential—Based on the ability of the matrices e.g. cell walls, starch etc. to be hydrated and bind water.

Maturation—Refers to qualitative changes which allow the plant or organ to express its full potential. This is accomplished by a gradual transition of morphology, growth rate, and flowering capacity.

Monoecious—Having both male and female flowers on the same plant, an example of this is maize.

Monoclonal antibody—An immunoglobulin produced by a single clone of lymphocytes (secreting plasma cell) and recognizing only a single epitope on an antigen.

Morphactin—A class of growth-retarding compounds which received their name because they have the ability to affect plant morphogenesis (morphologically active substances).

Myeloma cell line—An immortal antibody-secreting type of cell.

Negative curvature—Bending away from the tip.

Nonclimacteric—Refers to fruits which do not ripen in response to ethylene.

Ohne Wuchstoff, kein Wachstum—Without growth substances no growth.

Osmotic potential—Solute concentration.

Paradormancy—Dormancy due to the physical factors or biochemical signals originating external to the affected structure for the initial reaction, as in apical dominance or bud scale effects. In seeds control would come from any of the enclosing structures surrounding the embryo, not restricted to biochemical signals. This category could be identified by more rapid germination and normal seedling growth following excision of the embryo.

Parthenocarpy—Development of fruits without pollination or fertilization resulting in seedlessness.

Perfect flower—Containing both pistils and stamens.

Perennial plants—Plants which flower and set seeds year after year without dying.

Pesticide—A chemical used to control a pest.

Photodormancy—State of plants which require either light or dark to germinate.

Photoperiodism—State of plants which flower in response to day length conditions.

Phototropism—The movement of a plant organ in response to directional fluxes or gradients in light.

Plant growth regulators—Largely been used by agrichemical companies to designate synthetic regulators. The commonly used definition is that they are organic compounds other than nutrients (materials which supply either energy or essential mineral elements), which in small amounts promote, inhibit, or otherwise modify any physiological process in plants.

Plant growth retardant – An organic compound that retards cell division and cell elongation in shoot tissues and thus regulates plant height physiologically without causing malformation of leaves and stems.

Plant growth substance—(1) It must be a chemically characterized compound which is biosynthesized within the plant and is broadly distributed within the plant kingdom; (2) it must show specific biological activity at extremely low concentrations; and (3) it must be shown to play a fundamental role in regulating physiological phenomena in vivo in a dose-dependent and/or due to changes in sensitivity of the tissue during development.

Plumule—The growing point of the shoot which occurs at the upper end of the embryonic axis above the cotyledons.

Pollination—The transfer of pollen from the anther to the stigma.

Polyclonal antiserum—A serum sample containing antibodies against a specific antigen. Since most antigens have a large numbers of epitopes, an antiserum will contain many different antibodies against a given antigen, each antibody having been produced by a single clone of plasma cells.

Potato tuber—Morphologically a modified stem with nodes and internodes.

Pressure potential—Turgor.

Proximal stem cuttings—Roots form nearest the junction between the shoot and root.

Proximally—Located on the stem side of the abscission zone.

Q_{10}—Values which are calculated when reaction rates are known at any two temperatures (see Figure 9.2).

Quiescence—A condition where the seed or bud is under exogenous control (conditions such as water supply, temperature, or other environmental conditions may be limiting).

Radioimmunoassay (RIA)—An immunoassay which utilize isotopes such as^{125}I, ^{3}H and ^{14}C as a tracer.

Receptors—Molecules which specifically recognize and bind the hormone and as a consequence of this recognition can lead to other changes or series of changes which ultimately result in the biological response.

Rest—The condition where the seed or bud is under endogenous control (internal factors prevent growth even though environmental conditions are favorable).

Safener—Also called antidotes or protectants which have been developed to broaden the range of crop selectivity for particular herbicides. They interfere with the activity of an herbicide in a crop where the herbicide is not normally selective, thereby protecting the crop from injury.

Salicylates—A class of compounds having activity similar to salicylic acid (orthohydroxybenzoic acid) which is a plant phenolic. Phenolics are defined as substances that possess an aromatic ring bearing a hydroxyl group or its functional derivative.

Seed—A seed is a ripened ovule, which when shed from the parent plant consists of an embryo and a stored food supply, both of which are enclosed in a seed coat or covering.

Seed germination—A series of events which take place when dry quiescent seeds imbibe water resulting in an increase in metabolic activity and the initiation of a seedling from the embryo.

Senescence—The general failure of many synthetic reactions that precede cell death and is the phase of plant growth which extends from full maturity to death and is characterized by chlorophyll, protein, or RNA degradation as well as other factors.

Short-day plants—Plants which flower only when the dark period is greater then a certain critical length.

Short-long-day plants—Plants which flower only when subjected to short days followed by long days.

Skotodormancy—Dormancy caused by prolonged darkness.

Solvent partitioning—A purification step which involves partitioning between an aqueous phase and an immiscible organic solvent.

Spherosomes—Lipid bodies.

Statocytes—Cells which contain statoliths.

Statoliths—Greek *lithos*, stone. Are starch-containing plastids such as amyloplasts or chloroplasts which perceive changes in gravity.

Stimulative parthenocarpy—Requires pollen as the stimulus without subsequent fertilization in order to set fruit, e.g., Black Corinth grape.

Stolons—Potato tubers are initiated on stolons which are lateral shoots.

Summer annuals—Plants which do not require low temperatures in order to flower.

Thermodormancy—Inhibition caused by high temperatures.

Thermogenicity—Heat production.

Thermoinductive—Referring to plants which will flower in response to low temperatures.

Triple response—A bioassay based on the ability of ethylene to suppress stem elongation, increase radial expansion (lateral expansion), and promote bending or horizontal growth in response to gravity.

Tuber initiation—The process by which the tuber originates just below the stolon tip.

Vegetative parthenocarpy—Characterized by the development of fruit without pollination, e.g., pineapple and Washington Navel orange.

Vernalization—Plant species grown in temperate regions which will flower in response to low temperature treatments.

Viable—Referring to the embryo within the seed being alive and capable of germination.

Vivipary—A phenomenon of precocious germination before maturity.

Weed—A plant growing where it is not desired.

Winter annuals—The cold requirement in this group of plants is facultative, flowering will eventually occur without low-temperature treatment, however, its takes longer.

Chemical Names
and Abbreviations

2,4-D	(2,4-dichlorophenoxy)acetic acid
2,4-DB	4-(2,4-dichlorophenoxy)butanoic acid
2,4-DP	see Dichloroprop
2,4,5-T	2,4,5-trichlorophenoxyacetic acid
ABA	Abscisic acid
AC 94, 377	1-(3-chlorophthalimido)cyclohexanecarboxamide
ACC	1-aminocyclopropane-1-carboxlic acid
AdoMet	S-adenosylmethionine
Alachlor	2-chloro-N-(2,6-diethylphenyl)-N-(methoxymethyl)acetamide
Alar	see Daminozide
Alsol	(2-chloroethyl)tris(2-methoxyethoxy)silane
Amidochlor	N-(acetylamino)methyl-2-chloro-N-(2,6-diethylphenyl)acetamide
Amitrole	1H-1,2,4-triazol-3-amine
AMO-1618	2'isopropyl-4'-(trimethylammonium chloride)-5'-methylphenyl piperidine-1-carboxylate

Ancymidol	α-cycloproyl-α-(4-methoxyphenyl)-5-pyrimidienemethanol
AOA	aminooxyacetic acid
Atrazine	6-chloro-N-ethyl-N'-(1-methylethyl)-1,3,5-triazine-2,4-diamine
AVG	aminoethoxyvinylglycine
BA	6-benzylaminopurine
BPA	6-(benzyulamino)-9-(2-tetrahydropyranyl)-9H-purine
BAS 111	1-phenoxy-5,5-dimethyl-3-(1,2,4-triazole-1-yl)hexan-5-ol
Bensulfuron methyl	Methyl 2[[[[[4,6-dimethoxy-2-pyrimidinyl)amino]carbonyl]amino]sulfonyl]methyl]benzoate
Bonzi	see Paclobutrazol
BR	see Brassinosteroids
Brassinolide	2α,3α,22α,23α-tetrahydroxy-24α-methyl-B-homo-7-oxa-5α-cholestano-6-one
Bromacil	5-bromo-6-methyl-3-(1-methylpropyl)-2,4(1H,3H)pyrimidinedione
Bromoxynil	see Hydroxybenzonitrite bromoxymil
CCC	see Chlormequate chloride
Chloramben	3-amino-2,5-dichlorobenzoic acid
Chlorflurenol	2-chloro-9-hydroxy-9H-fluorene-9-carboxylic acid
Chlormequate chloride	2-chloroethyl-trimethyl-ammonium chloride
Cimetacarb	4(cyclopropyl-a-hydroxy-methylene)-3,5-dioxocyclohexane carboxylic acid ethyl ester
Clopyralid	3,6-dichloro-2-pyridinecarboxylic acid
CMP	2-(p-chlorophenoxy)-2-methylpropionic acid
Cycocel	see Chlormequate chloride
Daminozide	Butanedioic acid mono-(2,2-dimethylhydrazide)
DCMU	see Diuron
Dicamba	3,6-dichloro-2-methoxybenzoic acid
Dichlorprop	(±)-2-(2,4-dichlorophenoxy)propanoic acid
Diclofop	(±)-2-[4-(2,4-dichlorophenoxy)phenoxy]propanoic acid
Difenzoquat	1,2-dimethyl-3,5-diphenyl-1H-pyrazolium
Dikegulac	2,3:4,6,bis-O-(1-methylethylidene)-X-L-xylo-2-hexulofuranosonic acid
Dinoseb	2-1(1-methylpropyl)-4,6-dinitrophenol
Diuron	N'-(3,4-dichlorophenyl)-N,N-dimethylurea
Droop	see thidiazuron
Endothall	7-oxabicyclo(2,2,1)heptane-2,3-dicarboxylic acid
EPTC	S-ethyl dipropylcarbamothioate
Ethephon	(2-chloroethyl)phosphonic acid
Ethrel	see Ethephon
Fenoprop	(2,4,5-trichlorophenoxy)propanoic acid

Fluorene	Fluorene
Fluorene-9-carboxylic acid	Fluorene-9-carboxylic acid
Flurazole	phenylmethyl 2-chloro-4-(trifluoromethyl)-5-thiazolecarboxylate
Fluridone	1-methyl-3-phenyl-5-[3-(trifluoromethyl)phenyl]-4(1H)-pyridinone
Flurprimidol	α-(1-methylethyl)-α-(4-trifluoromethoxyphenyl)-5-pyrimidine methanol
GA3	Gibberellic acid
Glyphosate	N-(phosphonomethyl)glycine
Hydroxybenzonitrite bromoxymil	3,5-dibromo-4-hydroxybenxonitrile
IAA	Indole-3-acetic acid
IAAld	Indole-3-aldehyde
IBA	Indole-3-butyric acid
Imazapyr	(±)-2-[4,5-dihydro-4-methyl-4-(1-methylethyl)-5-oxo-1H-imidazol-2-yl]-3-pyridinecarboxylic acid
Imazethapyr	(±)-2-[4,5-dihydro-4-methyl-4-(1-methylethyl)-5-oxo-1H-imidazol-2-yl]-5-ethyl-3-pyridinecarboxylic acid
Inabenfide	4-chloro-2-(α-hydroxybenzyl)isonicotinanilide
IPA	isopentenyladenine
JA	Jasmonic acid
Kinetin	6-furfurylaminopurine
LAB 150 978	1-(4-trifluro-methyl)-2-(1,2,4-triazolyl(1)-3-(5-methyl-1,3-dioxan-5-yl)-propen-3-ol
MAA	methylarsonic acid
Maleic hydrazide	1,2-dihydro-3,6-pyridazinedione
MCPA	(4-chloro-2-methylphenoxy)acetic acid
MCPB	(4-chloro-2-methylphenoxy)butanoic acid
MCPP	see Mecoprop
Mecoprop	(±)-2-4(-chloro-2-methylphenoxy)propanoic acid
Mefluidide	N-[2,4-dimethyl-5-[[(trifluoromethyl)sulfonyl]amino]phenyl]acetamide
Mepiquate chloride	(1,1-dimethyl-piperidiumium chloride
Metribuzin	4-amino-6-(1,1-dimethylethyl)-3-(methylthio)-1,2,4-triazin-5(4H)-one
MTA	methyl thioadenosine
MTR	methyl thioribose
NAA	1-naphthaleneacetic acid
NOA	naphthoxyacetic acid
Norflurazon	4-chloro-5-(methylamino)-2-(3-(trifluoromethyl)phenyl)-3(2H)-pyridazinone
Oryzalin	4-(dipropylamino)-3,5-dinitrobenzenesulfonamide

Oxyfluorfen	2-chloro-1-(3-ethoxy-4-nitrophenoxy)-4-(trifluoromethyl)benzene
Paclobutrazol	1-(4-chlorophenyl)-4,4-dimethyl-2-(1H-1,2,4-triazol-1-yl)pentan-3-ol
Paraquat	1,1'-dimethyl-4-4'-bibyridinium ion
Phosphon D	tributyl(2,4-dichlorobenzyl)phosphonium clhoride
Pichloram	4-amino-3,5,6-trichloropicolinic acid
Piperidium bromide	1-allyl-1,3,7-dimethyloctyl)-piperidium bromide
PP333	see Paclobutrazol
Prohexadione calcium	Calcium 3,5-dioxo-4-propionylcyclohexanecarboxylate
Propanil	*N*-(3,4-dichlorophenyl)propanamide
Propham	1-methylethyl phenylcarbamate
Pyrazon	5-amino-4-chloro-2-phenyl-3(2*H*)-pyridazinone
SA	see Salicylic acid
SADH	see Daminozide
Salicylic acid	orthohyroxybenzoic acid
Sethoxydim	2-[1-(ethoxyimino)butyl]-5-[2-(ethylthio)propyl]-3-hydroxy-2-cyclohexen-1-one
Sevin	1-naphthyl *N*-methylcarbamate
Silaid	(2-chloroethyl)methylbis(phenylmethoxy)silane
Silvex	(2,4,5-trichlorophenoxy)propanoic acid
Simazine	6-chloro-*N,N'*-diethyl-1,3,5-triazine-2,4-diamine
STS	silver thiosulfate
Tetcyclacis	5-(4-chlorophenyl)-3,4,5,9,10-pentaza-tetra-cyclo-5,4,10-O-doceca-3,9-diene
TIBA	2,3,5-triiodobenzoic acid
Tordon	see Pichloram
Triclopyr	[(3,5,6-trichloro-2-pyridinyl)oxy]acetic acid
Triapenthenol	(E)-(RS)-1-cyclohexyl-4,4-dimethyl-2-(1*H*-1,2,4-triazole-1-yl)pent-1-en-3-ol
Uniconazole	(E-1-(4-chlorophenyl)-4,4-dimethyl-2-(1,2,4-triazol-1-yl)penten-3-ol
Zeatin	6-(4-hydroxy-3-methyl-trans-2-butenyl-amino)purine

Index